Nelson Mathematics 4

Western Edition

Series Authors and Senior Consultants
Mary Lou Kestell • Marian Small

Senior Authors
Heather Kelleher • Kathy Kubota-Zarivnij • Pat Milot
Betty Morris • Doug Super

Authors
Carol Brydon • Andrea Dickson • Catharine Gilmour
Elizabeth Grill-Donovan • Jack Hope • Wendy Klassen • David Leach
Pat Margerm • Gail May • Scott Sincerbox • Debbie Sturgeon • Rosita Tseng Tam

Assessment Consultant
Damian Cooper

THOMSON NELSON

Australia Canada Mexico Singapore Spain United Kingdom United States

THOMSON
NELSON

Nelson Mathematics 4
Western Edition

Series Authors and Senior Consultants
Mary Lou Kestell, Marian Small

Senior Authors
Heather Kelleher, Kathy Kubota-Zarivnij, Pat Milot, Betty Morris, Doug Super

Authors
Carol Brydon, Andrea Dickson, Catharine Gilmour, Elizabeth Grill-Donovan, Jack Hope, Wendy Klassen, David Leach, Pat Margerm, Gail May, Scott Sincerbox, Debbie Sturgeon, Rosita Tseng Tam

Assessment Consultant
Damian Cooper

Director of Publishing
David Steele

Publisher, Mathematics
Beverley Buxton

Senior Program Manager
Shirley Barrett

Program Managers
Lynda Cowan
Colin Bisset

Developmental Editors
Brenda McLoughlin
Linda Nigro
Michael Tabor
Robert Templeton
Susan Woollam

Developmental Consultants
Jackie Williams
Mary Reeve

Editorial Assistants
Christi Davis
Darren McDonald
Matthew Sheehan

Executive Managing Editor, Development & Testing
Cheryl Turner

Executive Managing Editor, Production
Nicola Balfour

Senior Production Editor
Linh Vu

Production Editor
Sheila Stephenson

Copy Editor
Julia Cochrane
Margot Miller

Production Coordinator
Franca Mandarino

Manufacturing Coordinator
Sharon Latta Paterson

Creative Director
Angela Cluer

Art Director
Ken Phipps

Art Management
ArtPlus Ltd., Suzanne Peden

Illustrators
ArtPlus Ltd., Andrew Breithaupt, Steven Corrigan, Deborah Crowle, Vesna Krstanovic, Sharon Matthews

Interior and Cover Design
Suzanne Peden

ArtPlus Ltd. Production Coordinator
Dana Lloyd

ArtPlus Ltd. Composition
Valerie Bateman, Miriam Semple

Photo Research and Permissions
Vicki Gould

Photo Shoot Coordinators
ArtPlus Ltd., Trent Photographics

Printer
Transcontinental Printing Inc.

Printed and bound in Canada
1 2 3 4 06 05 04 03

For more information contact Nelson, 1120 Birchmount Road, Toronto, Ontario, M1K 5G4. Or you can visit our Internet site at http://www.nelson.com

National Library of Canada Cataloguing in Publication

Nelson mathematics 4 / Mary Lou Kestell ... [et al.].—Western ed.

Includes index.
For use in grade 4.
ISBN 0-17-627021-3

1. Mathematics—Textbooks.
I. Kestell, Mary Louise
II. Title: Nelson mathematics four.

QA135.6.N445 2003
510 C2003-904667-2

Advisory Panel

Senior Advisor

Doug Duff
Learning Supervisor
Thames Valley District School Board
London, Ontario

Advisors

Donna Anderson
Coal Tyee Elementary School
Nanaimo, British Columbia

Keith Chong
Principal
School District #41
Burnaby, British Columbia

Anne Cirillo
Consultant
Toronto Catholic District
School Board
Toronto, Ontario

Attila Csiszar
Math Helping Teacher
Surrey School District #36
Surrey, British Columbia

David P. Curto
Principal
Hamilton-Wentworth Catholic
District School Board
Hamilton, Ontario

Marg Curto
Principal of Programs, Elementary
Hamilton-Wentworth Catholic
District School Board
Hamilton, Ontario

Edna Dach
Director, Instructional Services
Elk Island Public Schools

Wendy Dowling
Peel District School Board
Mississauga, Ontario

Lillian Forsythe, B.Ed., B.A., M.Ed.
Regina, Saskatchewan

Peggy Gerrard
Dr. Morris Gibson School
Foothills School Division
Okotoks, Alberta

Mary Gervais
Durham Catholic District
School Board

Janet Millar Grant
Ontario

C. Marie Hauk, Ph.D.
Consultant
Edmonton, Alberta

Rebecca Kozol
Maple Ridge School District #42
Maple Ridge, British Columbia

A. Craig Loewen, Ph.D.
Mathematics Education
University of Lethbridge
Lethbridge, Alberta

Frank Maggio
Mathematics Curriculum Consultant
Halton Catholic District
School Board
Burlington, Ontario

Moyra Martin
Principal
Msgr. O'Brien School
Calgary Catholic School District
Calgary, Alberta

Meagan Mutchmor
Consultant
Winnipeg School District #1
Winnipeg, Manitoba

Mary Anne Nissen
Consultant
Elk Island Public Schools
Sherwood Park, Alberta

Darlene Peckford, B.Ed., M.A.
Principal
Taber, Alberta

Kathryn Perry
Peel District School Board
Brampton, Ontario

Susan Perry
Consultant
Durham Catholic District
School Board
Oshawa, Ontario

Bryan A. Quinn
Consultant
Edmonton Public District
School Board
Edmonton, Alberta

Ann Louise Revells
Vice-Principal/Teacher
Ottawa-Carleton Catholic
School Board
Ottawa, Ontario

Evelyn Sawicki
Mathematics Consultant
Calgary, Alberta

Lorraine Schroetter-LaPointe
Program Facilitator
Durham District School Board
Whitby, Ontario

Nathalie Sinclair, Ph.D.
Simon Fraser University
Faculty of Education

Susan Stuart
Professor
Nipissing University
North Bay, Ontario

Joyce Tonner
Learning Coordinator
Thames Valley District
School Board
London, Ontario

Stella Tossell
Mathematics Consultant
North Vancouver, British Columbia

Sandra Unrau
Principal
Calgary Board of Education
Calgary, Alberta

Gerry Varty
Math Coordinator
Wolf Creek School Division #72
Ponoka, Alberta

Michèle Wills
Calgary Board of Education
Calgary, Alberta

Reviewers

Mary Adams
Thames Valley District School Board

Michael L. Babcock
Limestone District School Board

Anne Boyd
Campbell River School District #72
British Columbia

James Brake
Peterborough, Victoria,
Northumberland, Clarington,
Catholic District School Board

Nancy Campbell
Rainbow Board of Education

Deborah Colvin-MacDormand
Edmonton Public District School
Board

Anna Dutfield
Toronto District School Board

Linda Edwards, B.A., B.Ed.
Toronto District School Board

Susan Gregson
Peel District School Board

Susannah Howick
North Vancouver School District

Julie Keough
Waterloo Catholic District
School Board

Wendy King
Avalon West School District

Joan McDuff
Faculty of Education
Queen's University

Ken Mendes
Ottawa, Ontario

Gillian Rudge
Richmond School District

Rose Scaini

Naomi Young, B.P.E., B.Ed., M.Ed
Lewisporte Gander School
District #6
Newfoundland and Labrador

Aboriginal Content Reviewers

Brenda Davis
Education Consultant
Six Nations

Moneca Sinclaire, M.S.E., M.Sc.

Laura Smith
School District #34
Abbotsford, British Columbia

Perry Smith
District Teacher,
Aboriginal Education
School District #34
Abbotsford, British Columbia

Equity Reviewer

Arlene Campbell
Toronto District School Board
Educator
York University Doctoral Candidate

Literacy Reviewer

Dr. Roslyn Doctorow
Educational Consultant

Thank you to the following teachers for testing the Chapter Tasks.

Mary Lynne Alderdice
Niagara District School Board

Sonja Atwood
Central Okanagan School District 23

Rick Beetstra
Richmond School District 38

Kevin Corcoran
Peel District School Board

Anna Dutfield
Toronto District School Board

Lena Giorgio
Peel District School Board

Susan Gregson
Peel District School Board

Kelly Joel
Peel District School Board

Don Jones
Peel District School Board

Catherine Niven
Niagara District School Board

Kim Philip
Peel District School Board

Bryan Wood
Peel District School Board

Contents

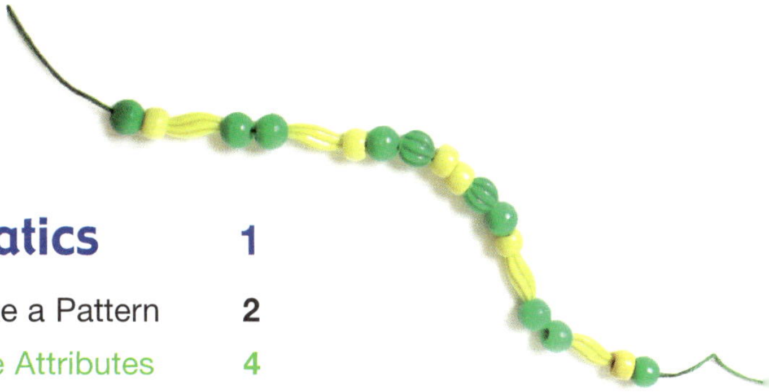

- Guided Activity
- Direct Instruction
- Exploration

CHAPTER 3 Data Management 59

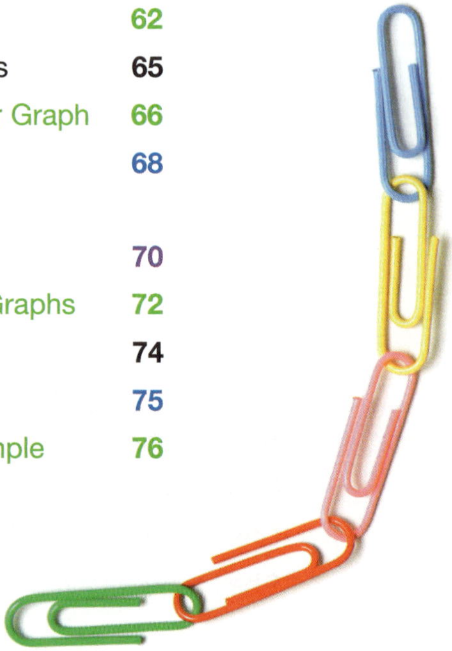

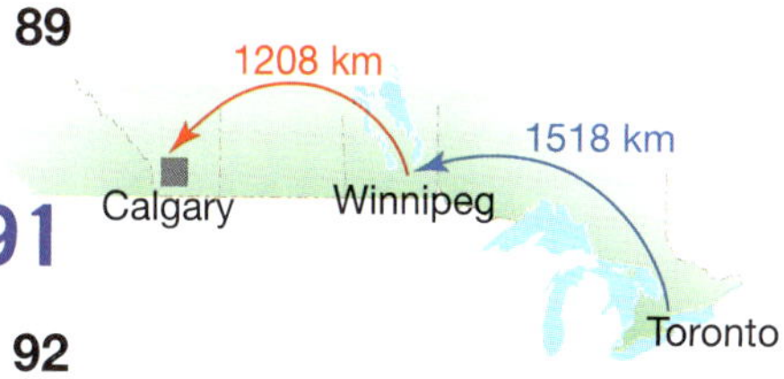

- Guided Activity
- Direct Instruction
- Exploration

CHAPTER 5

Measuring Length and Time 127

CHAPTER 6

Multiplication and Division Facts 155

CHAPTER 7

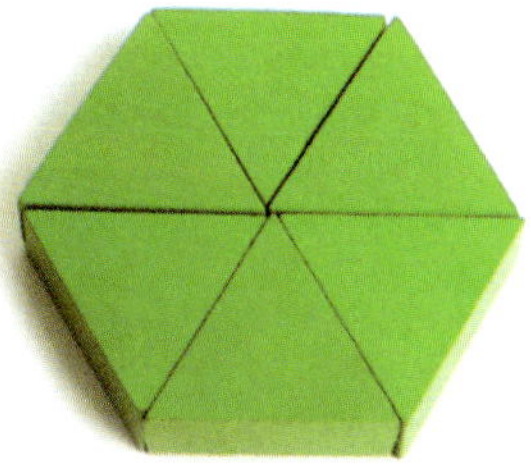

- Guided Activity
- Direct Instruction
- Exploration

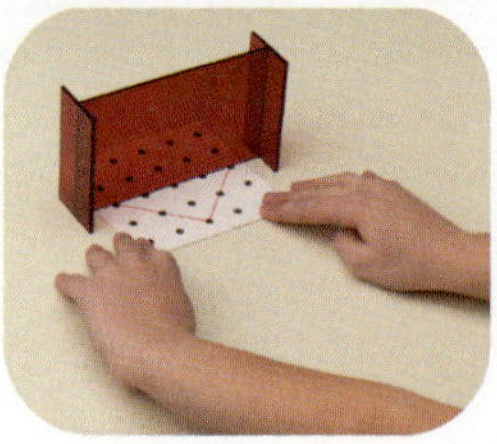

CHAPTER 9 Multiplying Greater Numbers 243

CHAPTER 10 Dividing Greater Numbers 271

- Guided Activity
- Direct Instruction
- Exploration

CHAPTER 11 3-D Geometry and 3-D Measurement 299

- Guided Activity
- Direct Instruction
- Exploration

- Guided Activity
- Direct Instruction
- Exploration

CHAPTER 1

Patterns in Mathematics

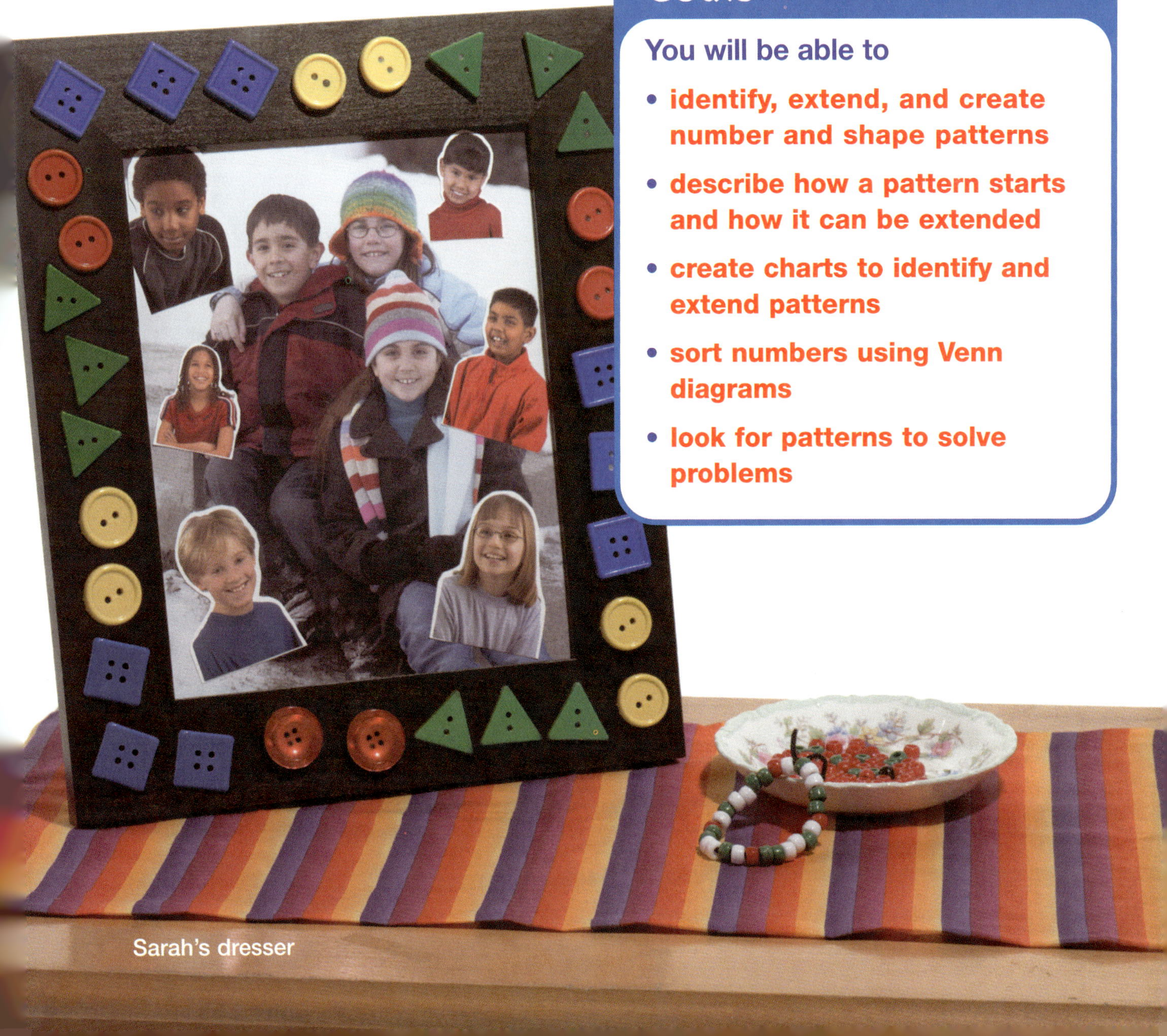

Sarah's dresser

Goals

You will be able to

- **identify, extend, and create number and shape patterns**
- **describe how a pattern starts and how it can be extended**
- **create charts to identify and extend patterns**
- **sort numbers using Venn diagrams**
- **look for patterns to solve problems**

Getting Started

Sorting to Make a Pattern

Sarah wants to make a bead necklace that shows a pattern.

You will need

- beads in different shapes, colours, sizes, and textures

- beading string

Sarah's Sorting

I sorted some of my beads into 4 groups.
I used 2 **attributes**: colour and texture.

Now I can find the beads I need to make my necklace.

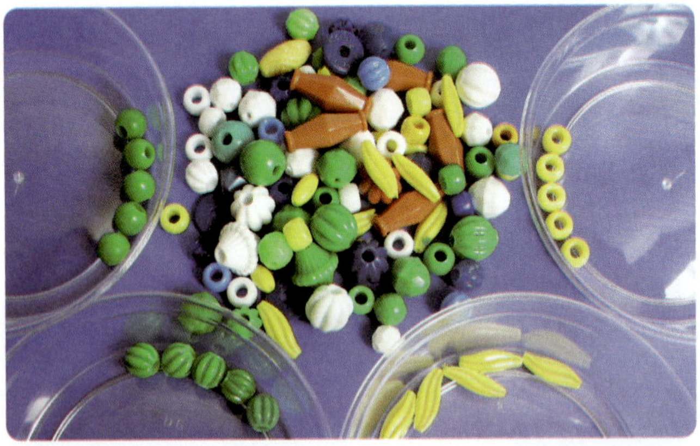

? How would you sort the beads so you could make a pattern?

A. List at least 4 attributes you could use to sort the beads.

B. Draw or describe a way to sort the beads using 2 or more attributes.
Label your work with the attributes you used.

C. Sarah made this bead necklace.
Describe the pattern by telling how each attribute changes.

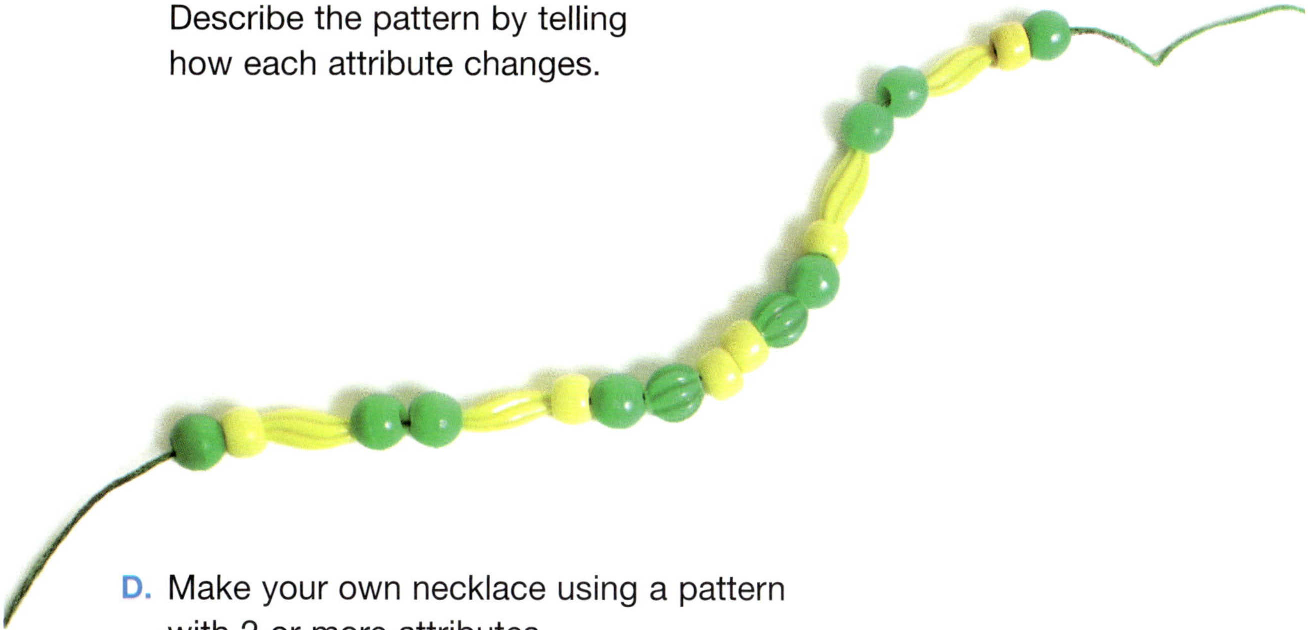

D. Make your own necklace using a pattern with 2 or more attributes.
Describe the pattern by telling how each attribute changes.

Do You Remember?

1. Draw or describe the next 3 shapes in each pattern.

   a) 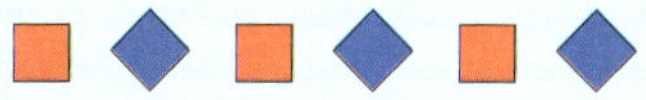

   b) 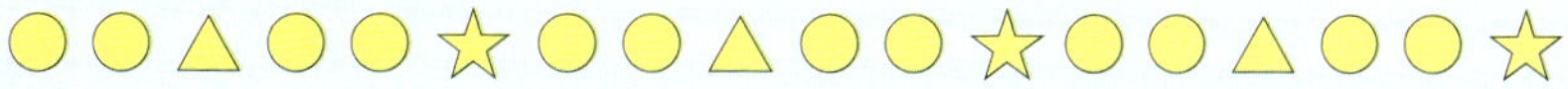

   c) 

2. Write the next 3 numbers or letters in each pattern.
   a) 25, 50, 75, 100, 125, ■, ■, ■
   b) 19, 28, 37, 46, 55, ■, ■, ■
   c) 100, 98, 96, 94, 92, ■, ■, ■
   d) A, C, E, G, ■, ■, ■

3. Write the number 5.
   Add the number 2 and write the sum.
   Keep adding 2 to create a pattern with 6 numbers.

CHAPTER 1

1 Patterns with Multiple Attributes

You will need
- pencil crayons

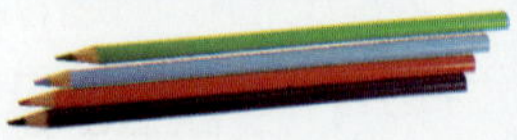

Goal **Describe, extend, and create patterns that change in many ways.**

Chris's Bracelet

I lost my favourite bead bracelet.

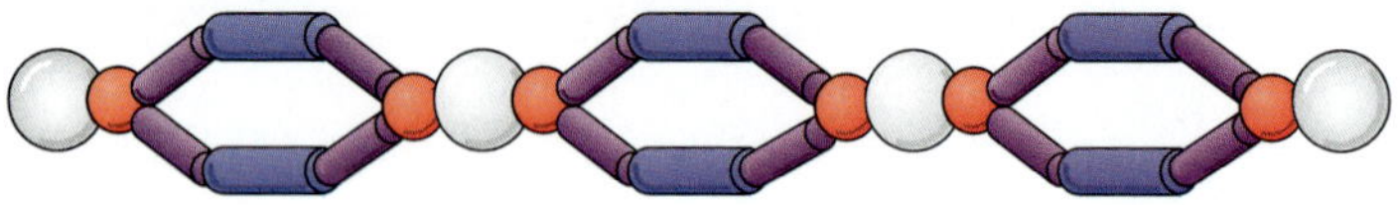

? How can Chris describe his bracelet so someone can make a copy for him?

A. Describe the pattern in Chris's bracelet by telling how the attributes of colour, shape, and size change.

B. Each type of bead in Chris's bracelet has been given a letter.
Complete the letter model to describe the pattern.
Repeat the pattern twice.

P	Q	R	S
large white ball	small red ball	small purple cylinder	large blue cylinder

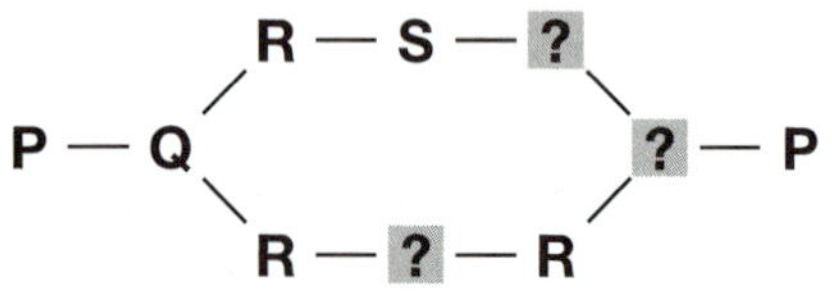

C. Draw a different bracelet using the same 4 types of beads that were in Chris's bracelet.

D. Describe the pattern in your bracelet by telling how each attribute changes.

E. Describe your bead pattern using a letter model.
Repeat the pattern twice.

Reflecting

1. The bead pattern in Chris's bracelet is made up of smaller patterns. What does this mean?
2. How does the letter model describe the pattern?

Checking

3. a) Describe the pattern by telling how each attribute changes.

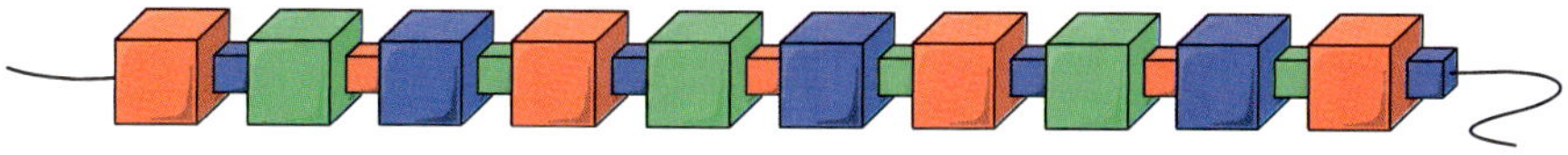

   b) Draw the next 3 beads.
   c) This letter model describes the pattern for the first 14 beads in part a). Describe the bead for each letter.

   A - B - C - D - E - F - A - B - C - D - E - F - A - B

Practising

4. Describe the pattern by telling how each attribute changes. Then draw the next 3 beads.

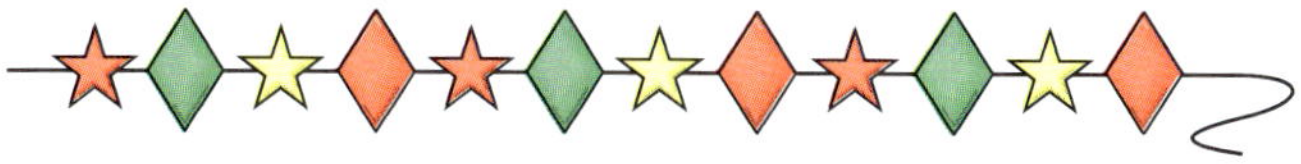

5. Create a letter model for this bead pattern.

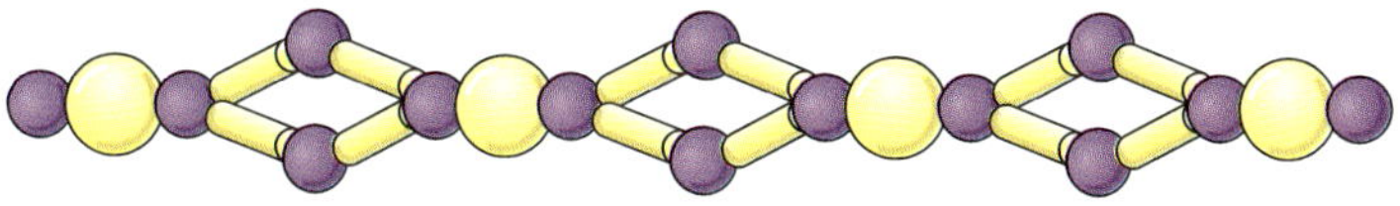

6. a) Draw your own bead pattern with 2 or more attributes that change.
   b) Describe the pattern by telling how each attribute changes.
   c) Describe your bead pattern using a letter model.

2 Number Patterns

Goal **Create, describe, and extend number patterns.**

Pedro and 4 other students each created a number pattern. Then they described their number patterns.

Pattern A: 42, 46, 50, 54, 58, ...

Pattern B: 42, 24, 42, 24, 42, 24, ...

Pattern C: 45, 44, 41, 40, 37, 36, ...

Pattern D: 100, 95, 90, 85, 80, ...

Pattern E: 10, 12, 16, 22, 30, 40, ...

? Which student made each pattern?

A. Match each student's pattern description with 1 of the number patterns above. Then write the next 3 numbers in the pattern.

B. Create 5 new number patterns, 1 pattern for each of the student descriptions. Include at least 5 numbers in each pattern.

Pedro's Pattern

The numbers decrease by the same amount each time.

Calvin's Pattern

The numbers decrease by different amounts.

Carmen's Pattern

The same 2 numbers repeat.

Josef's Pattern

The numbers increase by a different amount each time.

Mandy's Pattern

The numbers increase by the same amount each time.

Reflecting

1. a) Why is it possible to have more than 1 pattern for each student description?
   b) How could you change each description so that only 1 pattern is possible?

Checking

2. a) Create a number pattern with at least 5 numbers.
   b) Describe your number pattern. Make sure only 1 pattern is possible for your description.

Practising

3. Describe each pattern. Then write the next 3 numbers.
   a) 100, 98, 96, 94, 92, 90, ...
   b) 100, 99, 97, 94, 90, ...
   c) 3, 8, 13, 18, 23, ...
   d) 3, 6, 9, 3, 6, 9, 3, 6, 9, 3, ...

4. a) Create a number pattern that increases by a different amount each time.
   b) Explain why more than 1 pattern is possible for the description in part a).
   c) Change the description so that it describes only the pattern you created.

5. Terry has 8 pairs of socks.
   The pairs of socks made this pattern: 2, 4, 6, 8, …
   a) Write the next 3 numbers in Terry's pattern.
   b) Describe Terry's pattern.

6. Zola is watching this clock.
   She sees this pattern in the hours: 2, 4, 6, 8, …
   a) Write the next 3 numbers in Zola's pattern.
   b) Describe Zola's pattern.
   c) Compare Zola's pattern and Terry's pattern from Question 5. How are the patterns the same? How are they different?

3 Patterns in T-Charts

You will need

- pattern blocks

Goal Use t-charts to identify and extend patterns.

An inukshuk is a marker or signpost made of rocks built in the shape of a person.
The plural of inukshuk is inukshuit.
Manitok collected 15 small rocks and 26 large rocks.

? Does Manitok have enough small rocks to make 6 inukshuit like this one?

Manitok's Inukshuk Project

Step 1 I made models of the inukshuk using pattern blocks.
I used green blocks for the small rocks.

Step 2 I created a **t-chart** to record how many green blocks I used in each model.

Step 3 After making 3 inukshuit, I found a pattern in the t-chart: 3, 6, 9, ….

Step 4 I extended the pattern to complete the t-chart.

Inukshuk	Total number of green blocks
1	3
2	6
3	9
4	12
5	15
6	18

For 6 inukshuit, I need 18 small rocks.
I only have 15 small rocks, so I don't have enough.

Reflecting

1. Explain how the t-chart helped Manitok solve the problem.
2. Describe another strategy Manitok could have used to solve the problem.

Checking

3. a) Manitok has 26 large rocks. Does he have enough large rocks to make 6 inukshuit? Use this t-chart to solve the problem.
   b) Look at the numbers in the 2nd column of your t-chart. Describe the pattern.

Inukshuk	Total number of large rocks
1	
2	
3	

Practising

4. There is an inukshuk on the Nunavut flag.
   a) Create a t-chart to find the total number of rocks needed to make 7 of these inukshuit.

Inukshuk	Total number of rocks
1	
2	
3	

   b) Describe the pattern.

5. a) Build your own inukshuk model. Use 2 sizes of blocks. Use a different number of each size.
   b) How many blocks of each type do you need to make 7 inukshuit? Use t-charts.
   c) How many inukshuit can you make if you have 30 small blocks and 15 large blocks? Explain your thinking.

4 Measurement Patterns

You will need

- a calculator

Goal Extend time patterns in t-charts.

You spend about 195 days in school each year.

? How many days will you spend in school by the end of grade 4?

Jon's Calculator Pattern

In kindergarten, I spent 195 days in school.
In grade 1, I spent another 195 days in school, so I add:

195 [+] 195 [=] 390 [=] [=] [=]

I use the constant feature of the calculator to extend the pattern.

Grade	Total number of school days
K	195
1	390
2	
3	
4	

A. What is the **pattern rule** for the numbers in the 2nd column?

B. Complete Jon's t-chart.

C. How many days will Jon spend in school by the end of grade 4?

D. Look at the digits in the 2nd column of the t-chart. Describe the pattern.

E. Predict the number of school days by the end of grade 5 and by the end of grade 6. Use a calculator to check your predictions.

pattern rule
A description of how a pattern starts and how it can be extended

Reflecting

1. Explain how the pattern in the digits helped you to predict the number of days you spent in school by the end of grades 5 and 6.
2. Could 1456 ever appear in the 2nd column of Jon's t-chart? Explain your thinking.

Checking

3. There are about 19 holidays in a school year.
   a) How many holidays are there by the end of grade 4? Use the t-chart.
   b) Write the pattern rule.
   c) Describe any patterns you see in the digits in the 2nd column of the t-chart.
   d) Predict the number of holidays by the end of grade 6. Then check your prediction.

Grade	Total number of holidays
K	19
1	38
2	
3	

Practising

4. Complete 7 rows of each t-chart.

a)

Days	Total number of hours
1	
2	
3	

b)

Years	Total number of months
1	
2	
3	

5. Look at the 2nd column of each t-chart in Question 4. Write the pattern rule for each t-chart.
6. How old are you in months? Show your work.
7. Chantal has 45 minutes of computer time each day at home. How much computer time does she have in 1 week? Use a t-chart.

LESSON

Mid-Chapter Review

1 **1. a)** Describe the pattern by telling how each attribute changes.

b) Draw or describe the next 3 shapes in the pattern.

c) Describe the pattern using a letter model.

2 **2.** Write the next 3 numbers in each pattern.

a) 6, 12, 18, 24, ... **c)** 5, 15, 25, 35, ...

b) 49, 42, 35, 28, ... **d)** 1, 3, 5, 1, 3, 5, 1, 3, 5, 1, ...

3. Describe each pattern. Write the next 3 numbers.

a) 25, 27, 29, 31, ...

b) 0, 1, 3, 4, 6, 7, 9, ...

c) 22, 19, 16, 13, ...

d) 0, 15, 30, 45, 0, 15, 30, 45, 0, 15, 30, 45, ...

4. A pattern starts at 60 and then decreases by 1, then 2, then 3, and so on.
What is the 10th number in the pattern?
Show your work.

4 **5.** The library gets 16 new books each month.

a) Complete the t-chart to show how the number of new books grows over 1 year.

b) Write a rule for the pattern in the 2nd column of the t-chart.

c) Describe any patterns in the digits. Predict the next 3 numbers.

Month	Total number of new books
1	

6. It takes Suki about 45 minutes to make a bracelet. Make a t-chart that shows how many bracelets she can complete in 4 hours.

Calculator Patterns

You will need

- a calculator

Number of players: 2
How to play: Player 1 enters a secret pattern into a calculator.
Player 2 tries to figure out how the numbers in the pattern change.

Step 1 Enter a 3-digit start number.
Press [+].
Enter a change number.

Step 2 Press [=].
Give the calculator to your partner.

Step 3 Your partner presses [=] and tries to figure out how the numbers are changing.
After your partner figures out how the numbers change, it is your turn to figure out a secret pattern.

Josef's Secret Pattern

start number	change number	
134	[+] 4	[=] 138

Sarah's Solution

I pressed [=] twice.

138 [=] 142 [=] 146

I think the change is [+] 4.

Other ways to play:

- Use [−] instead of [+].
- Figure out the start number of the pattern.

5 Solve Problems Using a Patterning Strategy

You will need

- a 100 chart

1	2	3	4
11	12	13	14
21	22	23	24
31	32	33	34

Goal Look for a pattern to solve a problem.

Here comes a parade of 100 clowns!
Every 2nd clown has a red nose.
Every 3rd clown wears glasses.

? How many clowns have a red nose *and* glasses?

Miki's Solution

Understand

I need a way to count every 2nd and 3rd clown in a line of 100 clowns. That way I can see how many clowns have red noses *and* glasses.

Make a Plan

I can mark every 2nd and 3rd number on a 100 chart. Then I'll count the marks.

Carry Out the Plan

In the first 3 rows of the chart, I mark every 2nd number with \ and every 3rd number with /.

1	2	3	4	5	6	7	8	9	10
11	12	13	14	15	16	17	18	19	20
21	22	23	24	25	26	27	28	29	30
31	32	33	34	35	36	37	38	39	40
41	42	43	44	45	46	47	48	49	50
51	52	53	54	55	56	57	58	59	60
61	62	63	64	65	66	67	68	69	70
71	72	73	74	75	76	77	78	79	80
81	82	83	84	85	86	87	88	89	90
91	92	93	94	95	96	97	98	99	100

I see a pattern! Every 6th number has both marks.
I can now circle every 6th number and then count.

16 clowns have a red nose and glasses.

Reflecting

1. How did looking for a pattern make the problem easier to solve?

Checking

2. Every 5th clown in the parade of 100 clowns wears a hat. Every 2nd clown has a red nose. How many clowns have a hat *and* a red nose?

Practising

3. a) Every 3rd clown in the parade of 100 clowns wears glasses. How many clowns have a hat *and* glasses?
   b) How many clowns have a hat, glasses, *and* a red nose?

4. a) Describe a pattern in the digits of the badge numbers.

   b) Make a badge with a 7-digit number that has the same pattern.

5. Britney's family went skating together on a Tuesday. After that, Britney and her brother Liam skated every 2nd day. Their mom skated every 3rd day and their dad skated every 4th day. On what day of the week did Britney's family skate together again?

6. Look for a pattern in the numbers to find the answer.

   $20 - 19 + 18 - 17 + 16 - 15 + 14 - 13 + 12 - 11 + 10 - 9 + 8 - 7 + 6 - 5 + 4 - 3 + 2 - 1$

7. Create and solve a problem that can be solved by looking for a pattern.

6 Finding Missing Terms

You will need

- base ten blocks

Goal **Find the missing number in a pattern and in a number sentence.**

You can find a missing number in a pattern if you know how the numbers in the pattern change.

? What is the missing number in this pattern?
15, 21, 27, 33, ■, 45, ...

Natalie and Rami figured out that the numbers in the pattern change by 6 each time. Then they each wrote a number sentence with a missing number to find ■.

Natalie's Addition Sentence

+6

15, 21, 27, 33, ■, 45, ...

■ + 6 = 45

To find ■, I started at 45 on a number line.

Then I thought, "Where on the number line would I have started, if I added 6 and ended up at 45?"

So I subtracted 6 to find out.

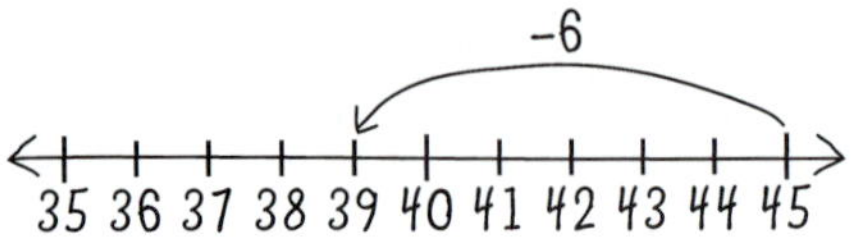

The missing number, ■, is 39.

Rami's Subtraction Sentence

−6

15, 21, 27, 33, ■, 45, ...

■ − 6 = 33

To find ■, I modelled 33 with blocks.

Then I thought, "What would I have started with, if I took away 6 and had 33 left?"

So I added 6 to find out.

39 − 6 = 33

The missing number, ■, is 39.

Reflecting

1. Explain how you would use a number line to find ■ in the subtraction sentence ■ − 6 = 33.
2. Explain how you would use blocks to find ■ in the addition sentence ■ + 6 = 45.
3. Explain how you might use guess and test and mental math to find ■ in the number sentences ■ − 6 = 33 and ■ + 6 = 45.

Checking

4. What is the missing number in this pattern?
   6, 11, ■, 21, 26, 31, …
   Use ■ + 5 = 21.

5. What is the missing number in this pattern?
   27, 23, 19, 15, ■, 7, …
   Use ■ − 4 = 7.

Practising

6. Use the number sentence to find the missing number in each pattern.

   a) 16, 25, 34, 43, ■, 61, ... ■ + 9 = 61
   b) 31, ■, 25, 22, 19, 16, ... ■ + 3 = 31
   c) ■, 77, 73, 69, 65, 61, 57, ... ■ − 4 = 77

7. Write the pattern rule for each pattern in Question 6.

8. Find ■ in each number sentence.

   a) ■ + 7 = 16
   b) 21 − ■ = 5
   c) ■ − 9 = 35
   d) ■ + 12 = 19
   e) 27 − ■ = 11
   f) 36 + ■ = 52
   g) 15 + ■ = 44
   h) ■ − 14 = 17
   i) 74 − ■ = 39

7 Venn Diagrams

Goal Sort and classify numbers using Venn diagrams.

Paulette is using a **Venn diagram** to sort some numbers by 2 attributes.

Venn diagram
A diagram that uses circles and a rectangle to show relationships

Paulette's Venn Diagram

1 circle has numbers with more than 1 digit. The other circle has numbers that are odd.

35 belongs in both circles. It belongs in the circle for numbers with more than 1 digit and in the circle for odd numbers.

194 belongs in 1 circle. It belongs in the circle for numbers with more than 1 digit, but not in the circle for odd numbers.

6 belongs outside the circles. It does not belong in the circle for numbers with more than 1 digit or in the circle for odd numbers.

Sorting Numbers

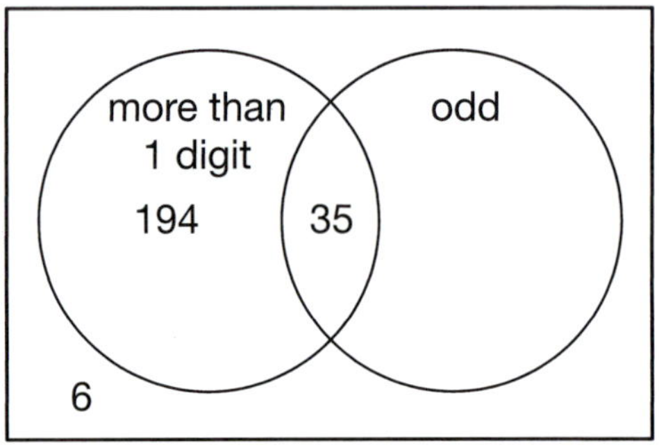

? How can Paulette sort these numbers in her Venn diagram?

650 7 48 283 8 5 71

A. What is Paulette's **sorting rule**?

B. Finish sorting Paulette's numbers in the Venn diagram. Explain how you know where to put each number.

C. Name 2 other numbers for each of the 4 parts of Paulette's Venn diagram. Explain how you used Paulette's sorting rule.

Reflecting

1. How are the numbers outside the circles the same?
2. Suppose the circle labelled odd was changed to even. How would you sort numbers differently?

Checking

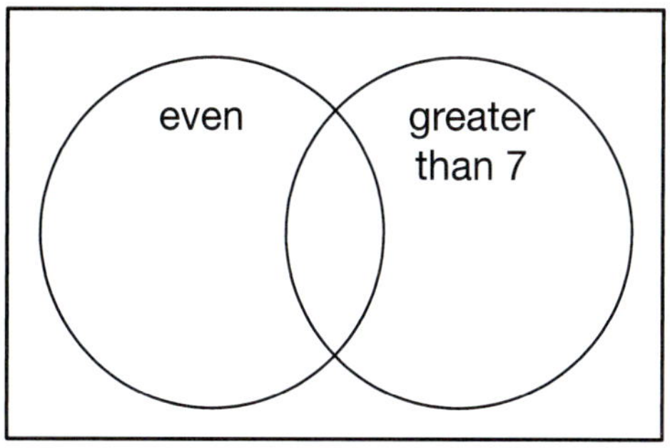

3. a) Sort these numbers using the Venn diagram.
   16, 21, 7, 10, 1, 8, 123, 2, 9
   b) Describe the sorting rule.
   c) Name 2 more numbers for each of the 4 parts of the Venn diagram.

Practising

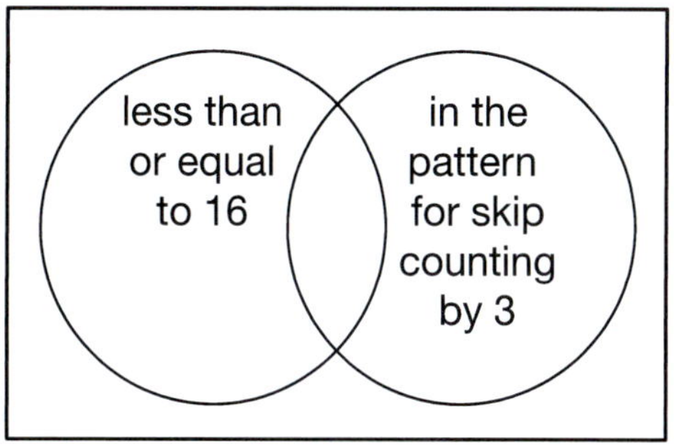

4. a) Sort these numbers using the Venn diagram.
   6, 7, 12, 24, 17, 20, 36, 99, 16, 32, 10
   b) Describe the sorting rule.
   c) Write 2 more numbers in each part of the Venn diagram.

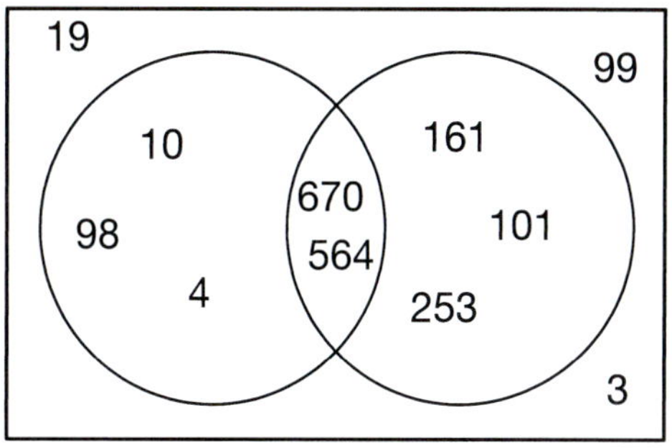

5. a) What labels could you use for the Venn diagram?
   Explain why your labels make sense.
   b) Write 2 more numbers in each part of the Venn diagram.
6. a) Make up a sorting rule. Sort these numbers in a Venn diagram. Describe your sorting rule.
   5, 18, 24, 162, 73, 2, 60, 254, 300, 45, 6
   b) Sort the same numbers in a different Venn diagram. Describe your sorting rule.

8 Multiple Number Patterns

Goal **Extend and describe special number patterns.**

Allison made a number chain. She wrote a number pattern inside the triangles: 3, 6, 9, 12, ...

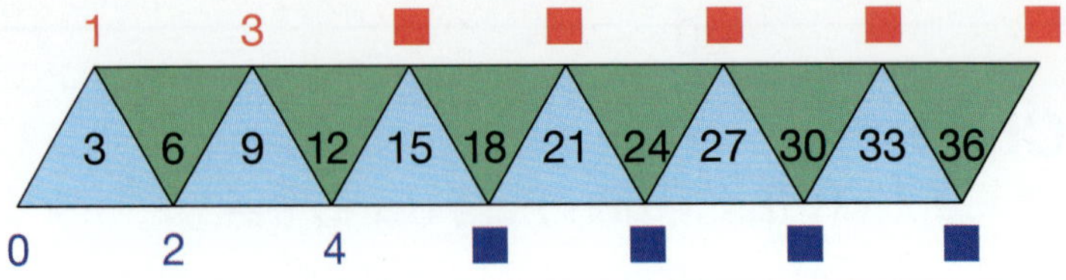

? What are the missing numbers in Allison's number chain?

Allison's Number Chain

On the corners of each triangle, I wrote numbers that add up to the number inside.

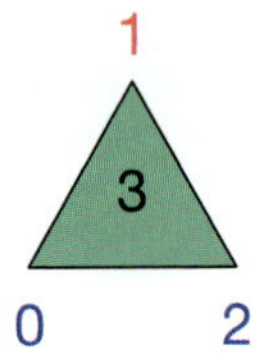

A. Complete Allison's number chain.

B. Write each number pattern. Then predict the next number.

a) the red corner numbers above the chain: 1, 3, ...

b) the numbers inside the chain: 3, 6, 9, 12, ...

c) the blue corner numbers below the chain: 0, 2, 4, ...

d) the zig-zagging blue and red numbers: 0, 1, 2, 3, 4, ...

C. Describe each number pattern in part B.

Reflecting

1. Compare the 4 number patterns in the chain.
How are the patterns the same?
How are they different?

Mental Math

Adding with 5s

Think of two 5s when you want to add a number ending in 5.

$15 + 7 = 15 + 5 + 2$

$\qquad\qquad 20 + 2 = 22$

A. Why is it easy to add 2 numbers ending in 5?

Try These

1. **a)** 15 + 8 **b)** 25 + 6 **c)** 45 + 9 **d)** 95 + 7

2. **a)** 5 + 18 **b)** 69 + 5 **c)** 38 + 45 **d)** 36 + 6

Curious Math

Pascal's Triangle

In 1653, Blaise Pascal used this triangular pattern of numbers to solve a problem.

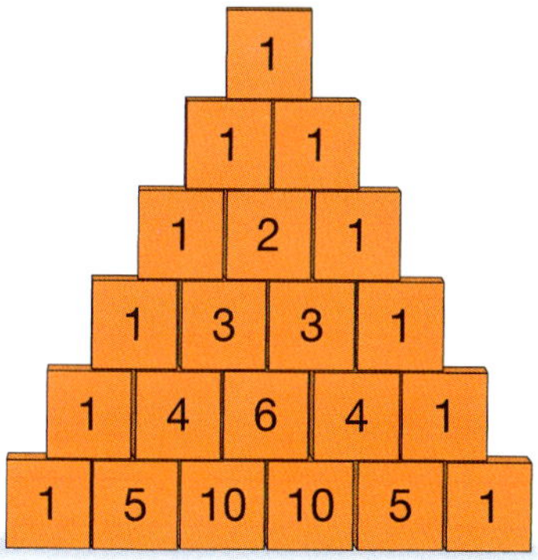

1 Choose any number inside the triangle. Look at the 2 numbers above it. How did Pascal make his triangle?

2 Write the next row of numbers in Pascal's triangle.

LESSON

Skills Bank

1 **1.** Describe each pattern by telling how each attribute changes.
Then draw the next 3 shapes or symbols.

a) ♣♦♠♥♣♦♠♥♣♦♠♥

b) 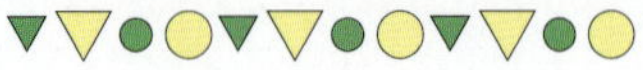

c) 

d) 

e) ↑→↓←↑→↓←↑→↓←

f) 

g) ▲▼◆ ◆▼▲ ▲▼◆ ◆▼▲ ▲▼◆ ◆▼▲

h) 

2 **2.** Describe each pattern. Then write the next 3 numbers.

a) 99, 96, 93, 90, 87, …
b) 4, 9, 14, 19, 24, …
c) 5, 10, 15, 5, 10, 15, 5, 10, 15, …
d) 98, 91, 84, 77, 70, …
e) 100, 98, 94, 88, 80, 70, …
f) 4, 9, 5, 10, 6, 11, 7, 12, 8, …

3 **3.** There is a maple leaf on the Canadian flag.

a) Create a t-chart to find the total number of points on 6 maple leaves.
b) Describe the pattern in the 2nd column of the t-chart.

Number of leaves	Total number of points
1	11
2	
3	

4. Vinh started with the middle hexagon in this quilt pattern. Each day he added another ring of hexagons.

a) Create a t-chart to find the total number of hexagons in the quilt after 5 days.
Use a calculator.
b) What is the first day there will be more than 200 hexagons in the quilt?

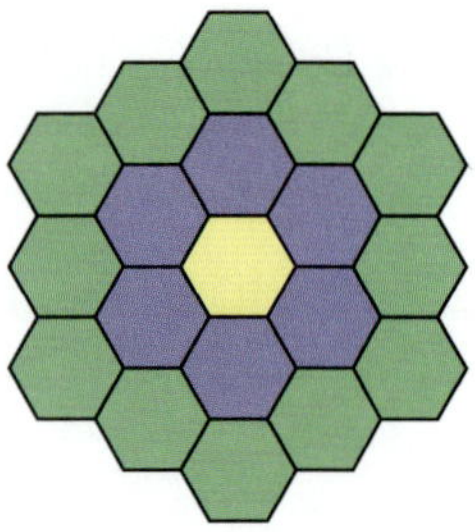

 NEL

4 **5.** Complete 7 rows of each t-chart. Use a calculator. Write a pattern rule for the pattern in the 2nd column of the t-chart.

a)

Years	Total number of weeks
1	52
2	■
3	■

b)

Weeks	Total number of hours
1	168
2	■
3	■

6 **6.** Use the number sentence to find the missing number in each pattern.

a) 8, 15, 22, ■, 36, 43, … ■ − 7 = 22

b) ■, 65, 59, 53, 47, 41, … ■ − 6 = 65

c) 17, ■, 23, 26, 29, 32, … 3 + ■ = 23

7. Find ■ in each number sentence.

a) 11 + ■ = 36

b) ■ + 26 = 54

c) 73 − ■ = 49

d) ■ − 12 = 56

e) 23 + ■ = 42

f) ■ − 8 = 76

7 **8.** **a)** Sort these numbers in the Venn diagram.
20, 11, 8, 145, 3, 14, 10, 374, 17, 86

b) Write 2 more numbers in each part of the Venn diagram.

Sorting Numbers

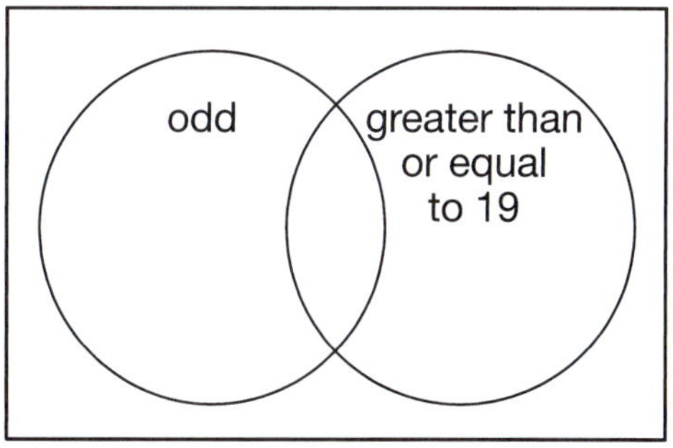

8 **9.** Complete each number chain. Predict the next 2 numbers in each pattern.

a)

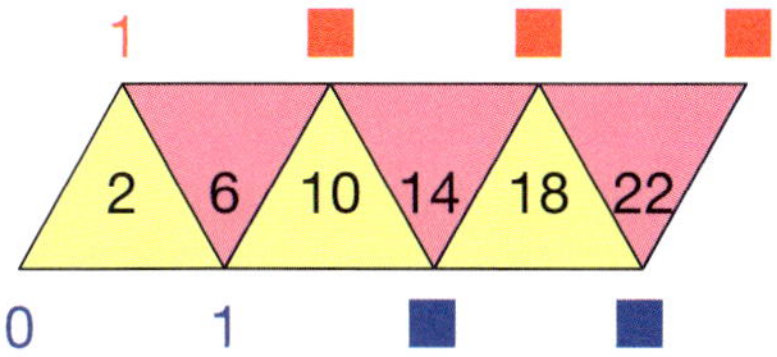

b)

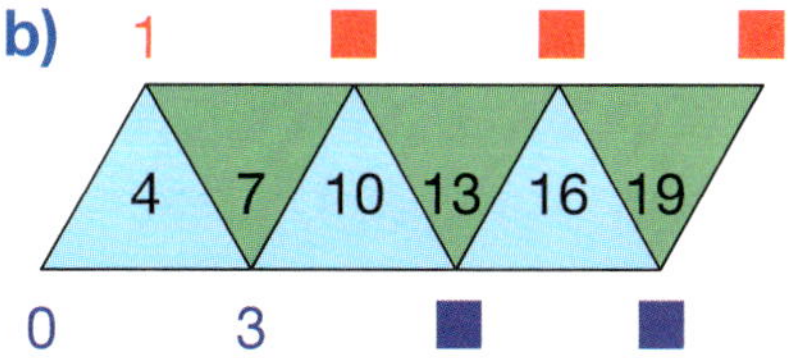

LESSON

Problem Bank

5 **1.** Jennifer put markers along a running track.
She put the first marker at 100 m.
Then she put a marker at every 50 m.
How many markers does she need to reach 400 m?

2. Kelly ran a 10 km race.
She ran the 1st kilometre in 5 minutes.
She then slowed down, and each kilometre after that took 1 minute longer to run than the kilometre before.
How long did it take Kelly to run the race?

3. Corey walks past doors 1 to 12 and opens every 2nd door. On his 2nd trip past the doors, he stops at every 3rd door. If the door is closed, he opens it. If the door is open, he closes it. Which doors are open after Corey's 2nd trip?

4. Brian just had his 5th birthday.
He gets a new haircut every 8 months.
How many haircuts has Brian had in his life?

8 **5.** Anil is helping his aunt around the house for 5 days.
He can choose between these 2 ways to be paid:

- $10 each day
- $1 for the 1st day and, on each day after that, double the amount of the day before

a) Which payment method should Anil choose? Why?

b) Would your answer change if he worked for 1 week? Explain your thinking.

LESSON

Chapter Review

1 **1.** Describe the pattern by telling how each attribute changes. Draw or describe the next 3 shapes.

a)  ...

b) 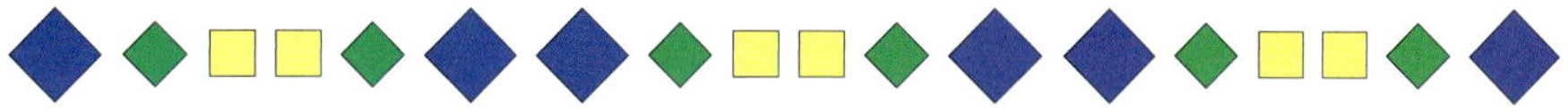 ...

2 **2.** Write the next 3 numbers in each pattern.

a) 7, 11, 15, 19, ... **b)** 24, 21, 18, 15, ...

3. A pattern starts at 51 and decreases by 8. Write 6 numbers in this pattern.

4 **4.** Complete each t-chart and describe the pattern rule.

a)

Hours	Total number of minutes
1	60
2	120
3	180
4	
5	
6	

b)

Cats	Total number of claws
1	18
2	36
3	54
4	
5	
6	

5 **5.** Create a problem that could be solved using 1 of the t-charts in Question 4. Solve your problem.

6 **6.** Find ■ in each number sentence.

a) ■ + 13 = 21 **b)** 5 + ■ = 34 **c)** 100 − ■ = 91 **d)** ■ − 22 = 66

7 **7.** Make up a sorting rule. Sort these numbers in a Venn diagram. Describe your sorting rule.
65, 743, 600, 35, 96, 5, 72, 981, 80

8 **8.** Complete the number chain. Describe each number pattern.

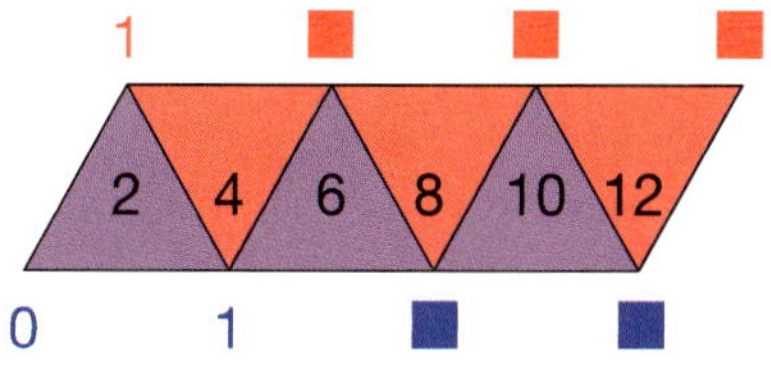

Chapter Task

Finding Number Patterns in Shape Patterns

Pedro created this shape pattern using pattern blocks.

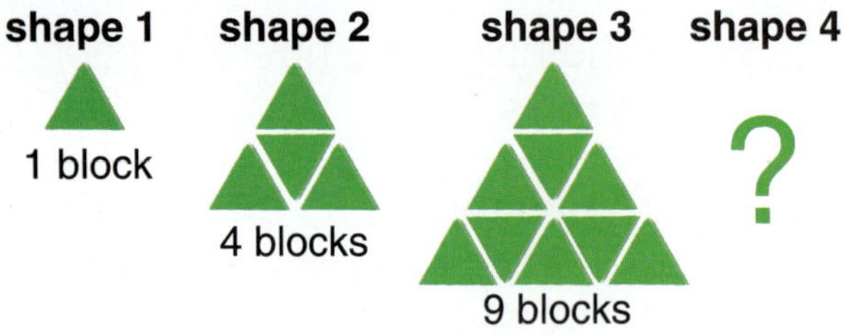

? What number patterns can you see in the shape pattern?

Part 1

A. Draw or make a model of shape 4. Describe the shape pattern in words.

B. Complete a t-chart to show the number pattern for Pedro's shape pattern.

C. Extend the t-chart to find out how many blocks would be in shape 5 and shape 6.

D. Write a rule for the number pattern.

Shape	Total number of blocks
1	1
2	4
3	9
4	
5	

Part 2

E. Create your own shape pattern using pattern blocks. Draw and describe your shape pattern.

F. Describe a number pattern in your shape pattern.

G. How are your pattern and Pedro's pattern the same? How are the patterns different?

Task Checklist

- ☑ Did you include a diagram?
- ☑ Did you use t-charts to show number patterns?
- ☑ Did you describe the number patterns by writing pattern rules?

CHAPTER 2

Numeration

Goals

You will be able to

- understand and compare numbers to 10 000
- represent 4-digit numbers in different ways
- explore place value number patterns
- communicate about ordering numbers
- sort using Carroll diagrams
- estimate the number of objects in a set
- count and estimate money amounts

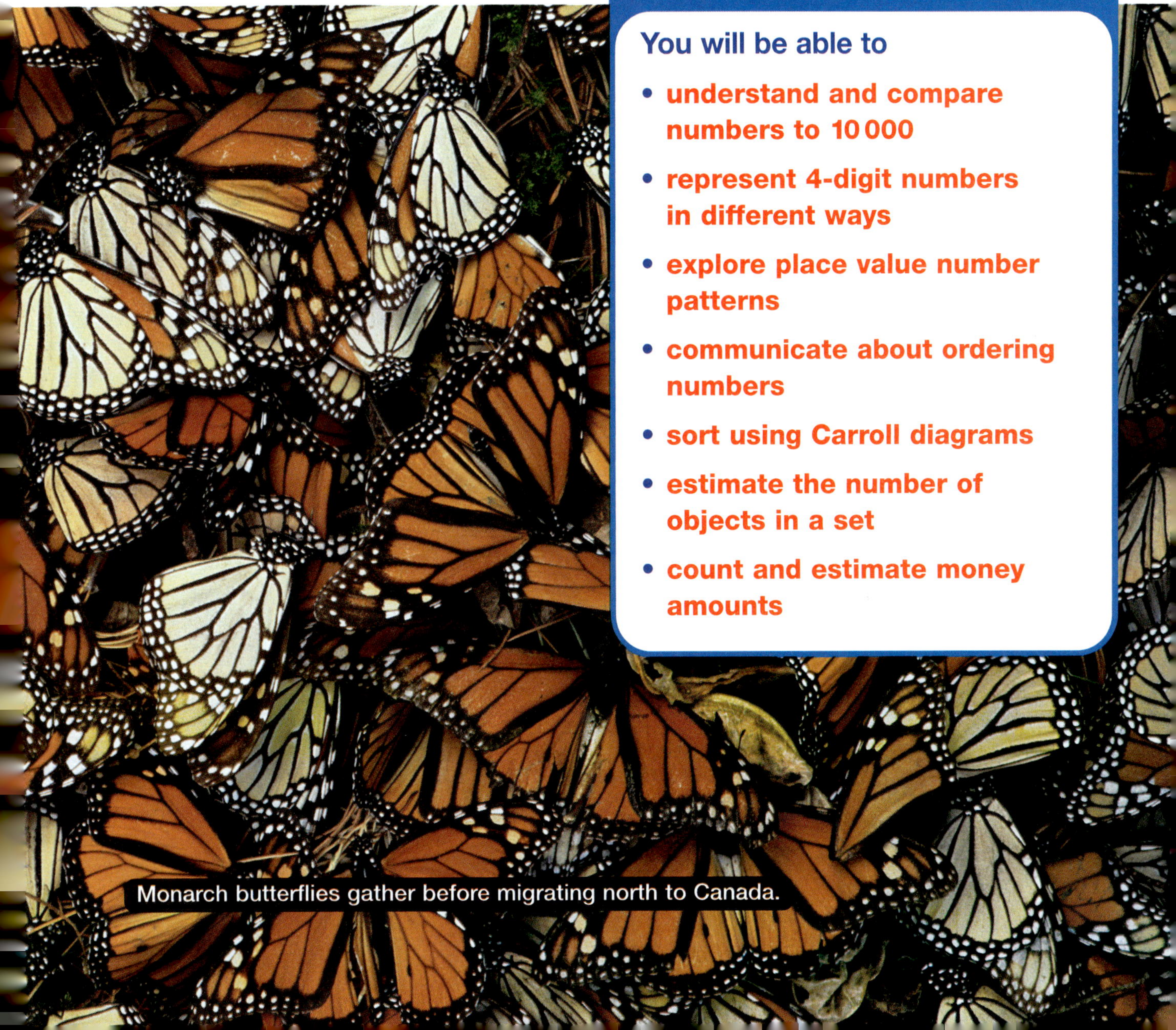

Monarch butterflies gather before migrating north to Canada.

CHAPTER 2

Getting Started

You will need

- base ten blocks

- play money

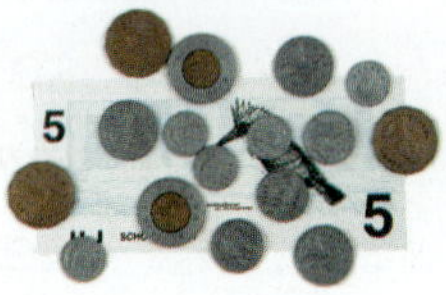

Modelling Numbers

A man in India has learned how to mimic the calls of 326 animals and birds.

Here is 1 way to represent 326 using base ten blocks.

11 blocks were used.

A. How do you know that 326 is between 300 and 400?

B. How could you model 326 if you had no tens blocks?

C. How could you model 326 using exactly 20 blocks?

For each task below, choose 11 blocks.
You may choose different blocks each time, but you must always have at least 1 hundreds block, 1 tens block, and 1 ones block.

D. What is the greatest number you can make with the 11 blocks you chose?
How do you know that this number is the greatest number?

E. What is the least number you can make with the 11 blocks you chose?

F. Use 11 blocks to make 2 numbers that are close to 300. Make 1 number that is less than 300 and 1 number that is greater than 300.

Do You Remember?

1. Represent each number using base ten blocks.
Write each number as ■ hundreds + ■ tens + ■ ones.
   a) 169 b) 961 c) 320 d) 507

2. Which number in each pair is greater?
   a) 639 or 714 b) 495 or 475 c) 306 or 360

3. Write each set of numbers in order from least to greatest.
   a) 384, 389, 364, 368 b) 870, 780, 817, 728

4. Describe 2 ways to make each amount of money using coins.
   a) \$4.50 b) \$9.75

CHAPTER 2

1 Place Value

You will need
- base ten blocks

Goal **Model numbers up to 10 000.**

Chantal is studying spiders.
She has found a spider that can lay 1200 eggs.

? How can Chantal model 1200 with the least number of base ten blocks possible?

Chantal's Model

I can model 1200 with 1200 ones blocks.

If the ones blocks are regrouped as tens blocks, I only need 120 blocks to model 1200.

A. If the tens blocks are regrouped as hundreds blocks, how many blocks will there be? Make the new model.

B. 10 hundreds blocks can be regrouped as a thousands block. Regroup the hundreds blocks. How many blocks are there now?

Reflecting

1. a) Which model uses the least number of blocks?

b) How is this model related to the digits in a place value chart?

Thousands	Hundreds	Tens	Ones

2. How is building a thousands block from hundreds blocks the same as building a hundreds block from tens blocks?

Checking

3. 3 spiders laid a total of 2065 eggs.
   a) Model 2065 with hundreds, tens, and ones blocks.
   b) Model 2065 using the least number of blocks possible. You can use thousands, hundreds, tens, and ones blocks.
   c) Did you need all 4 types of blocks in part b)? Explain why or why not.

Practising

4. A monarch butterfly migrated 3256 km. Find 3 ways to model 3256. Draw each model.

5. Write the number.
   a) 
   b) 

6. If you used only 1 type of block to model each number, how many hundreds blocks would you need? How many thousands blocks would you need?
   a) 2000 b) 5000 c) 9000

7. A fast-food restaurant has a sign that reads 4562 hamburgers sold. Each week for the next 5 weeks, the restaurant sells 100 more hamburgers.
   a) Model 4562 using the least number of blocks.
   b) Add blocks to your model to include the additional 100 hamburgers sold each week. Regroup blocks so the model uses the least number of blocks.
   c) What should the sign read at the end of 5 weeks?
   d) Which blocks changed in your model? Why?

2 Expanded Form

Goal **Write numbers up to 10 000 in expanded form.**

A professional baseball player played 3562 games in his career.

? In what ways can 3562 be represented?

You will need

- base ten blocks

- a place value chart

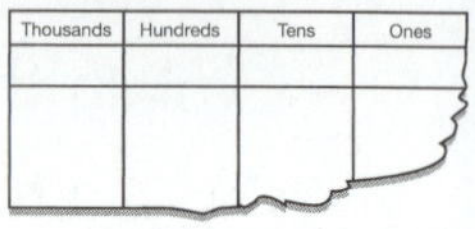

Shani's Ways

I can model 3562 using base ten blocks in a place value chart.

Thousands	Hundreds	Tens	Ones
3	5	6	2

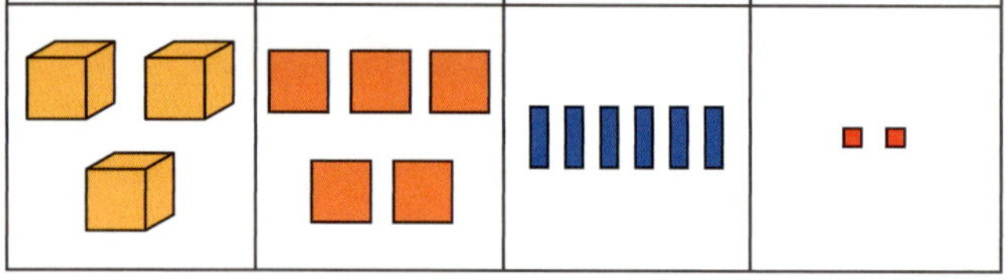

I can represent 3562 as
3 thousands, 5 hundreds, 6 tens, 2 ones.

I can read 3562 using the number words three thousand five hundred sixty-two.

When I write the number as 3562, it is in **standard form**.

I can also show a number in **expanded form**.
3562 in expanded form using words is
3 thousands + 5 hundreds + 6 tens + 2 ones.

3562 in expanded form using numerals is
3000 + 500 + 60 + 2.

standard form
The usual way we write numbers

expanded form
A way to write a number that shows the value of each digit

Reflecting

1. When writing in expanded form, why is 3000 + 500 + 60 + 2 better than 500 + 3000 + 2 + 60?
2. How are the number words for 659 and for the number that is 100 greater than 659 the same?
3. Why are numbers usually written in standard form rather than expanded form?

Checking

4. Pele of Brazil scored 1281 goals as a professional soccer player.
   a) Write 1281 in a place value chart.
   b) Write 1281 as ■ thousands, ■ hundreds, ■ tens, ■ ones.
   c) How would you read this number?
   d) Write 1281 in expanded form using numerals.
   e) Write the standard form and then the number words for the number that is 1000 less than 1281.

Practising

5. Write each number in standard form and as number words.
   a) 10 greater than 483
   b) 2001 less than 2701
6. Rob Peterson holds a record with 8000 + 10 + 7 good tennis serves in a row.
   a) Write this number in expanded form using words.
   b) Write this number in standard form.

7. Write in expanded form using a place value chart, numerals, and words.
   a) 7845
   b) 8050
   c) 6066

3 Comparing and Ordering Numbers

You will need

- base ten blocks

- a place value chart

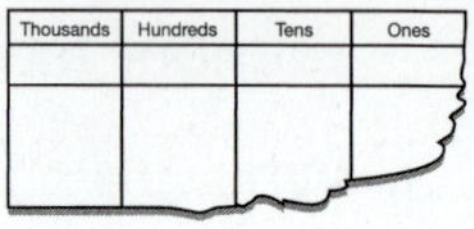

Goal Compare and order numbers up to 10 000.

In a Tour de France bicycle race, the cyclists rode 1368 km of mountain terrain and 1982 km of flat terrain.

? Which distance is longer, the mountain terrain or the flat terrain?

Chris's Solution

I record the digits in place value charts.
Then I model the numbers with base ten blocks.

Mountain Terrain			
Thousands	Hundreds	Tens	Ones
1	3	6	8
	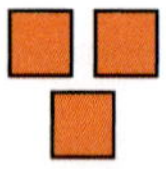	?	?

Flat Terrain			
Thousands	Hundreds	Tens	Ones
1	9	8	2
	?	?	?

A. Finish modelling the numbers.

B. Compare the numbers to find which distance is longer.

C. When comparing numbers, you can use the symbols **<** and **>**.
Show how the numbers compare by replacing ■ with <, =, or >.

1368 ■ 1982 1982 ■ 1368

< and >
The symbol > means the 1st number is greater than the 2nd number.
The symbol < means the 1st number is less than the 2nd number.

8 > 5 5 < 8

Reflecting

1. Write a number between 1368 and 1982. What digits can be in the thousands place and hundreds place?
2. When you compare numbers like 1368 and 1982, you look at each number from left to right. Explain what you do when you get to the 1st place value with a different digit.

Checking

Dogsleds cover 1795 km in Arctic race

Cyclists ride 3350 km in Tour de France

3. Look at the 2 headlines. Use < or > to show which race has the greater distance. Explain how you know.
4. Use the digits 3, 3, 5, 0 to make 4 new distances. Order the distances from shortest to longest.

Practising

5. Ashley lives in Vancouver. She made a table showing the distances between 6 Canadian cities and Vancouver.
   a) Which city is closest to Vancouver? Explain.
   b) Which city is farthest from Vancouver? Explain.
   c) Which cities are between 3000 km and 6000 km away from Vancouver?
   d) Rewrite Ashley's table with the cities in order from farthest from Vancouver to closest to Vancouver.

Distances from Vancouver

City	Distance (km)
Whitehorse, Yukon	2700
Toronto, Ontario	4500
Montreal, Quebec	4800
Moncton, New Brunswick	5825
Calgary, Alberta	1050
St John's, Newfoundland	7675

6. Complete by using >, =, or <. Explain your thinking.
   a) 9981 ■ 654 b) 6772 ■ 7276 c) 2365 ■ 7942

CHAPTER 2

4 Exploring 10 000

Goal Explore place value patterns to 10 000.

You will need
- base ten blocks

Josef's Place Value Patterns

I made place value puzzles using number patterns. Each pattern has 4 numbers in it.

Pattern 3 starts with 10 ones blocks. Replace all of the ones with tens. Continue replacing blocks with the next bigger size block.

Pattern 1 starts with 9 thousands + 9 hundreds + 7 tens. The number of tens blocks increases by 1 for each number.

Pattern 2 starts with 9 thousands + 9 hundreds + 9 tens + 1 ones. The number of ones blocks increases by 3 for each number.

Pattern 4 starts with 4 thousands blocks. The number of thousands blocks increases by 2 for each number.

? How are Josef's patterns alike?

A. Use base ten blocks to model the 4 patterns.

B. What do you notice about the 4th number in each pattern?

C. Create an increasing number pattern with the same 4th number. Write a description for your pattern.

Reflecting

1. Draw or describe a base ten block model for the 4th number in Josef's patterns.

Mental Math

Adding Tens, Hundreds, and Thousands

Natalie and Manitok played Spin to Win.
Natalie spun 3000 and 5000.
Manitok spun 4000 and 2000.

? Who wins the point?

Natalie's Solution

I know 3 + 5 = 8, so I can add 3000 + 5000 in my head.

3	3000	3 thousand
+ 5	+ 5000	+ 5 thousand
8	8000	8 thousand

Spin to Win

How to play:

Step 1 Spin twice.

Step 2 Add your spins to get a score.

Step 3 The person with the greatest score wins 1 point.

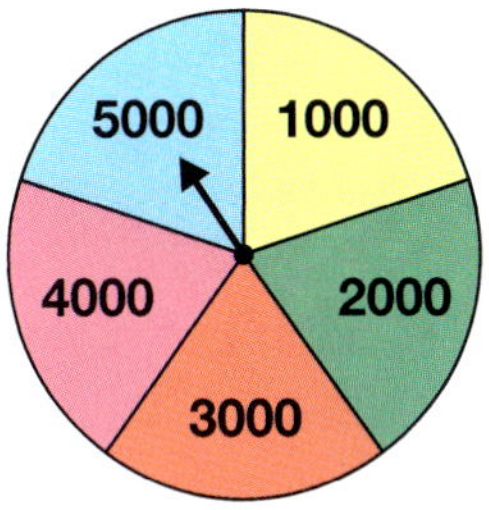

A. How does knowing 3 + 5 = 8 help Natalie calculate her score?

B. What score did Manitok get? Who wins the point?

Try These

1. Use 3 + 2 = 5 to find each sum.
   a) 30 + 20 **b)** 300 + 200 **c)** 3000 + 2000
2. Use 7 − 4 = 3 to find each difference.
   a) 70 − 40 **b)** 700 − 400 **c)** 7000 − 4000
3. Find 4 different scores you can get with 2 spins.
4. What is the best score you can get with 2 spins? Explain how you found the answer.

CHAPTER 2

5 Multiplying by 10, 100, and 1000

You will need

- base ten blocks

Goal **Multiply by 10, 100, and 1000.**

Jon can multiply by 10, 100, and 1000.
He has found a pattern that helps him to multiply.

? What pattern did Jon use to multiply by 10, 100, and 1000?

Jon's Pattern

I made a table to help me see the patterns.

I started by multiplying numbers by 10.

2 × 10 is like making 2 groups of 10.

There are 2 tens and 0 ones, so the value is 20.

Multiplication by 10	Standard form of the number
2 x 10	20
3 x 10	30

A. Choose 4 numbers. Use base ten blocks to multiply each number by 10. Write your results in the table.

B. Look for a pattern in the products.

C. Create a new table for multiplying by 100.
Multiply 4 numbers by 100.
Look for a pattern in the products.

D. Create another table to find a pattern for multiplying by 1000.

Reflecting

1. What pattern do you see when you multiply by each number?
   a) 10   b) 100   c) 1000
2. Why does it make sense that there will be 0 ones when you multiply a number by 10, 100, or 1000?

Checking

3. Use the patterns you found to multiply.
   a) 14×10 b) 61×100 c) 2×1000
4. What is the missing number?
   a) $2500 = ■ \times 100$ b) $2500 = ■ \times 10$

Practising

5. Multiply.
   a) 99×10 c) 1000×10 e) 15×100
   b) 5×1000 d) 10×1000 f) 100×100
6. For 10 years, Darren's parents have made him a photo album for his birthday. Each photo album has 144 photographs. How many photographs does Darren have in total?

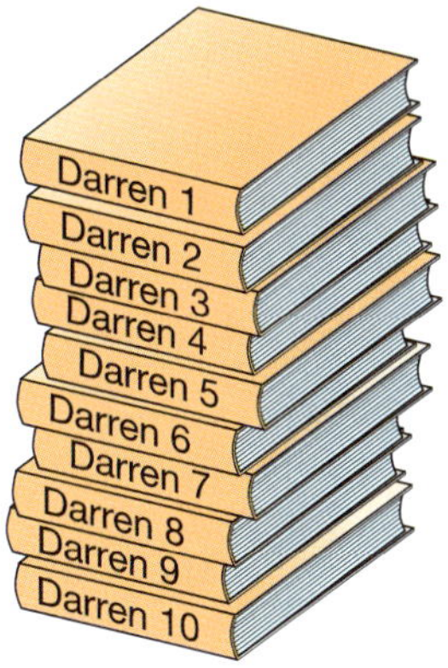

7. What is the missing number?
   a) $5000 = ■ \times 1000$ d) $8000 = ■ \times 100$
   b) $5000 = ■ \times 100$ e) $10\,000 = ■ \times 1000$
   c) $8000 = ■ \times 1000$ f) $10\,000 = ■ \times 100$
8. What is the missing number?
   a) $2090 = ■ \times 10$ c) $5000 = ■ \times 10$
   b) $1100 = ■ \times 10$ d) $4780 = ■ \times 10$
9. Keiko's family bought a used car with 45 one-hundred dollar bills. How much did the car cost?

Mid-Chapter Review

LESSON

1 **1.** A blue whale ate 3750 kg of plankton in 1 day. Model 3750 with the least number of blocks possible. Draw the model.

2 **2.** There are 4809 species of jumping spiders in the world. Write 4809 in expanded form using words and then using numerals.

3. Write each number in standard form.

a) 2000 + 80 + 5

b) 6 thousands + 2 hundreds + 5 tens + 6 ones

4. Use number words to write the number that is 3000 less than 3566.

5. Write the number that is three hundred greater than one thousand five hundred twenty-nine. Use standard form.

3 **6.** Complete each number sentence by using <, =, or >. Explain your thinking.

a) 654 ■ 7843

b) 9823 ■ 9832

c) 5478 ■ 8962

7. This table shows the number of dogs of different breeds owned by Canadians. Order the numbers from least to greatest.

Dog	Number
poodle	2817
Yorkshire terrier	2185
boxer	1444
golden retriever	6047
German shepherd	4576

4 **8.** Complete these number patterns.

a) 5, ■, 500, 5000

b) 8 ones, 8 tens, ■, 8 thousands

c) 4281, 4381, 4481, ■

d) 1, 10, ■, 1000, ■

5 **9.** What is the missing number?

a) 2000 = ■ × 100

b) 2000 = ■ × 10

c) 5600 = ■ × 100

d) 5600 = ■ × 10

e) 10 000 = ■ × 100

f) 10 000 = ■ × 10

Math Game

Getting to 10 000

Number of players: 2 or more
How to play: Get as close as you can to 10 000, but don't go over!

You will need

- a die

- game sheet

Getting to 10 000

Turn	Number rolled

- a calculator

Step 1 Roll a die and record the number rolled on your game sheet.

Step 2 Decide if you will multiply the number rolled by 1, 10, 100, or 1000.
Record your decision on your game sheet.

Step 3 Multiply and then record the product.

Step 4 Use a calculator to add the product to your previous running total. Record the new running total on your game sheet.

After every player has had 10 rolls, the player closest to, but not over, 10 000 wins.

Sarah's Game

On my first 3 turns, I rolled a 5, a 6, and another 5.

Getting to 10 000

Turn	Number rolled	× 1, × 10, × 100, or × 1000	Product	Running total
1	5	x 1000	5000	5000
2	6	x 100	600	5600
3	5	x 100	500	6100

6 Rounding to the Nearest 10, 100, or 1000

Goal **Round numbers to the nearest 10, 100, or 1000.**

Allison, Calvin, and Pedro report on school sports for the local newspaper.
2943 spectators came to a track and field meet.
They decide to round the number of spectators for the headline in the newspaper.

? How can Allison, Calvin, and Pedro round the number 2943?

Allison's Rounding

I'll round 2943 to the nearest thousand.

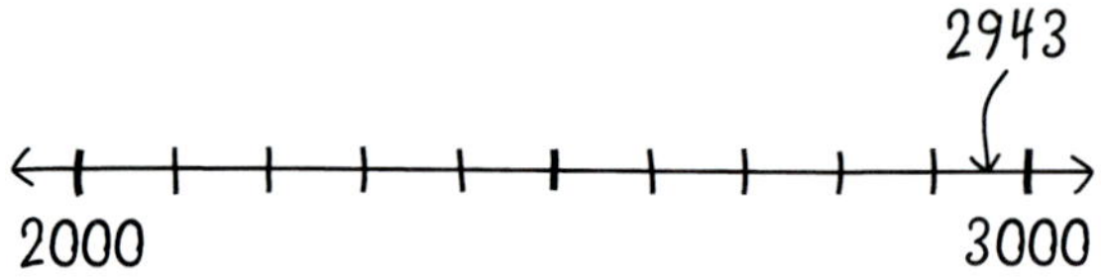

2943 is between 2000 and 3000.

It is closer to 3000,
so 2943 rounds to 3000.

Calvin's Rounding

I'll round 2943 to the nearest hundred.

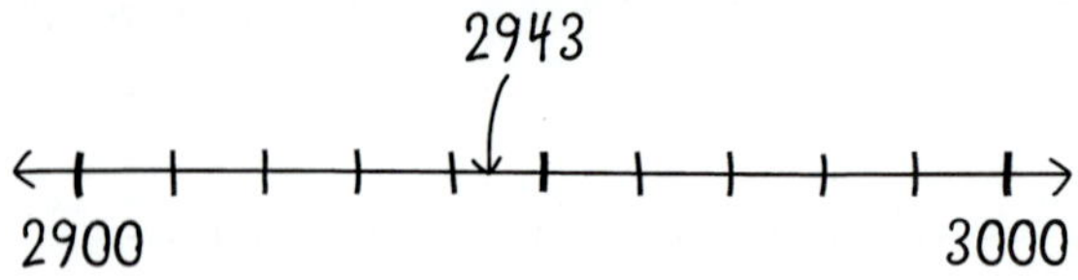

2943 is between 2900 and 3000.

It is closer to 2900,
so 2943 rounds to 2900.

Pedro's Rounding

I'll round 2943 to the nearest ten.

2943 is between 2940 and 2950.

It is closer to 2940,
so 2943 rounds to 2940.

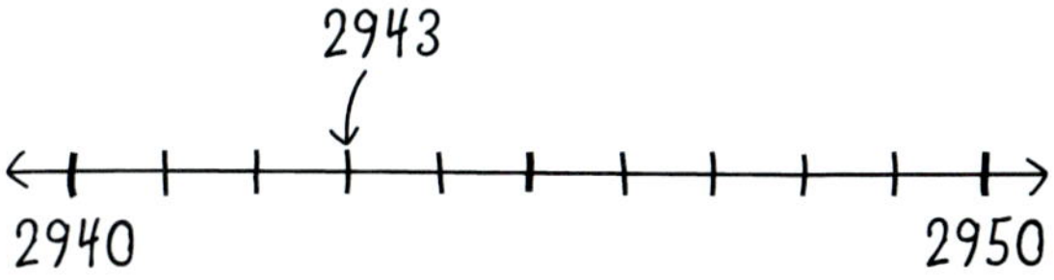

2940 spectators attend track meet

Reflecting

1. Which of the 3 rounded numbers do you think is best for the newspaper headline? Why?
2. In a 4-digit number like 1729, how is rounding to the nearest thousand different from rounding to the nearest hundred?

Checking

3. There were 4365 spectators at a provincial softball championship. Round 4365 to the nearest thousand, the nearest hundred, and the nearest ten.

Practising

4. a) Draw a number line to round the population of Drumheller to the nearest thousand.
   b) What community has about 3000 people?
   c) Explain why the populations of La Ronge, Jasper, and Humboldt are all about 5000 when rounded to the nearest thousand.

Place	Population
Drumheller	7833
Cowley	273
Inuvik	3451
La Ronge	4519
Jasper	4691
Humboldt	5163

5. Round the population of Cowley. Write the number words.
   a) Round to the nearest hundred.
   b) Round to the nearest ten.

7 Communicate About Ordering Numbers

Goal Explain how to order a set of numbers in a complete, clear, and organized way.

Miki organized a computer-game contest for her sister's friends.
When she awarded 1st, 2nd, 3rd, and 4th prizes, the players asked how she ordered the scores.

Top Scores

909	9009
9909	999

? How can Miki explain the steps she followed to order the numbers?

Miki's Rough Copy

I wrote the numbers with 4 digits first.

I wrote the numbers with 3 digits last.

I compared the digits to put them in order.

Miki's Good Copy

The winning scores were 909, 9009, 999, and 9909.

I wrote the numbers with 4 digits first.

Then I wrote the numbers with 3 digits, because 4-digit numbers are greater than 3-digit numbers.

9009 9909 909 999

I compared the hundreds digits in the 4-digit numbers to see which number was greater.

9909 9009 909 999

I compared the tens digits in the 3-digit numbers to see which number was greater.

9909 9009 999 909

Reflecting

1. Find and describe differences between Miki's rough copy and her good copy. Use the Communication Checklist.

Communication Checklist

- ☑ Did you show all the steps?
- ☑ Did you explain your thinking?

Checking

2. Rami was asked to put these game scores in order from greatest to least.
   865 1876 1540 86 1000
   Here is the rough copy explaining his steps.
   Use the Communication Checklist to write a good copy.

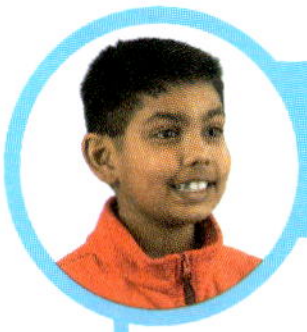

Rami's Rough Copy

I wrote 1000 in the middle.

I wrote the 4-digit numbers first.

I wrote the numbers less than 1000 last.

1876 1540 1000 86 865

They were all in order except 86 and 865, so I switched them.

1876 1540 1000 865 86

Practising

3. a) Order these numbers from greatest to least.
   3867 3869 392 473 450
   b) Write an explanation of how you did the ordering. This is your rough work.
   c) Use the Communication Checklist to find ways to improve your rough copy. Then write a good copy.

8 Carroll Diagrams

Goal **Sort and classify numbers using Carroll diagrams.**

Vinh is using a **Carroll diagram** to sort numbers by 2 attributes.

Carroll diagram
A diagram that uses rows and columns to show relationships

Vinh's Carroll Diagram

I wrote the attributes on the left and across the top of the Carroll diagram.

1 row has numbers that are less than 5000.

The other row has numbers that are equal to or greater than 5000.

1 column has numbers that can be modelled with hundreds blocks only.

The other column has numbers that cannot be modelled with hundreds blocks only.

	Can be modelled with hundreds blocks only	Cannot be modelled with hundreds blocks only
Less than 5000	3200	
Equal to or greater than 5000		7050

? How can Vinh sort these numbers in his Carroll diagram?

2782 6444 10 000 6400 90 8930 907 1000 400 5000

A. Finish sorting Vinh's numbers in the Carroll diagram. Explain how you know where to put each number.

B. Name 2 other numbers for each part of Vinh's Carroll diagram. Explain your thinking.

Reflecting

1. Describe how the numbers in each part of Vinh's Carroll diagram show the attributes Vinh used for sorting.
2. Is there any number that does not fit in Vinh's Carroll diagram? Explain how you know.

Checking

3. a) Sort the numbers in the Carroll diagram.
   5020, 51, 6000, 702, 90, 8599,
   8, 310, 643, 4100

   b) Name 2 more numbers for each of the 4 parts of the Carroll diagram.

	Has the digit 0	Does not have the digit 0
Fewer than 4 digits		
4 digits		

Practising

4. a) Sort the numbers in the Carroll diagram.
   5624, 8360, 7565, 1721, 6402, 4199,
   999, 9087, 3600

   b) Write 2 more numbers in each part of the Carroll diagram.

	Even	Not even
Rounds up to the nearest 1000		
Rounds down to the nearest 1000		

5. a) What labels could you use for the Carroll diagram?
   Explain why your labels make sense.

   b) Write 2 more numbers in each part of the Carroll diagram.

6315 3000	1110 20 5
1704 8676 5888	356 1052

6. a) Make up a sorting rule. Sort the numbers in a Carroll diagram.
   300, 6405, 5300, 9024, 1070, 4221,
   856, 3130

   b) Make up another sorting rule.
   Sort the same numbers in a different Carroll diagram.

9 Estimating the Number of Objects

Goal **Estimate, then count the number of objects.**

? How can you estimate how many objects are in a set?

You will need

- counters such as beads or small cubes

- clear containers of different sizes

A. Take a handful of counters.

a) Estimate how many counters you took.

b) Count the counters.
Is your estimate reasonable? Explain.

c) Repeat parts a) and b).

B. Fill a small container with the same counters you used in part A.

a) Without counting, take out about 20 counters.

b) Count the counters you took.
Is your estimate of 20 counters reasonable? Explain.

c) Put the counters back in the container.
Estimate how many of these counters fill the container.

d) Count to find out.
Is your estimate reasonable? Explain.

C. Use the same counters again.
Fill a container that is larger than the container you used in part B.

a) Estimate how many counters are in this container.

b) Count to find out.
Is your estimate reasonable? Explain.

D. Look around your classroom.
Find a set of objects you can count.

a) Estimate the number of objects in the set.

b) Count the objects. Is your estimate reasonable? Explain.

E. Manitok counted about 220 white beads on this medallion.
Estimate each number of beads on the medallion.

a) red beads

b) green beads

c) all the beads

F. Count the approximate number of beads for each estimate in part E.
Are your estimates reasonable? Explain.

Reflecting

1. a) How might the shape of a container make a difference when you estimate?

   b) How might the sizes of the counters make a difference when you estimate?

   c) What else could make a difference?

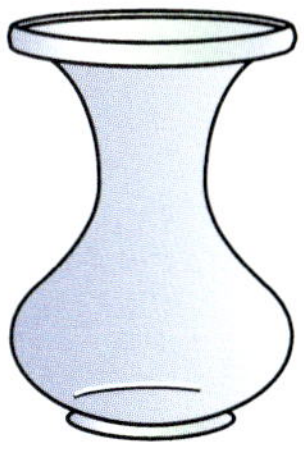

2. Rey and Mandy estimated the number of beads in a jar.
   Rey's estimate was 380.
   Mandy's estimate was 390.
   Then they counted. There are 381 beads.

   a) Mandy's estimate is as reasonable as Rey's. Explain why.

   b) What would be an unreasonable estimate? Explain your thinking.

10 Counting Money Collections

Goal **Estimate, count, and write money amounts up to $50.00.**

You will need

- play money

- a calculator
- a cup

The class is fund-raising for a camping trip.
Paulette records how much money is collected each day.

Day 1	Day 2	Day 3	Day 4
3 five-dollar bills	1 toonie	1 five-dollar bill	1 five-dollar bill
5 loonies	2 quarters	2 toonies	1 toonie
3 quarters	10 dimes	4 quarters	5 loonies
5 dimes	3 nickels	5 dimes	18 nickels
2 nickels	10 pennies	2 nickels	50 pennies
5 pennies		80 pennies	

? How much money has Paulette's class collected?

Paulette's Estimate

First, I will estimate the total for each day.

When I write the estimates, the number before the decimal tells how many dollars.
The number after the decimal tells how many cents.

Day	Estimate
1	$21.00
2	$4.00
3	$11.00
4	$13.00

A. Find the actual amount for each day.
Compare each amount with Paulette's estimate for the day.

B. Use a calculator to find the total.

Reflecting

1. Explain how you could estimate the total for each day to get the same estimates as Paulette.
2. How can listing the amounts for each day from greatest to least make estimating easier?

Checking

3. On day 5 of fund-raising, Paulette's class collected this amount.
   a) Estimate how much money the class raised.
   b) Calculate the actual amount for day 5.
4. Use bills and coins to make $10.00 in 3 different ways.

Practising

5. Estimate each total. Calculate the actual total.
   a) 2 twenty-dollar bills, 1 quarter, 2 dimes, 4 nickels
   b) 1 five-dollar bill, 2 toonies, 1 loonie, 5 pennies
   c) 4 ten-dollar bills, 4 loonies, 4 quarters, 4 pennies
6. Jeff has 25 quarters, 10 dimes, 5 nickels, and 1 penny in his piggy bank. How much money does Jeff have?
7. Use play money. Fill a cup with pennies.
   a) Estimate the number of pennies in the cup.
   b) Count the pennies. Record the value.
   c) Explain how knowing the value of the pennies in a cup could help you estimate the value of money in a large jar of pennies.
8. a) Repeat Question 7 with nickels.
   b) Repeat Question 7 with loonies.

LESSON

Skills Bank

1

1. Write the number.

a) 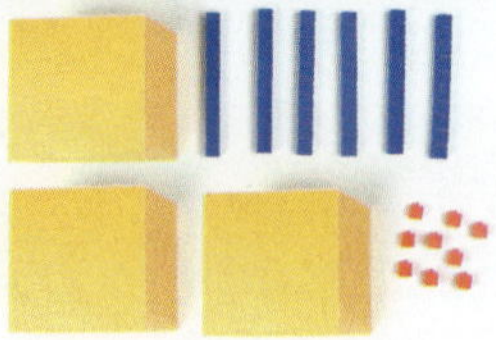

c) 

b) 

d) 

2. Model each number using the least number of blocks. Draw each model.

a) 1873 c) 6037 e) 4000 g) 621
b) 3604 d) 2080 f) 1004 h) 9999

3. Suppose you used only hundreds blocks to model each number. How many blocks would you need for each number?

a) 1000 b) 6000 c) 7500 d) 8300

2

4. Write each number in expanded form using numerals and words.

a) 9803 c) 1030 e) 7777 g) 573
b) 7007 d) 9999 f) 9876 h) 7500

5. Write each number in standard form.

a) one thousand six hundred fifteen
b) 7 thousands, 8 hundreds, 3 ones
c) 8000 + 300 + 50 + 4
d) 5000 + 200 + 7

6. Write the number words.

a) 1010 less than 1634 b) 101 greater than 872

3 7. Complete each using <, =, or >.

a) 986 ■ 953
b) 2234 ■ 2432
c) 7629 ■ 983
d) 10 000 ■ 1000
e) 9909 ■ 9990
f) 7685 ■ 7658
g) 559 ■ 5590
h) 1342 ■ 1351
i) 3980 ■ 4995

8. Order from least to greatest.

a) 8561, 7982, 8642, 693
b) 9805, 3248, 653, 3379, 3241
c) 7982, 7984, 7992, 7899
d) 543, 5672, 9870, 5070, 9930

9. Use each set of digits to write four 4-digit numbers. Order the numbers from least to greatest.

a) 7, 9, 9, 0
b) 2, 4, 4, 8
c) 1, 3, 5, 0

10. The table shows the number of people in Victoria who speak each language as their first language.

a) Which language is spoken by the greatest number of people?
b) Which language is spoken by the least number of people?
c) Write the numbers from greatest to least.

First language	Number of people
Dutch	2245
German	5155
French	4930
Chinese	6900
Portuguese	1575
Punjabi	2395

4 11. Complete each number pattern.

a) 7, ■, 700, 7000
b) 4997, 4998, 4999, ■, ■
c) 8719, 8723, 8727, ■, ■
d) 4682, 5682, 6682, ■, ■
e) 683, 783, 883, ■, ■
f) 8970, 8980, 8990, ■, ■

5 12. Find each product.

a) 12×100
b) 25×100
c) 6×100
d) 8×1000
e) 3×1000
f) 4×1000
g) 10×10
h) 10×100

6 13. Round each number to the nearest thousand.

a) 8245
b) 9079
c) 789
d) 6378
e) 3333
f) 7690
g) 2954
h) 6193

14. Round each number in Question 13 to the nearest ten.

15. Write the number words for 623 rounded to the nearest hundred.

8 **16. a)** Sort the numbers in the Carroll diagram.
3764, 3099, 4499, 4092, 4500, 3500, 4935, 3498, 4863, 4256

b) Write 2 more numbers in each part of the Carroll diagram.

	Greater than 4000	Not greater than 4000
To the nearest 1000, rounds to 4000		
To the nearest 1000, does not round to 4000		

9 **17.** Take 2 handfuls of counters such as beads or ones blocks.

a) Estimate the number of counters.

b) Count the counters you took.
Is your estimate reasonable?

10 **18.** Estimate each total. Then calculate the actual total.

a)

c)

b)

d)

19. a) Sort these amounts in the Carroll diagram.
\$5.00, \$10.25, \$6.55, \$9.95, \$0.75, \$28.90, \$12.50, \$37.54, \$0.20

b) Write 2 more amounts in each part of the Carroll diagram.

	Greater than \$10.00	Equal to or less than \$10.00
Can be made with quarters only		
Cannot be made with quarters only		

LESSON

Problem Bank

1 **1.** Ravi wants to build a model of 2232.
He has these blocks. Can Ravi build the model?
Use pictures, numbers, and words to explain.

2. A number is represented by 10 hundreds blocks and some tens blocks. What could the number be?

2 **3.** Write the number in standard form that is 100 less than the least 5-digit number.

4. Enter the number seven thousand eighty-four on a calculator.
- a) Change the eight to one.
- b) Change the zero to five.
- c) Change the calculator display to show the first number you entered.

5. Create your own problem like Question 4.

3 **6.** a) Write as many different 4-digit numbers as you can using the digits 7, 9, 7, and 0.
b) Which of your 4-digit numbers is the greatest? Which of your numbers is the least?

6 **7.** 3 towns each have a population that rounds to 8000. Use the digits 2, 3, 7, and 8 to create possible populations for the 3 towns.

9 **8.** a) Estimate the number of holes in this ceiling tile.
b) Predict how many tiles you need for a total of 1000 holes.
c) Count the holes in this tile. Should you change your prediction? Explain why or why not.

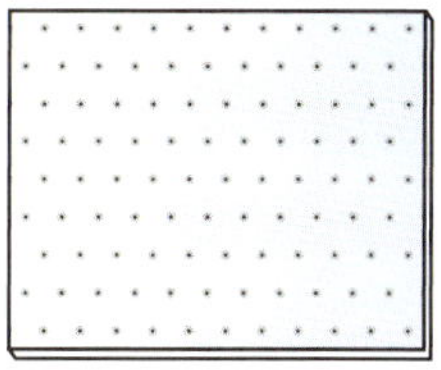

10 **9.** Lily has 2 bills and 8 coins worth $42.60.
What bills and coins could she have?

LESSON

Chapter Review

2 **1.** There are 2230 organic farms in Canada.
- **a)** Model 2230 using the least number of blocks. Draw the blocks.
- **b)** Use number words to write 2230.
- **c)** Write 2230 in expanded form using numerals and then using words.

2. Write each number in standard form.
- **a)** 1000 + 90 + 6
- **b)** 4 thousands + 2 hundreds + 6 ones
- **c)** six thousand one hundred twenty-nine

3. What number is two hundred less than nine thousand one hundred twenty-four? Write the number in standard form.

4. Use number words to write the number that is 300 greater than 649.

3 **5.** Find 3 ways to make this number sentence true. ■295 > 15■4
Explain your thinking.

6. This table lists the number of performances for some Broadway shows.
- **a)** Which shows had between 2000 and 4000 performances?
- **b)** List the shows in order from the least number of performances to the greatest number of performances.

Broadway show	Number of performances
Chicago	1891
Grease	3388
Cats	7485
Beauty and the Beast	2887
The Phantom of the Opera	5566

4 **7.** Extend each pattern.
- **a)** 2, 20, 200, ■
- **b)** 5481, 6481, 7481, ■
- **c)** 9920, 9940, 9960, ■

5 **8.** What is the missing number?
- **a)** 6000 = ■ × 100
- **b)** 4700 = ■ × 100
- **c)** 9900 = ■ × 100

6 9. In one year there were six thousand four hundred seventy-eight earthquakes worldwide. What is the number of earthquakes rounded to the nearest thousand? Use standard form.

10. A giant squid weighed 2946 kg. How much did it weigh to the nearest thousand kilograms, nearest hundred kilograms, and nearest ten kilograms?

11. What is 895 rounded to the nearest hundred? Write the number words.

8 12. a) Make up a sorting rule. Sort the numbers in a Carroll diagram. 682, 5012, 1940, 963, 400, 9, 6229, 815, 56, 9312, 3020
   b) Write 2 more numbers in each part of the Carroll diagram.

9 13. A 14 g bag of raisins contains 40 raisins. How many raisins would you expect to find in a 42 g box? Explain your thinking.

14. Look for a set of small objects that are all the same and that you can count.
   a) Estimate the number of objects. Explain how you estimated.
   b) Then count. Is your estimate reasonable? Explain.
   c) How does having all the objects the same help you estimate?

10 15. Estimate how much money is shown. Then calculate the actual total.

a)

b)

Chapter Task

Creating a Puzzle

Carmen has created a population puzzle. She created 6 make-believe towns.

- All of the towns have 4-digit populations.
- Each population can be modelled using exactly 16 base ten blocks.

? What is the population of each town?

A. Use Carmen's clues to figure out the population of each town. Explain your thinking for each answer.

Everytown
The population is made using the same number of each type of block.

Centuria
The population is made using only hundreds blocks.

Littleville
The population is the least number you can make.

Roundtree
The population rounds down to 8000 and is made using only thousands and tens blocks.

Middletown
The population is between 1200 and 1300 and is made using only hundreds and tens blocks.

Hugo
The population is the greatest number you can make less than 10000.

B. List the populations of the towns in order. Explain how you ordered the populations.

Task Checklist

- ✓ Did you explain how you figured out each population?
- ✓ Did you show all your steps when you explained your order?

CHAPTER 3

Data Management

Goals

You will be able to

- collect and organize data
- make pictographs and bar graphs by hand
- make graphs using a computer
- read and interpret tables, charts, and graphs
- explain how people use data

Planning a playground

Getting Started

Graphing Treasures of the Sea

You will need

- a sorting chart

Category	Number of objects

- grid paper

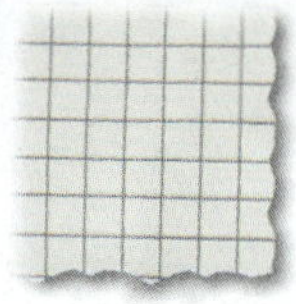

? How can you make a graph to represent the objects in the picture?

A. List categories you can use to sort the objects in the picture.

B. Make a **tally chart** to show how many objects there are for each category.

C. Use the **data** in your tally chart to make a **bar graph**. Number the **axis** by 2s or 5s.

D. Which bar or bars on your graph don't end exactly on a line? Explain how you decided where to end each bar.

E. Can you use your graph to find the total number of objects in the picture? Explain your answer.

Do You Remember?

1. Write 2 things you can tell from each data display.

a) bar graph

b) Venn diagram

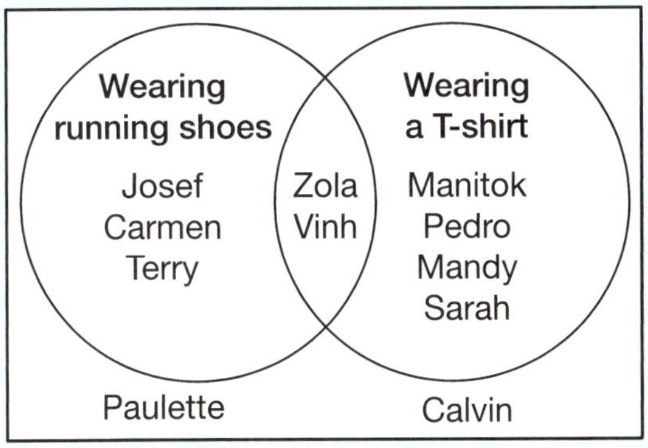

c) chart

How Many Pets Do You Have?

Name	Pets
Carmen	5
Manitok	2
Vinh	1
Sarah	7
Zola	0
Terry	2

d) pictograph

1 Constructing a Pictograph

Goal **Construct and interpret pictographs.**

You will need

- a pictograph frame

Title:

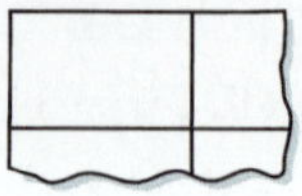

- graphing software
- a computer

Gotta Find a Footprint

Gotta find a footprint, a bone, or a tooth
To grab yourself a piece of dinosaur truth.
All the dino knowledge that we've ever known
Comes from a footprint, a tooth, or a bone.

JEFF MOSS

Pedro's Data

I asked people which dinosaur they like best. I will make a pictograph to display the data I collected.

Which Dinosaur Do You Like Best?

Dinosaur	Number of people
tyrannosaurus rex	60
triceratops	45
brachiosaurus	25
velociraptor	50
other	20

? How should Pedro display his data in a pictograph?

Pedro's Pictograph

I chose a footprint symbol to represent the number of people who like each dinosaur.

My **scale** is "Each  means 10 people."

I drew 6 footprints for tyrannosaurus rex, but I don't know how many to draw for triceratops.

A. How many symbols should triceratops have? Explain.

B. Is a footprint the best symbol to use or would a bone, a tooth, or another symbol be better? Explain.

C. Make a pictograph to show the dinosaur data. Use a different scale than Pedro. Write the title and show the scale you used.

Reflecting

1. What scale did you use for your pictograph? Explain your thinking.
2. How many symbols did you use for brachiosaurus? Why?
3. What do you need to know about the data before you can choose a scale for a pictograph?

Checking

4. The chart shows some data for you to display in a pictograph.
   a) What symbol will you use to represent the number of visitors? Why?
   b) How many people will 1 symbol represent on your pictograph? Explain your choice of scale.
   c) Make the pictograph.

Visitors to the Dinosaur Show

Day	Number of visitors
Monday	150
Tuesday	300
Wednesday	200
Thursday	225
Friday	450

Practising

5. Dinosaur bones were found at many sites. Use this pictograph to answer the questions.

Number of Dinosaur Bones Found

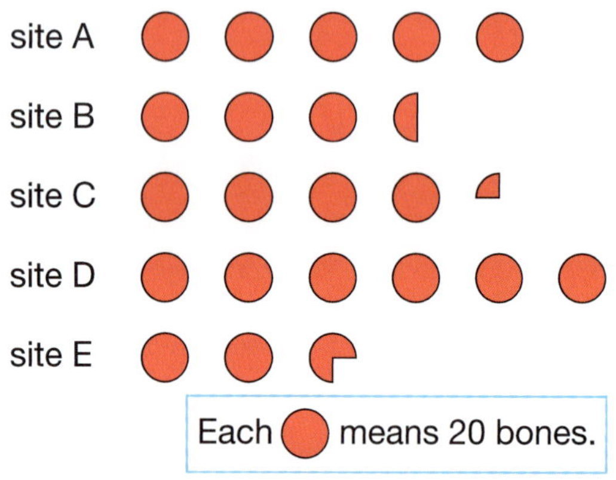

   a) Which site had the greatest number of bones? Which site had the least number of bones?
   b) Make a chart to show how many bones were found at each site.
   c) How many bones were found at site E? Explain how you know.
   d) Why do you think the person who made this graph used a circle symbol instead of a bone?
   e) Make a new pictograph of the data that uses a different scale. Explain your choice of scale.

Dot-Paper Diagrams

You will need
- dot paper

The sides of this rectangle touch 6 dots.
There are 0 dots inside.

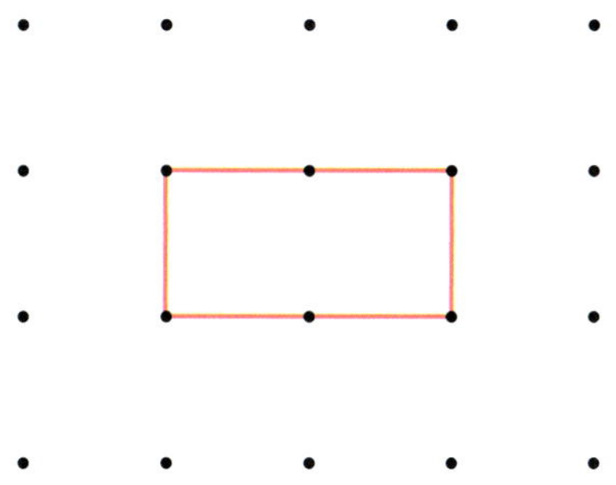

A. What other rectangles can you make with 0 dots inside?

B. What triangle can you make with sides that touch 6 dots and with 0 dots inside?

Try These

1. Draw each shape on dot paper.
   a) The sides of a rectangle touch 6 dots. There are 2 dots inside.
   b) The sides of a square touch 4 dots. There is 1 dot inside.
   c) The sides of a square touch 4 dots. There are 4 dots inside.

2. Find as many shapes as you can where the sides of the shape touch 4 dots and there is 1 dot inside.

2 Choosing a Scale for a Bar Graph

Goal **Explain how to choose a graph and a scale that are appropriate for the data.**

You will need

- 3 dice

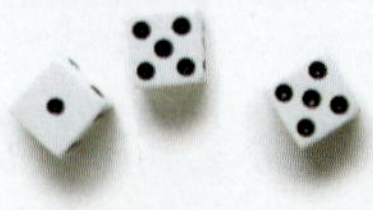

- grid paper

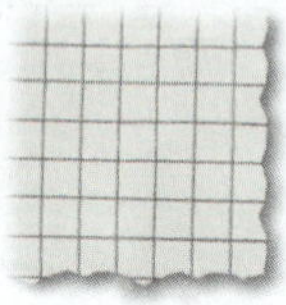

Mandy's Experiment

When I play board games, it seems like somebody rolls a 5 on almost every turn.

To see if this is true, I'm going to roll 3 dice and see if I get a 5 on at least 1 die.

I'll do 50 sets of rolls.

? How can Mandy use an appropriate graph to display the results of her experiment?

A. Try Mandy's experiment.
Make a tally chart to keep track of how many rolls have a 5. Roll 50 times.
Make only 1 tally mark for each roll, even if you roll more than one 5.

B. This bar graph shows Mandy's results.
About how many times did she roll at least one 5?

Did I Roll a 5?

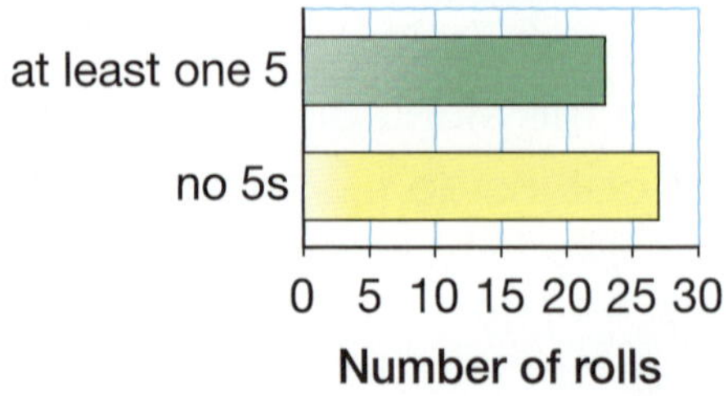

C. Make a bar graph to show the results of your experiment. Use the same scale as Mandy.

D. What does your graph show about how often you rolled a 5 compared to how often you didn't?

E. Make a chart with the combined results for your whole class. Show the results in a bar graph. Use a different scale than Mandy.

Reflecting

1. What scale did you use for the class graph? Why did you use this scale?
2. How did you decide where to graph numbers that were between the scale marks?
3. Mandy's graph shows if there is a relationship between the number of rolls and the number of times she rolled a 5. How does it show this?

Checking

4. Rami did an experiment with 2 dice. On each roll, he added the 2 dice to find the total. He rolled 500 times and graphed the results.
   a) Tell how many rolls each bar could represent.
   b) Why is there more than 1 possible answer for part a)?
   c) If you add the numbers that the bars represent, what should the sum be? Why?

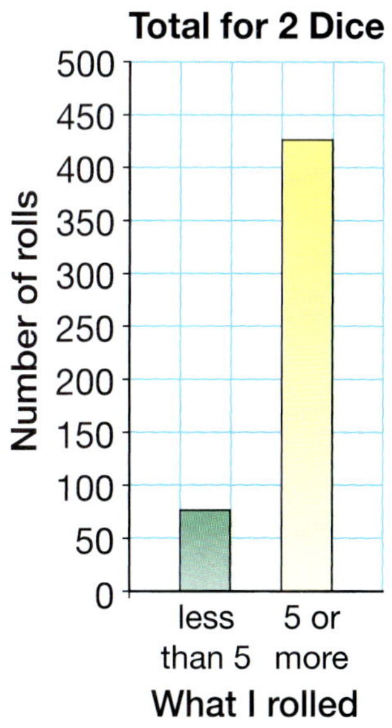

Practising

5. a) Try an experiment like Rami's. Roll 2 dice and tally whether the sum is "less than 5" or "5 or more." Repeat 100 times.
   b) Make a bar graph to show what happened. Choose a scale that is appropriate for the data.
   c) Explain how you chose your scale.
   d) What relationship does your graph show?

3 Collecting Data

Goal **Predict results, and collect and organize data.**

You will need

- paper clips

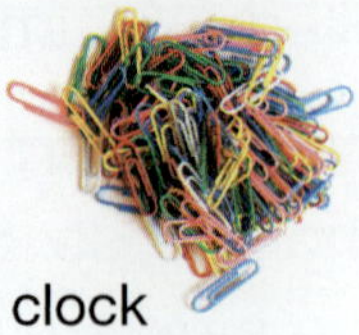

- a clock

? How many paper clips can you link in 2 minutes?

A. Predict how many paper clips you can link in 2 minutes. Record your prediction.

B. Make a paper-clip chain while your partner times 2 minutes.

C. Count and record the number of paper clips in your chain.
Compare the number with your prediction.

D. Now it's your partner's turn to make a chain while you time 2 minutes.

E. Make a chart that shows all the class chains. What is the least number of paper clips in a chain? What is the greatest number of paper clips in a chain?

Person	Number of paper clips in chain
1	
2	
3	
4	
5	
6	

F. How can you organize the numbers in groups to make the data easier to understand?

G. Work with a partner.
Make a table to record the groups.
Show every number in its appropriate group.
Record every number, including those that came up more than once.

Reflecting

1. a) What groups did you use to organize the numbers in your table?
   Explain your thinking.
   b) What other groups make sense?
   Why do they make sense?

2. a) What information is easier to see in the organized numbers than it was in the original chart?
   What makes it easier to see?
   b) What suggestion would you make for organizing the data?
   Explain advantages of your suggestion.

3. a) Create 2 questions about the data in your organized table.
   b) Tell how you would answer each question.
   Explain why your answers are reasonable.

4 Constructing a Bar Graph with Intervals

You will need
- chart of class chains from Lesson 3
- grid paper

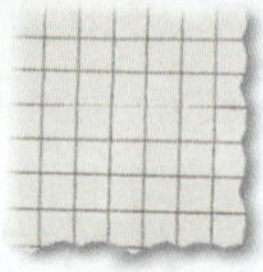

Goal Construct a bar graph using appropriate intervals.

? How can you make a bar graph to show how many paper clips can be linked in 2 minutes?

Natalie's Bar Graph

How Many Paper Clips Can We Link in 2 Minutes?

12	12	14	16	17	17	19	21	22	22	23
24	26	26	28	28	28	30	31	31	33	35
35	36	38	43							

Our class made a list of everybody's paper-clip chain lengths.

A graph with a bar for every number would have 18 bars! That's too many.

I decided to group the numbers in **intervals** so my graph would have only 3 or 4 bars.

The difference between the least and greatest number of paper clips linked is about 30.

An interval of 2 would mean about 15 bars.
An interval of 5 would mean about 6 bars.
I will use an interval of 10.
Each bar shows an interval of 10.

Number of paper clips	Number of people
12–21	8
22–31	12
32–41	5
42–51	1

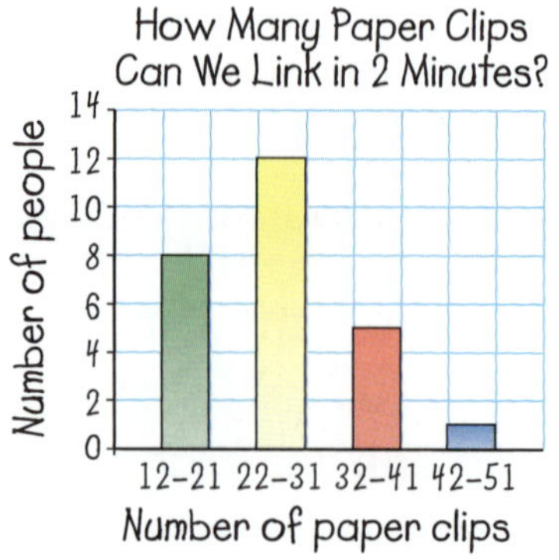

interval
The distance between 2 values on a graph scale. Intervals on a graph should be equal.

12–21 and 22–31 are intervals of 10.

Reflecting

1. Which bar on Natalie's graph includes all the people who linked 28 paper clips? Explain how you know.
2. How did Natalie use the difference between the greatest and the least number of chain lengths to help her make the bar graph?
3. Write a problem about Natalie's bar graph. Solve your problem.

Checking

4. This list shows how many small paper clips some students were able to hang from a big paper clip in 2 minutes.
   a) Choose intervals and organize the data in a table.
   b) Explain how you chose the intervals for your table.

Number of Paper Clips Attached

16	18	19	20	21
21	22	24	24	24
25	26	26	26	27
28	29	29	30	32
32	35	36	36	37

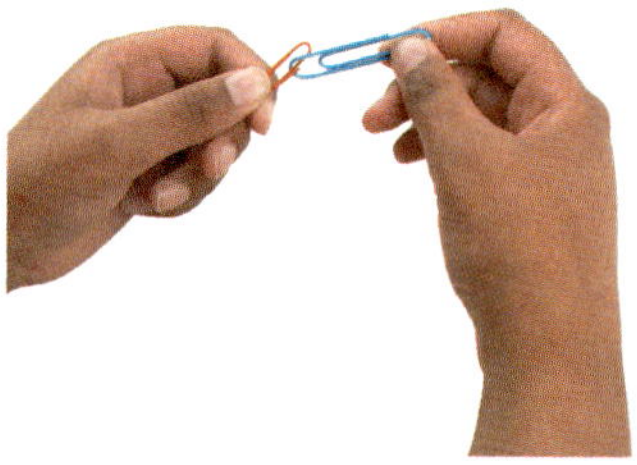

Practising

5. This list shows how many elastics some students were able to tie together in a chain in 4 minutes.
   a) Make a table and a bar graph with appropriate intervals.
   b) Explain how you chose the intervals for your graph.
   c) Explain how you chose the scale for the axis that shows the number of people.

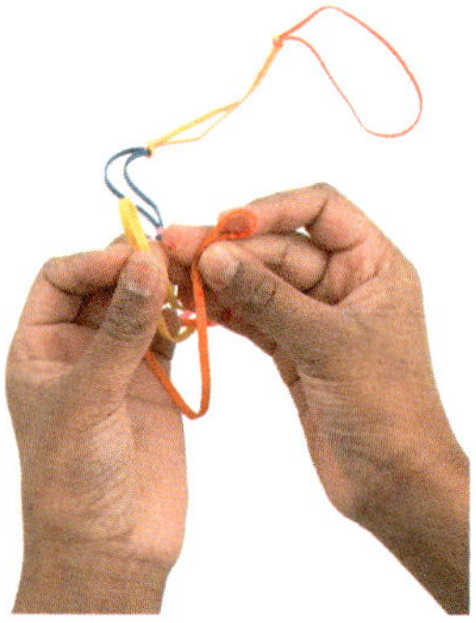

Number of Elastics Tied Together

31	32	32	33	35
36	36	36	37	37
37	39	39	41	41
41	42	42	42	42
43	43	44	44	45
45	45	47	48	48
49	49	49	50	50

5 Reading and Interpreting Graphs

Goal **Read and interpret graphs and identify their features.**

Manitok's Data

I collected these data about favourite lunchtime activities at my school.

Favourite Lunchtime Activities

Activity	Number of students
chess	40
floor hockey	75
volleyball	20
craft club	60
choir	25

These 2 graphs are supposed to show Manitok's data.

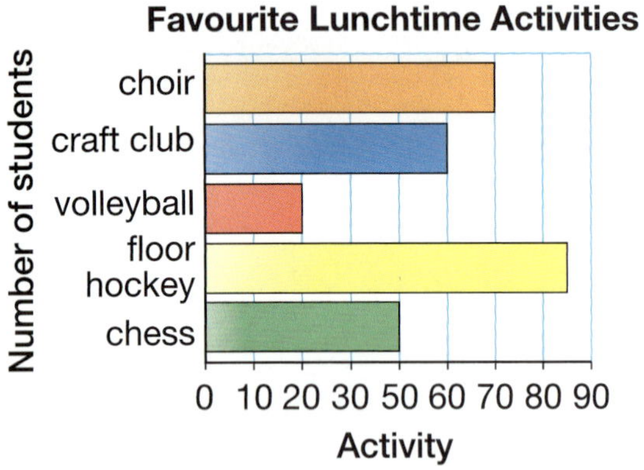

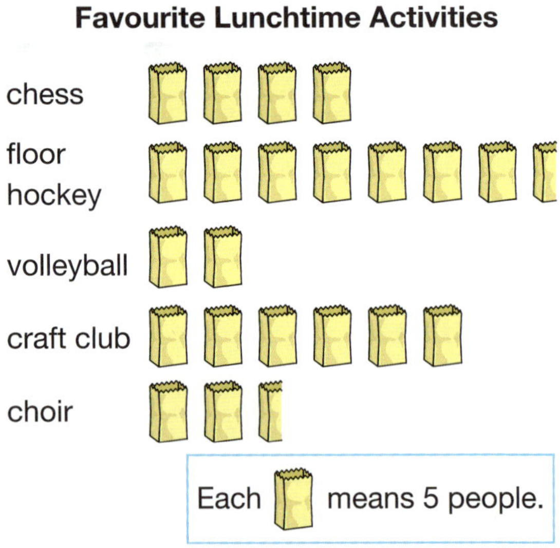

? What mistakes can you find in each graph?

A. Describe the mistakes you found on each graph. Use words such as *title*, *axis*, *label*, *scale*, and **legend**.

B. Describe a different mistake you think someone might make when showing Manitok's data in a graph.

legend
A feature on a graph that explains what the symbols mean

Reflecting

1. What would a correct bar graph show about the students' favourite lunchtime activities?
2. Explain how each type of graph helps you to compare data.
3. How might Manitok use the data he collected?

Checking

4. What mistakes can you find in each graph?

a) **Games Won by Each Floor Hockey Team**

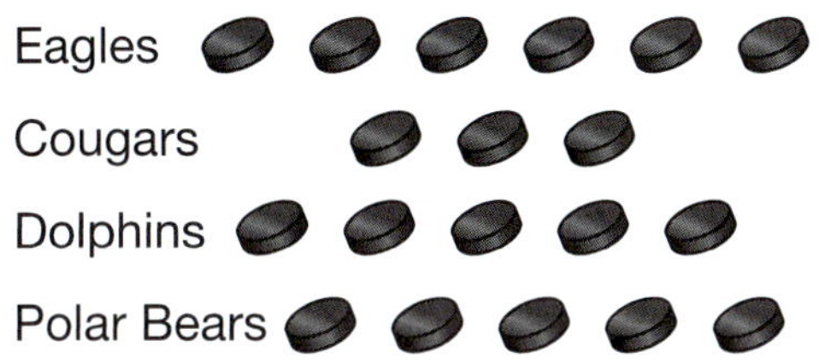

b)

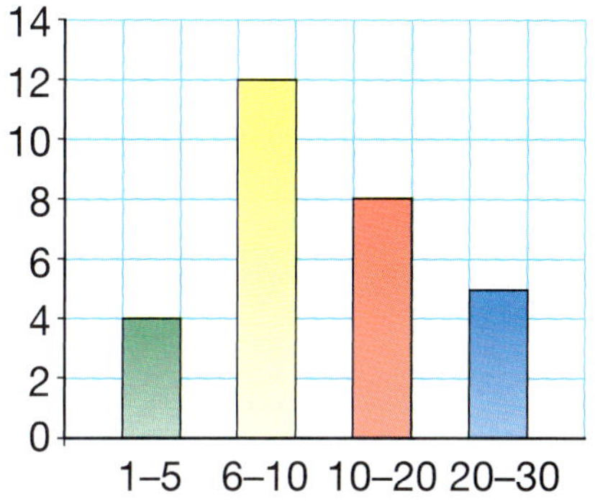

Practising

5. What's wrong with this bar graph?

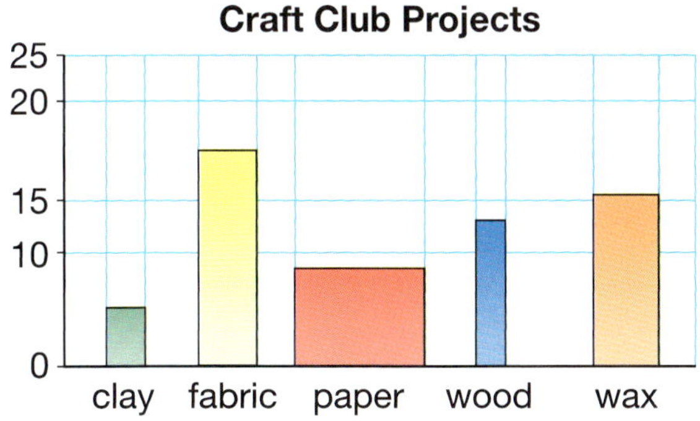

6. These 2 graphs are supposed to show the same information about how many games a floor hockey team has won and lost. Do the graphs show the same information? Explain your thinking.

LESSON

Mid-Chapter Review

1 **1.** The pictograph shows data about what students chose to read in the library.

- **a)** How many students chose comics?
- **b)** How many more students chose books than magazines?
- **c)** Draw a new pictograph of the data. Use a different scale.
- **d)** Explain your choice of scale.

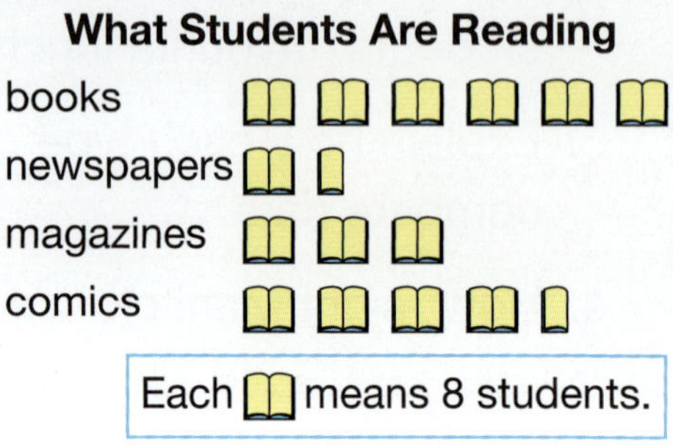

2 **2.** The Sub Shop kept track of sales for 1 week to find out which sandwiches are most popular.

Sandwich	turkey	vegetarian	roast beef	tuna	meatball
Number sold	168	137	206	155	124

- **a)** Make a bar graph to show the data. Remember to include the title and labels for the axes.
- **b)** What scale did you use on the Number sold axis? Why did you choose this scale?
- **c)** Which sandwich is most popular? Which sandwich is least popular?

5 **3. a)** List 3 things the graph shows about students' hot dog orders.

- **b)** Which bar includes students who ordered 18 hot dogs?
- **c)** About how many students ordered hot dogs this year? Explain how you know.

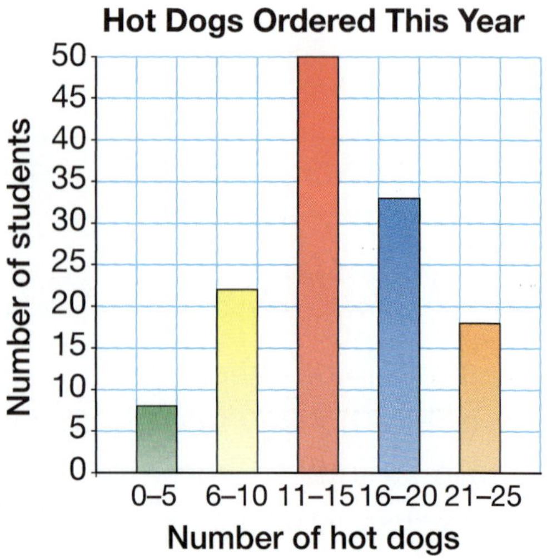

4. Write a question you can answer using the hot dog graph. Write the answer for your question.

CHAPTER 3

6 Graphing with Technology

Goal **Use spreadsheet software to organize and display data.**

You will need

- coloured candies

- spreadsheet software
- a computer

? What colour will you find most often in a package of coloured candies?

A. Predict the colour you will find most often.

B. Open a package of candies.
Sort the candies by colour.
Enter the results into a spreadsheet.

	A	B	C
1	blue	5	
2	brown	12	
3	green	4	
4	orange	5	
5	pink	8	
6	purple	6	
7	red	7	
8	yellow	9	
9			

C. Use the graphing function in your spreadsheet to make a bar graph of your data.

D. Change the data in your spreadsheet by doubling the number of candies for 1 colour.
What happened to the graph?
Explain why this happens.

Reflecting

1. What are some of the advantages of using software to make a graph?
What are some of the disadvantages?

2. Predict what would happen to your graph if you changed the data by halving the number of candies for 1 colour.
Explain the reason for your prediction.

CHAPTER 3

7 Collecting Data with a Sample

Goal **Choose a population and a sample.**

You will need

- small pieces of paper

Ice Castle is choosing 1 of these names for its new popsicle flavour. They need a name that will help sell popsicles. They want opinions of people who buy popsicles.

Ice Castle thinks that most people who buy popsicles are 3 to 17 years old. This is the **population**.

Ice Castle can't **survey** the whole population. They can only survey a **sample** of the population.

population
The whole group

sample
A part of the population or group

? How can Ice Castle choose a sample from the population and conduct a survey?

A. Create an imaginary sample and conduct a survey.

a) Imagine that you are an age from 3 to 5 years. Choose 1 of Ice Castle's names for the new popsicle flavour that someone this age might like. Write it on a small piece of paper. Collect the papers.

b) Repeat part a) for each age group. Use a different paper for each choice.

- 6 to 8 years
- 9 to 11 years
- 12 to 14 years
- 15 to 17 years

B. Calculate the total for each name for the new popsicle flavour.

C. Do you think Ice Castle should make a decision based on the data from your sample? Explain.

Reflecting

1. Why do you think adults were not included in the sample?
2. Suppose these were the results for the choices. What do you think Ice Castle should do? Explain why you think this.

Super Fancy	7
Bubble Chew	5
Blueberry	5
Sour Pop	8

3. Explain why your imaginary group is a sample for the population.

Checking

4. Josef is choosing a movie to show at his school. Explain why each group is not the population for his choice.
   a) people who shop at a nearby store
   b) children at a nursery school
5. Explain why each sample does not represent the population for Question 4.
   a) students on the soccer team at Josef's school
   b) grade 1 students in Josef's school
   c) high school students who went to Josef's school

Practising

6. a) What is the population for Question 4?
   b) How do you think Josef should choose a sample? Explain how this sample represents the population.
7. Choose 1 of these topics or make up your own.
   - Decide what new club to start at your school.
   - Plan the music to play at a school party.

   a) What is the population? Explain why.
   b) How would you choose a sample? Explain why.
8. Explain how a population and a sample from the population are the same. Explain how they are different.

8 Communicate About Collecting Data

You will need

- survey steps

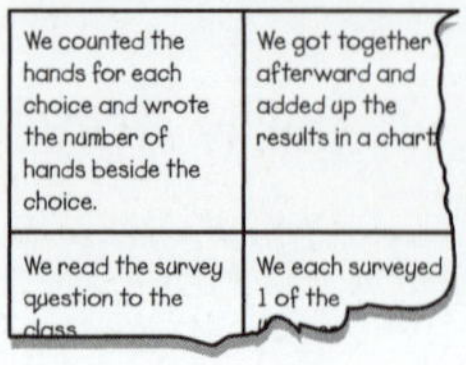

- scissors

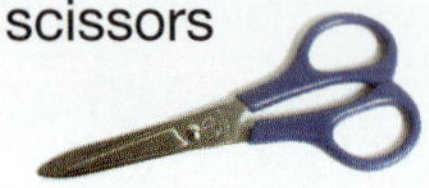

Goal **Describe the steps for collecting data in a clear and organized way.**

Miki's Survey

Our group planned a survey. We surveyed four grade 4 classes to find out about their favourite activities during indoor recess.

We made cards to show the steps we used to collect our data, but the cards got all mixed up.

Survey Question

What is your favourite activity during indoor recess?

- playing board and card games
- playing sports in the gym
- reading in the library
- other

? How can Miki put the steps in order?

We counted the hands for each choice and wrote the number of hands beside the choice.	We got together afterward and added up the results in a chart.	We wrote a survey question with 3 choices and "other."	We read the question again and asked students to raise their hands for their choice.
We read the survey question to the class.	We each surveyed 1 of the 4 classrooms.	My friend Chris picked sports because he likes basketball.	We told the students to raise their hands only once.

A. Which step do you think Miki's group did first? Which step did they do next?

B. Arrange the rest of the steps in order.

C. Read the Communication Checklist.
How would Miki's group answer the 3rd question?

D. Compare your order of the steps with another group's order. Were there any differences? If so, what were the differences?

Communication Checklist

- ✓ Did you show all the steps?
- ✓ Did you put the steps in order?
- ✓ Did you include only necessary information?

Reflecting

1. How did you decide on the order of the steps?

Checking

2. Allison's group collected data about favourite board and card games. Put their steps in order.

We made a survey form for each person. We put our survey question on the form.	We went to each class and gave each person a survey form.	We asked everyone to write an answer on the survey form.	We wrote our survey question, "What is your favourite board or card game?"	We sorted the survey forms to see how many people liked each game.

Practising

3. Sarah's group collected data about favourite sports.
   a) Put their steps in order.
   b) Which step is not necessary? Explain your thinking.

We asked the people who like sports to put a cube in one of 3 pails. We had 1 pail for each sport.	We decided to ask, "Which sport do you like best – basketball, floor hockey, or volleyball?"	The cubes were all the same colour.	We went to each class and asked the people who like sports to raise their hands.	We counted the cubes in each pail to see which sport was the favourite.

9 Conducting a Survey

Goal **Conduct a survey and make a graph to display the data.**

You will need

- grid paper

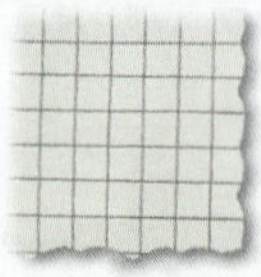

- pencil crayons

- spreadsheet software
- a computer

? How can you conduct a survey and display the data you collected?

A. Make a plan for collecting the data.
- What do you want to find out?
- What is the population for your survey?
- How will you choose a sample?
- Where and when will you do your survey?

B. Create the question you will ask in your survey.

C. What answer do you think you will get most often? Give reasons for your prediction.

D. Conduct your survey. Record your data on a tally chart.

E. Make a graph to show the data you collected. Decide whether you will use a spreadsheet or draw your graph by hand. Explain your choices.

Reflecting

1. How well does your graph help you answer the question for your survey?

2. **a)** Explain why the way you collected data makes sense.

b) Does your sample represent the population? Explain the reason for your answer.

3. Create 2 questions about the data in your graph. Tell how you would answer each question.

Race to the Top

Number of players: 2
How to play: Find the sum of your cards and place a tile on the appropriate bar in the graph.
Try to fill a bar of the graph.

You will need

- Race to the Top bar graph

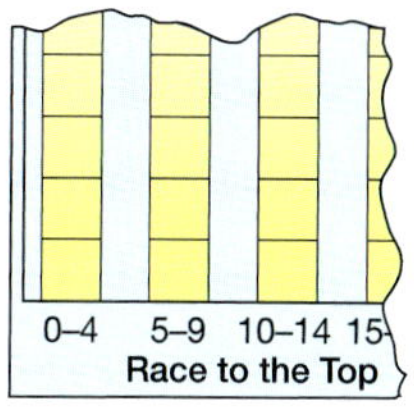

- 4 sets of number cards (1 to 10)

- tiles

Step 1 Shuffle the cards and deal 2 cards to each player. Place the other cards face down in a pile.

Step 2 Find the sum of your 2 cards.

Step 3 The dealer plays first.
Decide whether you will take 1 more card from the pile.
If you do, add the new card to your sum.

Step 4 Put a tile on the appropriate bar to show your sum.

Step 5 Switch dealers and deal again.

The person who fills the last square in any bar wins.

Calvin's Turn

I was dealt 2 and 6.
I drew 9.
My sum is $2 + 6 + 9 = 17$.
I put a tile on the 15–19 bar.

Skills Bank

LESSON

1

1. This pictograph shows the number of students that ride each bus.
   a) Which bus has the most students? How many students ride this bus?
   b) Which bus has the fewest students? How many students ride this bus?
   c) Make a chart to show how many students ride each bus.
   d) Make a new chart to show how many students will be on each bus after these changes:
      - 5 students from bus 3 move to bus 1.
      - 10 students from bus 2 move to bus 4.
   e) Make a new pictograph to show the data from part d). Use a different scale. Explain why your scale makes sense.

Students Who Ride the Bus

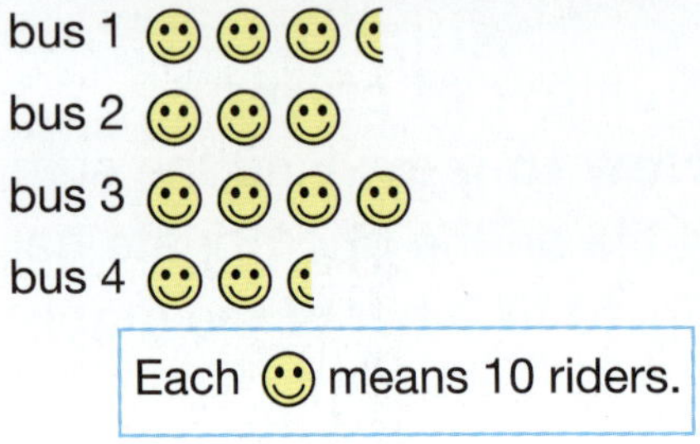

2. a) Write a question about the graph you drew for Question 1.
   b) Answer your question.

2

3. This table shows the cost of some food items at a restaurant in 1950.
   a) Make a bar graph to show the data. Choose an appropriate scale.
   b) Explain why your scale makes sense.
   c) Write 3 things the graph shows about the cost of the foods.

Cost of Some Foods in 1950

Food item	Cost
hamburger	45¢
soup	25¢
milk	15¢
cheese sandwich	40¢
banana split	60¢
apple pie	30¢

4. a) Write a question about your graph for Question 3.
   b) Answer your question.

4

5. This list shows the ages of 48 customers at a restaurant.

a) Organize the data in a table like this.

Customer Ages

2	5	6	7	7	8
11	15	16	16	16	19
20	21	21	21	21	23
24	27	28	29	29	30
31	32	32	35	36	36
37	37	37	38	41	45
46	48	50	51	51	52
54	56	60	62	73	81

Age of customer	Number of customers
1–10	
11–20	
21–30	
31–40	
41–50	
51–60	
61–70	
71–80	
81–90	

b) Make a bar graph to show the data from your table.

c) Write at least 1 question you could answer using your graph from part b). Answer your question.

5

6. Describe the mistakes on each graph.
Use words such as title, axis, label, scale, and legend.

a)

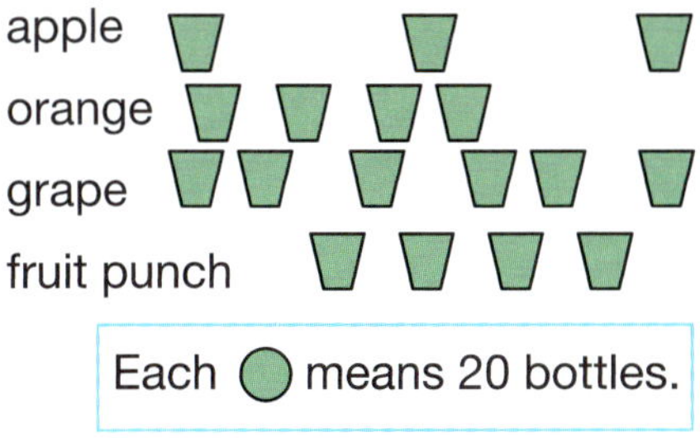

b)

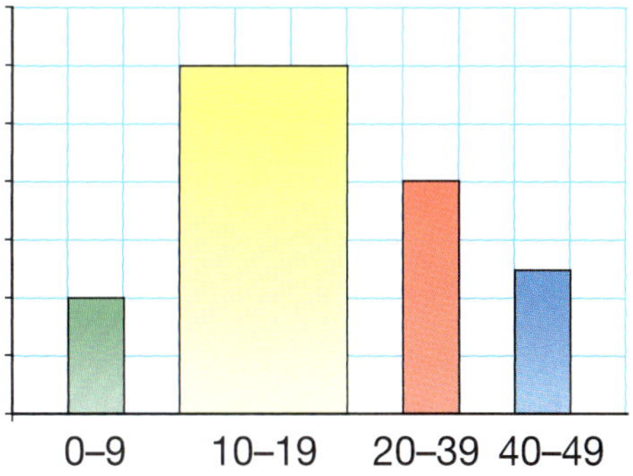

7. Match each feature of a graph with its purpose.

a) title
b) axis
c) scale
d) legend
e) label

V. tells what the symbols mean
W. tells what the graph is about
X. tells the value of each pictograph symbol
Y. tells what each bar represents
Z. where you show the scale on a bar graph

Problem Bank

LESSON

1 **1.** The pictograph shows the number of milk cartons some grade 4 students bought at the school in October.
The students bought 84 cartons altogether.
How many cartons does each  represent?
Explain your answer.

2 **2.** Use the graph to solve the problem.
The temperature in one city was 7°C warmer than in another city.
Which were the 2 cities?
Explain your thinking.

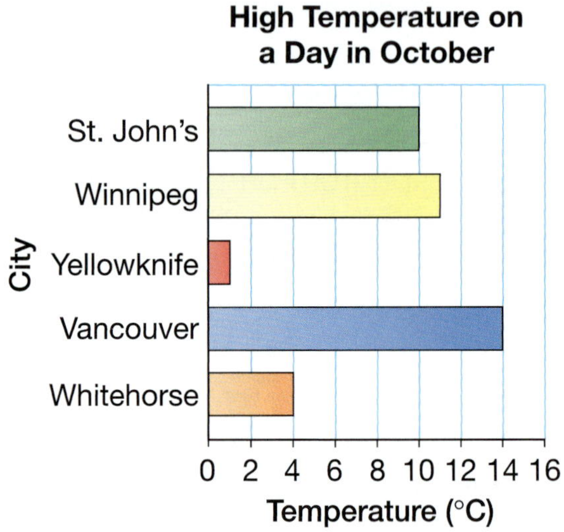

4 **3.** Greg surveyed 50 students to see how many CDs they owned. There were twice as many students who owned between 11 and 15 CDs as students who owned fewer than 6 CDs.
How many students owned fewer than 6 CDs? Explain your thinking.

Number of CDs	Number of students
fewer than 6	
6–10	8
11–15	
more than 15	18

5

4. This graph shows that there are 45 fish in Casey's tank. How many of each kind of fish are in Casey's tank? Explain your thinking.

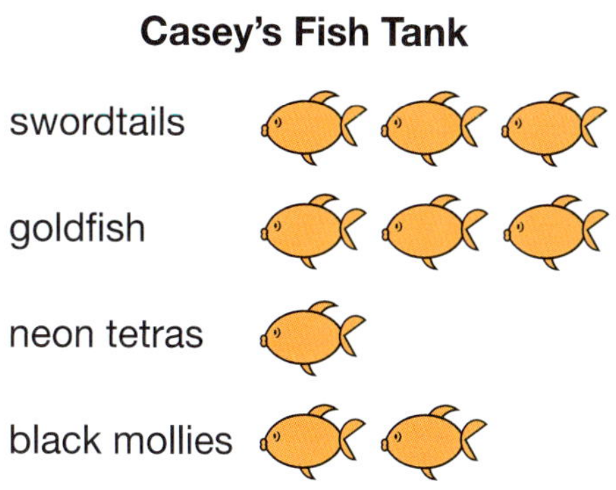

5. a) What could this graph represent? Explain your thinking.
   b) Label the title and axes and show the scale.

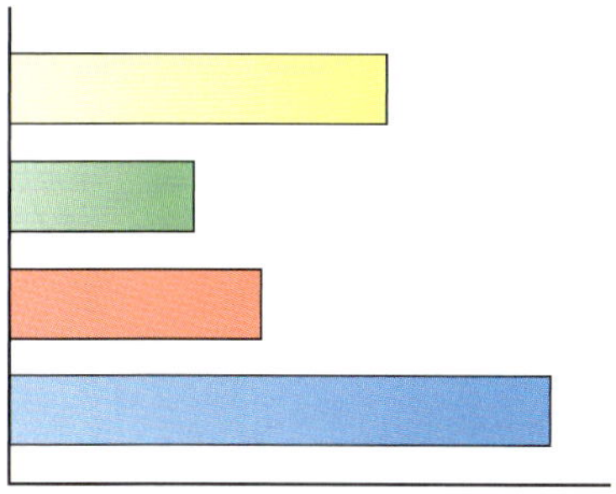

6. Create a problem with addition, subtraction, or both operations that someone could solve using the graph about favourite seasons.
   Write the answer to your problem.
   Show the steps you would use to solve your problem.

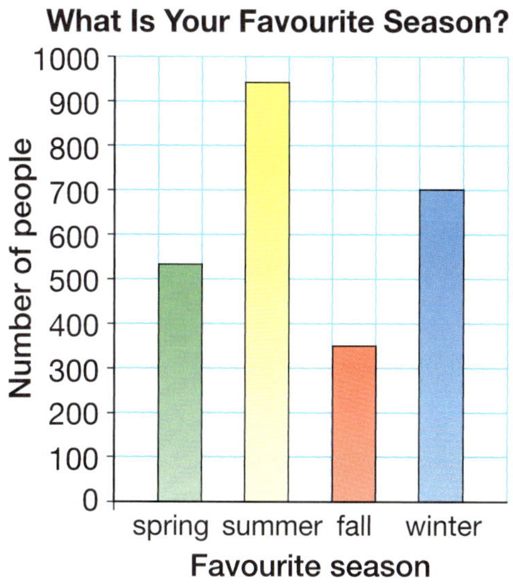

LESSON

Chapter Review

1 **1.** Use the pictograph to answer these questions.

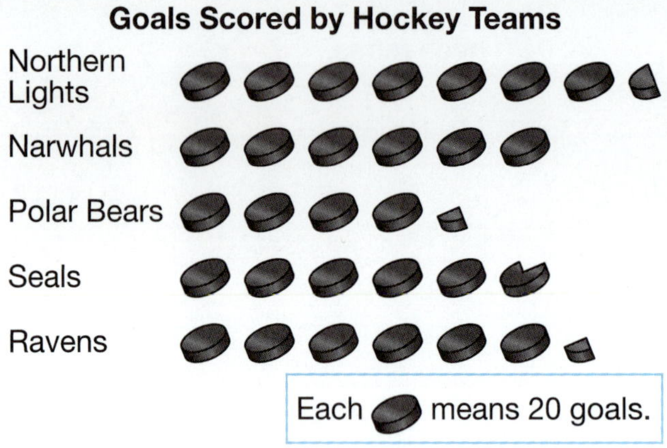

a) Is a hockey puck a good symbol to use for this pictograph? Explain your answer.

b) Is "Each means 20 goals" a good scale to use for this pictograph? What other scales could have been used? Explain your answer.

2 **2.** Use the bar graph to answer these questions.

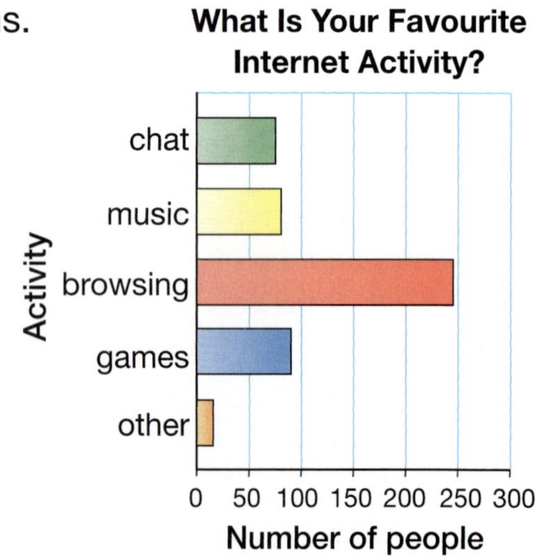

a) About how many more people prefer browsing to games?

b) About how many people in total are represented on the graph?

c) What scale was used to make the graph?

d) What are some of the advantages and disadvantages of this scale? Would another scale be better? Explain your answer.

4 3. These data show how long some students were able to stand on 1 foot.

Seconds on 1 Foot

12 19 20 21 21 21 22 28 29
32 34 36 37 37 37 41 41 42
43 45 46 48 50 51 57 58 59
60 63 64 65 67 70 72 78 80

Seconds	Number of students
1–10	
11–20	
21–30	
31–40	
41–50	
51–60	
61–70	
71–80	

a) Complete the table.
b) Make a bar graph with intervals.

7 4. Josef and Sarah asked students at their school whether they prefer watching movies or cartoons.

Josef asked the first 25 students he saw at school. Then, he said about half the students in the school prefer watching each.

Sarah asked 50 students in grades 3 and 4.

Then, she said all the students in the school prefer watching cartoons.

a) Did Josef or Sarah have a better plan for collecting data? Explain your thinking. Use the words population and sample.
b) How would you collect the data? Explain why your method is reasonable. Use the words population and sample.

9 5. Describe a survey you could conduct to find out how many families bake cookies at home.

a) Write a survey question.
b) What would be the population for the survey?
c) How would you choose a sample?
d) How would you organize the survey information?
e) Explain how well your sample would represent the population.

Chapter Task

Planning a Playground

Lakeview School will buy 4 new pieces of playground equipment. Lakeview students conducted a survey of all the students to find out what they think the school should buy.

? What 4 pieces of playground equipment should the school buy?

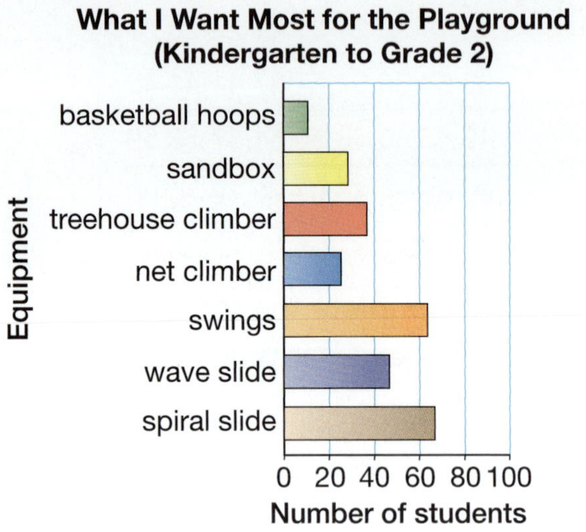

Part 1

Justin saw the results of the survey. He said the school should buy the spiral slide, the wave slide, the swings, and the treehouse climber.

A. Do you think Justin was looking at the graph or the table? Explain.

B. Why do you think the most popular suggestions are different in the graph and in the table?

C. Make a bar graph to show the data in the table. Explain how you chose your scale.

What I Want Most for the Playground (Grades 3 to 6)

Equipment	Number of students
basketball hoops	58
sandbox	2
treehouse climber	34
net climber	15
swings	51
wave slide	16
spiral slide	37

Part 2

D. What 4 pieces of playground equipment do you think the school should buy? Justify your decision.

Task Checklist

- ✓ Did you use math language?
- ✓ Did you explain your thinking?

Cumulative Review

Cross-Strand Multiple Choice

Cars	Wheels
1	4
2	8
3	12
4	16
5	20

1. Which rule describes the number pattern in the Wheels column?

A. start at 4 and multiply by 2 each time
B. start at 4 and multiply by 4 each time
C. start at 4 and add 1 each time
D. start at 4 and add 4 each time

2. Extend this pattern:
2 + 6 − 5, 3 + 7 − 5, 4 + 8 − 5, ■

E. 5 + 9 − 5 **G.** 5 + 9 − 7
F. 5 + 8 − 7 **H.** 5 + 8 − 5

3. Which set of numbers is in order from least to greatest?

A. 3508, 5081, 7125, 7095, 9091
B. 9109, 9108, 9107, 9105, 9103
C. 8236, 8335, 8434, 8533, 8632
D. 3919, 3818, 3717, 3616, 3515

4. Which is the expanded form of 8205?

E. 8200 + 200 + 5 **G.** 8000 + 200 + 5
F. 8000 + 20 + 5 **H.** 8200 + 5

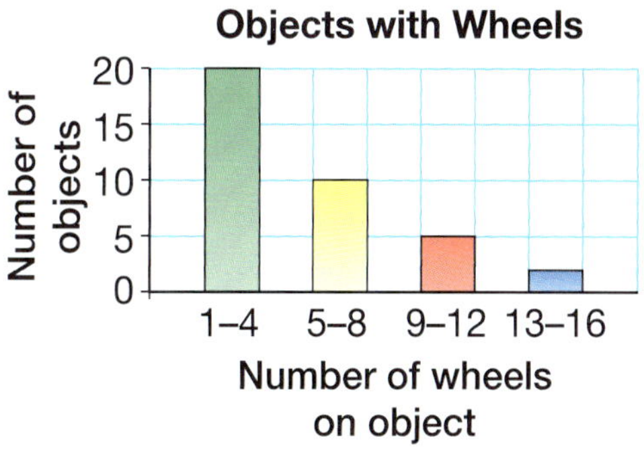

5. What colour bar includes the objects with 11 wheels?

A. green
B. yellow
C. red
D. blue

6. How many of the objects have fewer than 9 wheels?

E. 35 **G.** 30
F. 5 **H.** 15

Cross-Strand Investigation

Wheels in Our Lives

7. Some students picked their favourite toy with wheels.
   a) Create a bar graph of the data.
   b) How do you think the data were collected?
   c) Describe the data. Use math words.
   d) Bicycle wheels can have solid centres or spokes. Here is a pattern of wheels in a bicycle store.

Favourite Toy with Wheels

Favourite toy	Number of students
skateboard	30
bicycle	25
in-line skates	15
scooter	20
other	15

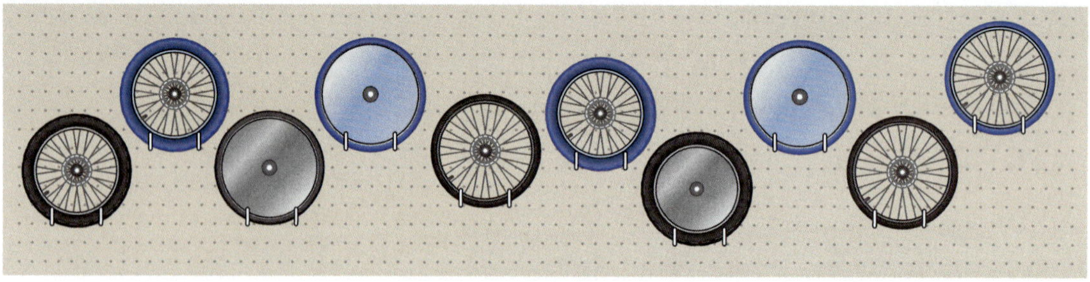

      i) What attributes change in the pattern?
      ii) Describe how each attribute changes.
   e) Make a different wheel pattern.
      i) List attributes that could change in your pattern.
      ii) Make a pattern that has 2 or more changing attributes.
      iii) Describe your pattern.

8. Vehicles are parked in a garage.
   a) Describe the number patterns in the chart.
   b) How is the number pattern for Vehicles related to the number pattern for Number of wheels?
   c) What will be the number of wheels on level 2? Show your work.
   d) Describe the patterning strategy you used in part c).

Level	Vehicles	Number of wheels
7	130	520
6	170	680
5	210	840

CHAPTER 4

Addition and Subtraction

Goals

You will be able to

- estimate sums and differences
- add and subtract with 4-digit numbers
- add and subtract numbers mentally
- create and solve addition and subtraction problems

A school bus yard

Getting Started

Counting Students

The bar graph shows the numbers of students who go to 3 different schools. South School has 205 students.

? About how many students go to the 3 schools?

A. Estimate the number of students at North School. Explain how you estimated.

B. Estimate the number of students at Lakeview School. Explain how you estimated.

C. Estimate the total number of students at the 3 schools. Explain how you estimated.

Do You Remember?

1. Explain how to use one number sentence to complete the other number sentence.

a) Use 18 + 20 = 38 to find 18 + 19 = ■.

b) Use 11 + 19 = 30 to find 30 − 19 = ■.

2. Estimate by **rounding**. Show your rounded numbers.

a) 185 + 188 **b)** 435 + 75 **c)** 811 − 378 **d)** 456 − 378

3. Calculate each **sum** or **difference**. Show your thinking.

a) 150 + 150 **b)** 456 − 200 **c)** 418 + 99 **d)** 456 − 66

4. **a)** Find the total cost of the 2 caps.

b) Describe or draw the coins and bills you would use to pay the exact amount for each cap.

1 Adding Mentally

Goal **Use mental math strategies to add 2-digit numbers.**

There are 28 people in the top section of an arena.
There are 22 people in the bottom section of the arena.

? What is the total number of people in the arena?

Mandy's Method

I imagine rows of 10 people.

When I think of moving people to other rows, it's easier to see how many people there are.

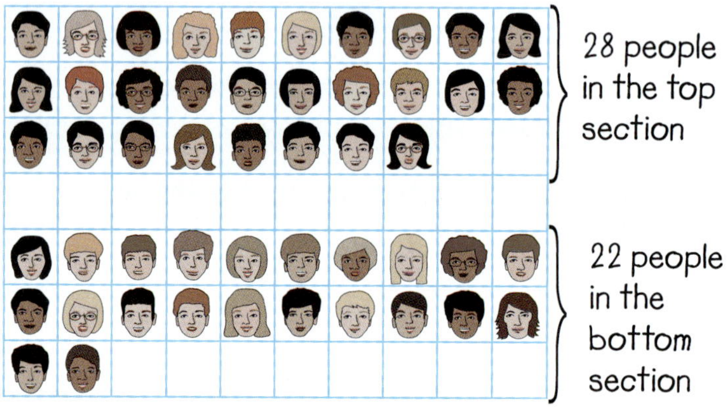

A. Which people might Mandy think of moving? Why?

B. What is the total number of people in the arena? Show your calculations.

Vinh's Method

I used a different method to add 28 and 22.

I drew a number line and used mental math to find the sum.

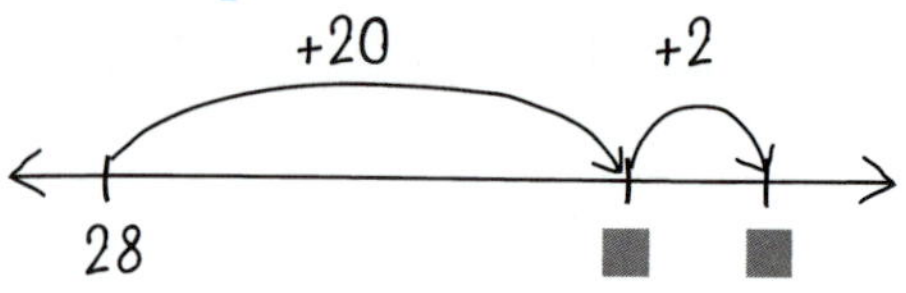

C. What numbers are missing from Vinh's number line?

D. Explain how Vinh used a number line and mental math to add 28 and 22.

Reflecting

1. Use Mandy's method and Vinh's method to add 47 and 23. Show your thinking.
2. What names would you give Mandy's method and Vinh's method? Explain your thinking.

Checking

3. In an arena, there are 49 people in the top section and 21 people in the bottom section.
   Show 2 ways to use mental math to find the total number of people in the arena.

Practising

4. Use mental math to calculate the total number of people in each arena.
   Choose a method that fits each situation.
   Explain what you did.

a)

Top section	Bottom section
18	42

b)

Top section	Bottom section
49	39

c)

Top section	Bottom section
37	23

d)

Top section	Bottom section
25	45

5. How can $40 + 40 = 80$ be used to calculate each sum? Show your work.
   a) $39 + 41$
   b) $38 + 42$
   c) $39 + 39$
   d) $45 + 45$

CHAPTER 4

2 Estimating Sums

You will need

- a calculator

Goal **Estimate sums by rounding.**

? How far is it from Vancouver to Toronto?

Manitok's Number Line

I used a calculator to calculate the distance from Vancouver to Toronto.

Sketching a number line can help me figure out if 4387 km is a reasonable answer.

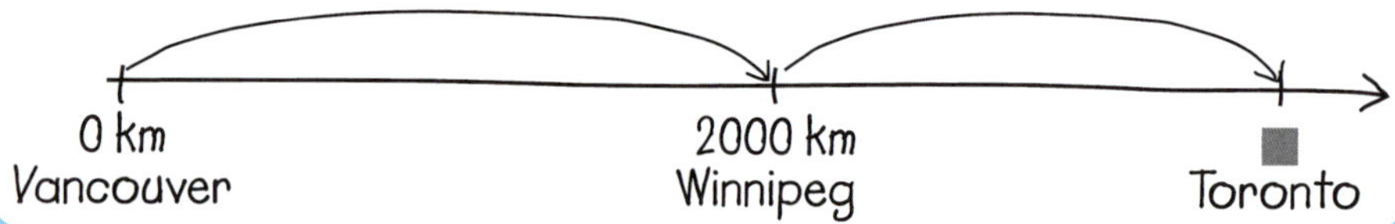

A. Why did Manitok round 1869 km to 2000 km instead of 1000 km?

B. What rounded number could Manitok use to estimate the distance from Winnipeg to Toronto?

C. Estimate the total distance from Vancouver to Toronto. Use rounded numbers.

D. Does Manitok's calculator show a reasonable answer? Explain your thinking.

E. Use a calculator to find the distance from Vancouver to Toronto.

Reflecting

1. How did the number line help Manitok estimate?
2. When you estimate, can you tell if your estimate will be greater or less than the answer from a calculator?
   Explain your answer using examples.

Checking

3. Estimate these total flight distances.
   Show your rounded numbers.

   a)

   b)

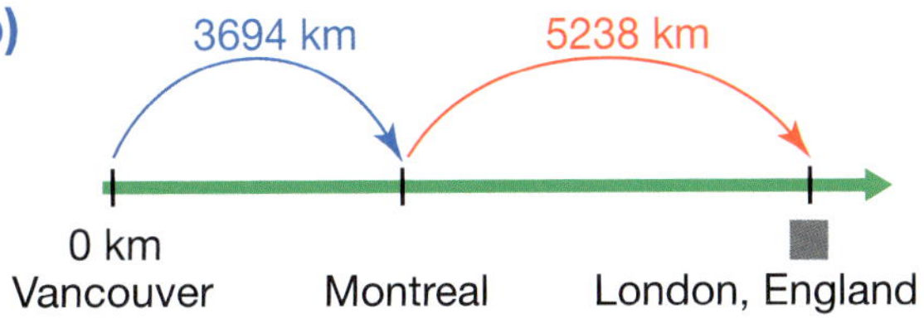

Practising

4. Estimate. Show your rounded numbers.
   a) 1567 + 813 b) 2611 + 1489 c) 4156 + 1722

5. Choose a trip you can make that combines at least 2 flights. Estimate the total distance.
   Show your work.

Trip	Flight distance
Toronto ➤ St. John's	2112 km
St. John's ➤ London, England	3774 km
London, England ➤ Cairo, Egypt	3530 km
Cairo, Egypt ➤ Athens, Greece	1130 km
Athens, Greece ➤ St. John's	6106 km

6. Josef estimates that 1437 + 1498 is about 2000.
   Allison thinks that the sum is close to 3000.
   Are the 2 estimates reasonable? Explain.

3 Communicate About Number Concepts and Procedures

You will need

- a calculator

Goal **Explain your thinking when estimating a sum.**

The local arena has 9000 seats for a hockey game.
3488 tickets for blue seats and 4834 tickets for green seats were sold.

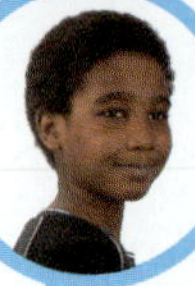

Rey's Sum

I used a calculator to find out how many tickets were sold altogether.

I pressed 3488 [+] 4834 [=] and got 8322.

? Is Rey's answer reasonable? Explain how you know.

Rami's Explanation

There are 9000 seats, so 8322 tickets seems reasonable.

I'll estimate to check. I'll use rounding because rounded numbers are easy to add.

3488 is about 3000 because it's closer to 3000 than 4000.

4834 is close to 5000.

I'll start at 3000 and add on 5 jumps of 1000.

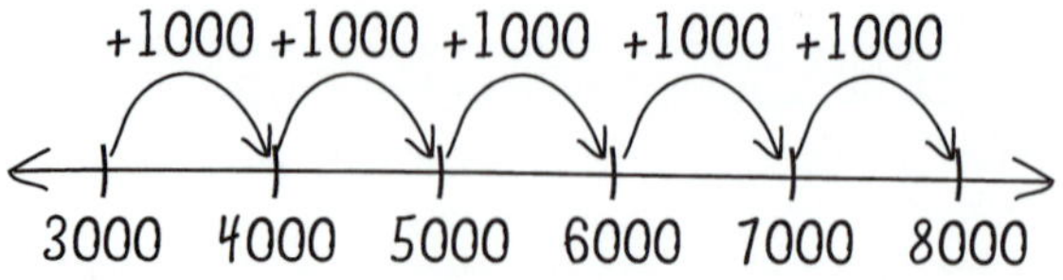

The answer should be close to 8000, so I think 8322 is reasonable.

Allison's Explanation

I'll use rounded numbers because they're easy to work with.

3488 rounds to 3000 and 4834 rounds to 5000.

The sum should be close to 3000 + 5000.

I know 3 + 5 = 8, so I can add the thousands.

8322 is close to 8000, so it's a reasonable answer.

	3 thousand
+	5 thousand
	8 thousand

Reflecting

1. Read the Communication Checklist. How would Rami and Allison answer each question? Give evidence from each person's work.
2. How does Rami's number line help you understand what he is thinking?

Communication Checklist

- ✓ Did you show the right amount of detail?
- ✓ Did you explain your thinking?

Checking

3. The arena has 6000 seats for the circus. 1631 adult tickets and 3712 children's tickets were sold. How many tickets were sold altogether? Use a calculator. Explain why your answer is reasonable.

Practising

4. The arena has 7500 seats for a basketball game. 2815 tickets for blue seats and 3947 tickets for green seats were sold. How many tickets were sold altogether? Use a calculator. Explain why your answer is reasonable.
5. Is the sum reasonable? Explain how you know.

$$\begin{array}{r} 4829 \\ +\ 2301 \\ \hline 6130 \end{array}$$

4 Adding 4-Digit Numbers

You will need

- place value chart

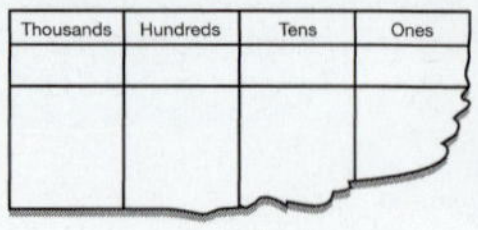

Goal Solve addition problems using regrouping.

? What is the total number of Girl Guides in the Lions and Mid Island areas?

Girl Guide Membership

Area	Number of Girl Guides
Lions	2029
Loughheed	1874
Mid Island	1418
Monashee	1675
Peace Arch	1690
Vancouver	1141

Natalie's Addition

There are 2029 Girl Guides in the Lions area.

There are 1418 Girl Guides in the Mid Island area.

2029 is between 2000 and 3000.

1418 is between 1000 and 2000.

That means there must be between 3000 and 5000 Girl Guides altogether.

$$\begin{array}{r} 2000 \\ +\,1000 \\ \hline 3000 \end{array} \longleftarrow \begin{array}{r} 2029 \\ +\,1418 \\ \hline \end{array} \longrightarrow \begin{array}{r} 3000 \\ +\,2000 \\ \hline 5000 \end{array}$$

I can find the actual number of Girl Guides by adding on a place value chart.

Step 1 I added the ones and regrouped them.

9 ones + 8 ones = 17 ones

17 ones = 1 ten and 7 ones

$$\begin{array}{r} {}^{1} \\ 2029 \\ +\,1418 \\ \hline 7 \end{array}$$

Thousands	Hundreds	Tens	Ones
		1	
2	0	2	9
1	4	1	8
			7

Step 2 I added the tens.

1 ten + 2 tens + 1 ten = 4 tens

$$\begin{array}{r} 1 \\ 2029 \\ +\ 1418 \\ \hline 47 \end{array}$$

Thousands	Hundreds	Tens	Ones
		1	
2	0	2	9
1	4	1	8
		4	7

Step 3 I added the hundreds.

0 hundreds + 4 hundreds = 4 hundreds

$$\begin{array}{r} 1 \\ 2029 \\ +\ 1418 \\ \hline 447 \end{array}$$

Thousands	Hundreds	Tens	Ones
		1	
2	0	2	9
1	4	1	8
	4	4	7

Step 4 I added the thousands.

2 thousands + 1 thousand = 3 thousands

$$\begin{array}{r} 1 \\ 2029 \\ +\ 1418 \\ \hline 3447 \end{array}$$

Thousands	Hundreds	Tens	Ones
		1	
2	0	2	9
1	4	1	8
3	4	4	7

There are 3447 Girl Guides in the Lions and Mid Island areas. My estimate was between 3000 and 5000 Girl Guides, so 3447 is reasonable.

Reflecting

1. In Step 1, why did Natalie **regroup** 17 ones as 1 ten and 7 ones?
2. Explain how Natalie could calculate the total using base ten blocks.

regroup

Trade 10 smaller units for 1 larger unit, or 1 larger unit for 10 smaller units

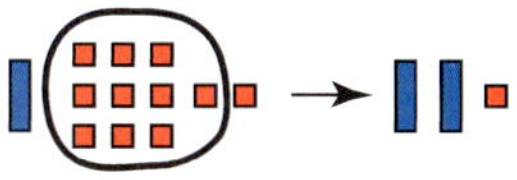

Checking

3. a) There are 1874 Girl Guides in the Lougheed area and 1690 Girl Guides in the Peace Arch area. What is the total number of Girl Guides in the Lougheed and Peace Arch areas?
   b) Explain how you know your answer is reasonable.

Practising

4. There are 3256 Cubs in the Chinook area and 1437 Cubs in the Pacific Coast area. Estimate and then calculate the total number of Cubs in the Chinook and Pacific Coast areas.

5. Estimate. Complete the calculation if the estimated sum of the numbers is between 4000 and 6000. Show your rounded numbers.

   a) $2987 + 2145$
   b) $3254 + 162$
   c) $2311 + 2499$
   d) $1300 + 2253 + 1701$

6. Is the answer reasonable? Explain how you know.
   a) $4566 + 2869 = 9435$
   b) $1859 + 2899 = 3758$
   c) $3532 + 4499 = 8031$

7. Create your own addition problem about the Girl Guide data. Show how to solve your problem.

8. Write two 4-digit numbers with a sum that is just less than 10 000. Show your work.

9. Would you calculate each sum using mental math, base ten blocks, pencil and paper, or a calculator? Give a reason for each choice.
   a) $26 + 24$
   b) $3456 + 239$
   c) $18\,264 + 11\,495$

10. Calculate each sum in Question 9 using your chosen method. Show your work.

Race to 150

Number of players: 2
How to play: Cross out numbers and find the total.

Step 1 Copy this array of numbers.

5	10	15	20	25	30
5	10	15	20	25	30
5	10	15	20	25	30
5	10	15	20	25	30

Step 2 Player 1 crosses out a number.

Step 3 Player 2 crosses out another number and adds to find the total of the crossed out numbers.

Step 4 Take turns crossing out and totalling all the numbers.

The player who reaches a total of exactly 150 wins.

Chris's Turn

Shani and I have crossed out a total of 120 so far.

I can win if I cross out a 30!

5	~~10~~	15	20	25	~~30~~
5	~~10~~	~~15~~	20	25	30
5	10	15	20	~~25~~	~~30~~
5	10	15	20	25	30

Mid-Chapter Review

LESSON

1 **1.** Use mental math to add.
Explain your steps for 2 of the questions.

a) 46 + 18
b) 38 + 22
c) 19 + 46
d) 39 + 56
e) 29 + 31
f) 75 + 25

2 **2.** Estimate. Explain your steps for 2 of the estimates.

a) 6455 + 2876
b) 999 + 1786
c) 3658 + 4444
d) 5110 + 3459

3 **3.** Is each answer reasonable?
Explain your thinking.

a) 4579 + 518 = 5097
b) 2367 + 2710 = 6077
c) 5145 + 4145 = 9290

4 **4.** Estimate and then add.

a) 4217 + 3572
b) 746 + 615
c) 3169 + 825
d) 7394 + 2527
e) 3876 + 5678
f) 1750 + 5250

5. Josef's address is 8419 Main St. Is 84 plus 19 equal to 8419? Explain your reasoning.

6. On the same day, a plane flew 5456 km to one city and then flew 3567 km to another city. Calculate how far the plane flew that day. Explain how you know that your answer is reasonable.

7. Pedro estimates that the sum of 2 numbers is close to 5000. One of the numbers is 2878. What might the other number be? Explain how you know that your answer is reasonable.

Subtract by Adding On

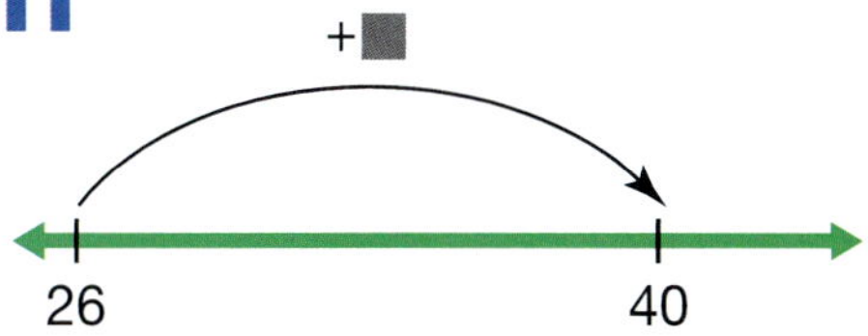

You can subtract by thinking about adding on.
To subtract 26 from 40, think 26 + ■ = 40.

Chantal's Method

I jumped 4 + 10.

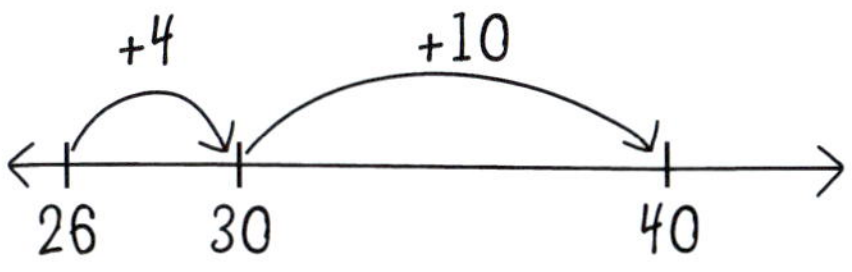

Carmen's Method

I jumped 10 + 4.

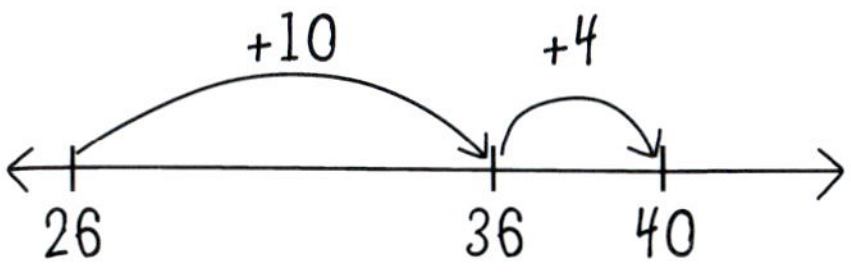

A. How much is 40 − 26?
Explain how you know.

B. How would you use your answer from part A to find 40 − 27?

C. Write another pair of numbers you could subtract by adding on.
Draw a number line to show the jumps you would use to find the answer.

Try These

1. Subtract by adding on.
Use a number line to show your steps.

a) 30 − 12	**c)** 50 − 12	**e)** 30 − 19	**g)** 40 − 25
b) 30 − 11	**d)** 80 − 45	**f)** 45 − 15	**h)** 55 − 35

5 Subtracting Mentally

Goal **Develop mental math strategies for subtracting 2-digit numbers.**

Pedro's brother is 19 and his father is 45.

? How much older is Pedro's father than Pedro's brother?

Pedro's Solution

I can show the problem on a number line.

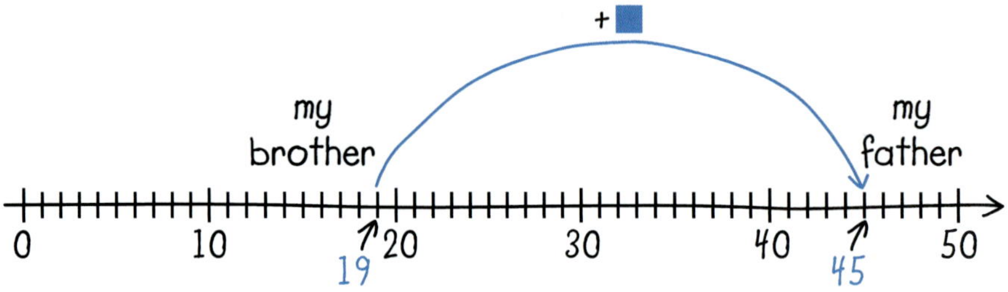

If I add 1 year to each age, the difference between the ages will still be the same, but the numbers will be easier to subtract.

Then 45 − 19 = ■ becomes 46 − 20 = ■.

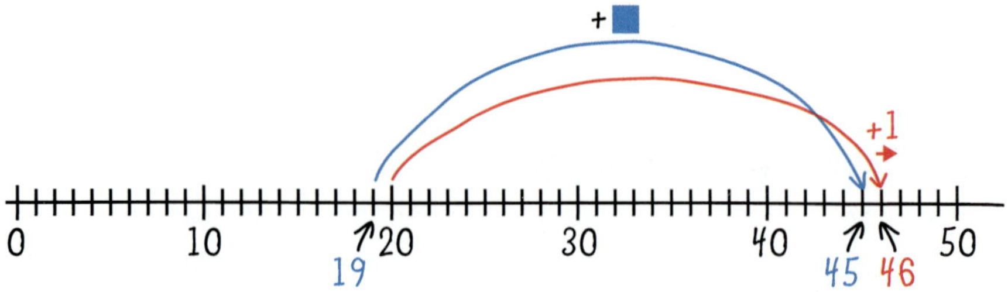

A. How much older is Pedro's father than Pedro's brother? Explain how you know.

B. Did you subtract by taking away or by adding on? Explain your method.

Reflecting

1. Explain why the red arrow and the blue arrow show the same size of jump on the number line.
2. Why is 46 − 20 easy to solve mentally?

Checking

3. Sarah's sister is 28 years old and Sarah's mother is 51 years old.
   How much older is Sarah's mother than Sarah's sister?
   Use mental math to find the answer.
   Explain your steps.

Practising

4. Use mental math to find each age difference.
   Use a different strategy to check your answers.
   a) 28 years and 53 years
   b) 49 years and 81 years

5. Use mental math to subtract.
   Explain your steps.
   a) 33 − 19
   b) 66 − 18
   c) 67 − 29
   d) 81 − 69

6. Write 2 different ages between 20 and 100.
   Show how to use mental math to find the difference between the ages.

6 Estimating Differences

Goal **Estimate differences by rounding.**

The height of Ishpatina Ridge in Ontario is 693 m. The height of Cypress Hills in Saskatchewan is 1468 m.

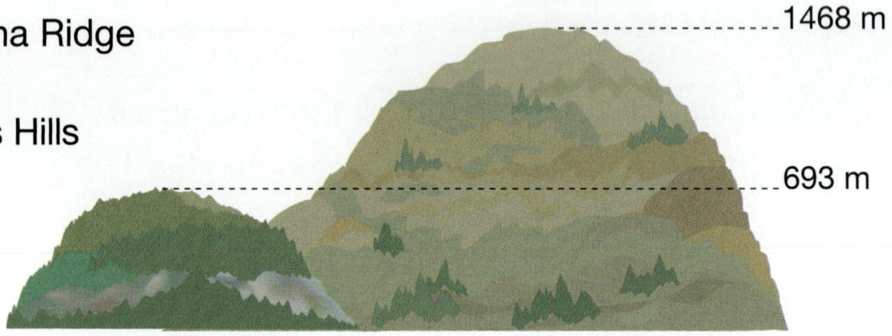

? How can you estimate the height difference between Cypress Hills and Ishpatina Ridge?

A. How can you use rounding to estimate? Show 2 different ways.

B. How can you use the drawing to estimate?

C. How else could you estimate? Show at least 2 ways.

Reflecting

1. Carmen said she estimated the height difference by subtracting 1468 − 693 and rounding the result to the nearest thousand. Is this a good way to estimate? Explain.
2. Manitok said he estimated by counting on by 100s from 700 to 1400. Is this a good way to estimate? Explain.
3. Find out 10 estimates for your class. Why are many different estimates possible?

River Crossing

Number of players: team of 2
How to play: Try to get the greatest total in your river crossing.

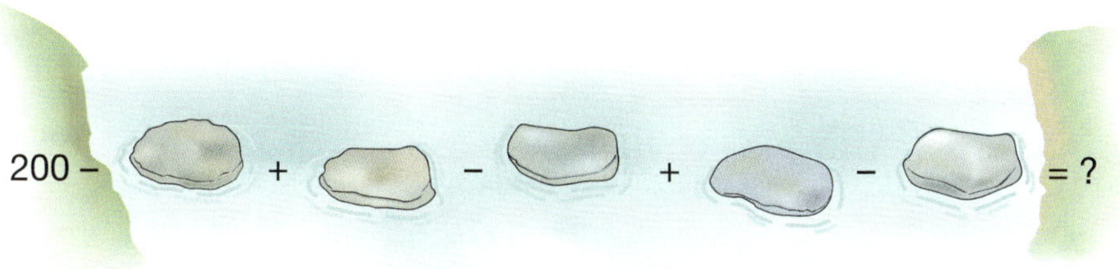

You will need

- 2 dice

- River Crossing game sheet

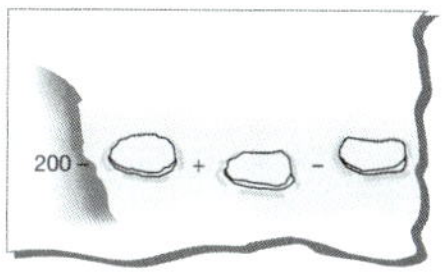

Step 1 Player 1 rolls both dice to make a 2-digit number. The player writes this number on the 1st stepping stone.

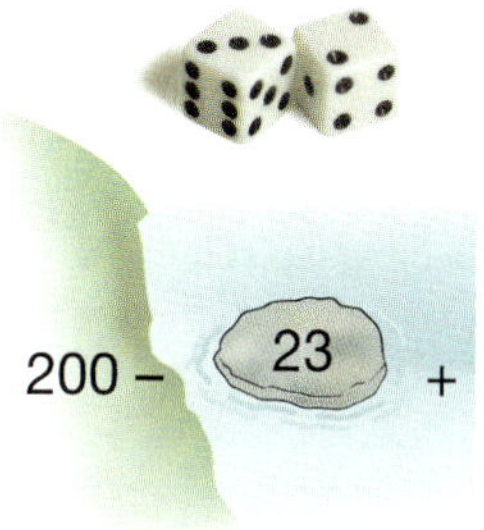

Step 2 Take turns rolling the dice and writing numbers on the stones until you have crossed the river.

Step 3 Once you have crossed the river, use subtraction and addition to calculate your score. Use the chart below to see how you did.

0 to 50	51 to 100	101 to 150	151 to 200	201 to 250	251 to 300
Watch out for the alligators!	I hope you brought some dry clothes!	Nice work! You made it all the way across without falling in!	Hey, you're pretty good! Let's see you do that again!	Didn't I see you doing stunts in that white-water rafting movie?	Would you like a job as a nature tour guide?

7 Subtracting from 4-Digit Numbers

Goal **Use a pencil and paper method to subtract from a 4-digit number.**

A video store has 1257 DVDs.
848 DVDs have already been rented.

? How many DVDs are left at the video store?

Jon's Subtraction

I think 1257 − 848 must be about 400 because
12 hundreds − 8 hundreds = 4 hundreds.
I can subtract to find the actual difference.

Step 1 I compare the numbers column by column.
I need more hundreds and more ones before I can subtract 8 **hundreds**, 4 **tens**, 8 **ones**.

$$\begin{array}{r} 1257 \\ -\ 848 \\ \hline \end{array}$$

Thousands	Hundreds	Tens	Ones

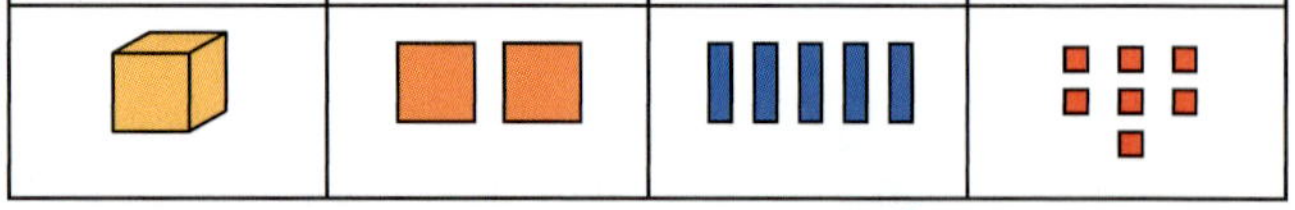

Step 2 If I regroup 1 **thousand** as 10 **hundreds**,
I'll have 12 **hundreds** altogether.
But I still need more ones before I can subtract.

$$\begin{array}{r} {\scriptstyle 0\,12} \\ \not{1}\not{2}57 \\ -\ 848 \\ \hline \end{array}$$

Thousands	Hundreds	Tens	Ones

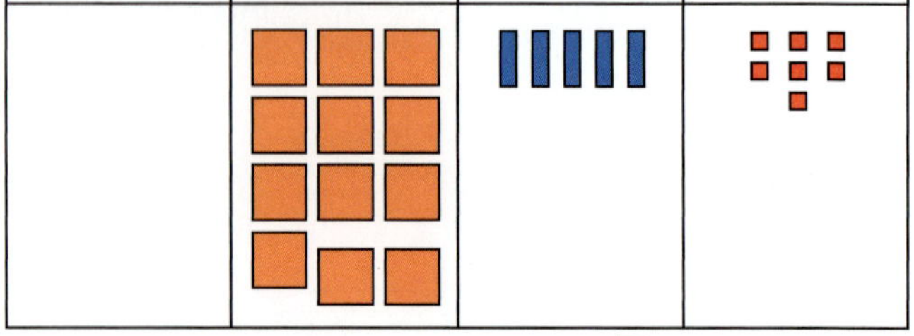

Step 3 I'll regroup 1 **ten** as 10 **ones**.
Now I have 0 **thousands**,
12 **hundreds**, 4 **tens**,
17 **ones**.

0 12 4 17
~~1257~~
− 848

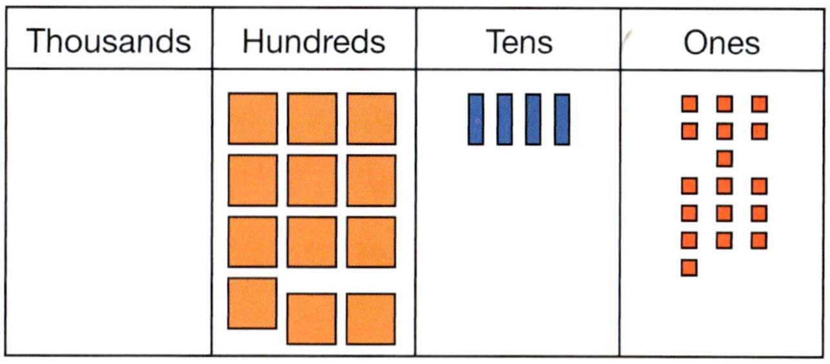

Step 4 Now I have enough in each place value column to subtract.

0 12 4 17
~~1257~~
− 848
409

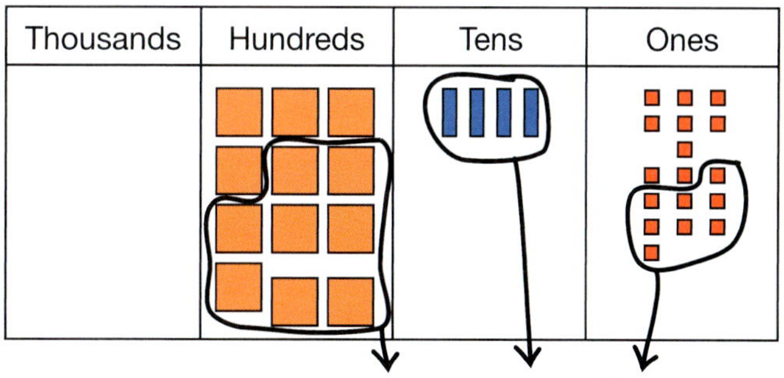

There are 409 DVDs left at the video store.

This answer seems reasonable because it's pretty close to my estimate of 400.

Reflecting

1. In Step 1, how did Jon know he would need more hundreds and more ones before he could subtract?
2. How does Step 4 show that 1257 was regrouped as 1200 + 40 + 17?
3. Explain how Jon could use addition to make sure he subtracted correctly.

Checking

4. On Saturday, 788 of the 1257 DVDs are rented.
   a) How many DVDs are left at the video store?
   b) Show how to add to check your answer.

Practising

5. The video store was giving away tickets. By Wednesday, 571 of the 4325 tickets had been given away. How many tickets were left?

6. Estimate each subtraction. Complete the subtraction if the estimated difference is greater than 2000.
   a) 2348 − 999   b) 3649 − 682   c) 4567 − 768

7. Estimate and then subtract. Show your work.

a)	b)	c)	d)
1234	5045	8129	1000
− 621	− 347	− 518	− 235

8. Is the answer reasonable? Explain.
   a) 6434 − 178 = 6256
   b) 2257 − 388 = 1689
   c) 4533 − 751 = 3782
   d) 5624 − 591 = 6215

9. Would you calculate each difference using mental math, base ten blocks, pencil and paper, or a calculator? Give a reason for each choice.
   a) 5324 − 324   b) 16 905 − 12 876   c) 1542 − 519

10. Use a 3-digit number and a 4-digit number to create a subtraction problem about renting DVDs from a video store. Solve your problem. Show your work.

11. What is the greatest difference you can get when you subtract a 3-digit number from a 4-digit number? Explain how you know.

Hidden Digits

Someone spilled ink on Paulette's notebook. Can you help her figure out the hidden digits?

When you replace the ink spots, the addition or subtraction should be correct.

1

```
   4 7
+  9 ■
 1 ■ 5
```

2

```
 1 ■ 9
+  3 6
 ■ 3 ■
```

3

```
 2 3 ■
−  9 8
 1 ■ 1
```

4

```
 1 4 ■ 5
−   ■ 6 9
 ■ 2 1 6
```

5

```
   ■ ■ ■
   2 6 4
+  3 0 3
 1 2 9 9
```

6

```
   4 8 7
   ■ ■ ■
+  4 9 1
 1 5 8 1
```

8 Subtracting in a Different Way

Goal **Use regrouping to make subtraction easier.**

In 1998, the population of Norman Wells was 879. For the year 2013, the predicted population of Norman Wells is 1000.

? How much would the population of Norman Wells grow during these years for this prediction?

Paulette's Subtraction

The answer must be between 100 and 200 because 879 is between 800 and 900. I know 1000 − 800 = 200 and 1000 − 900 = 100.

I will regroup the 1000 so it's easier to subtract 8 hundreds, 7 tens, and 9 ones.

1000 ←
− 879

Step 1 If I think of 1000 as 999 + 1, then I can do all the regrouping in 1 step.

```
 999 + 1
 ~~1000~~
- 879
```

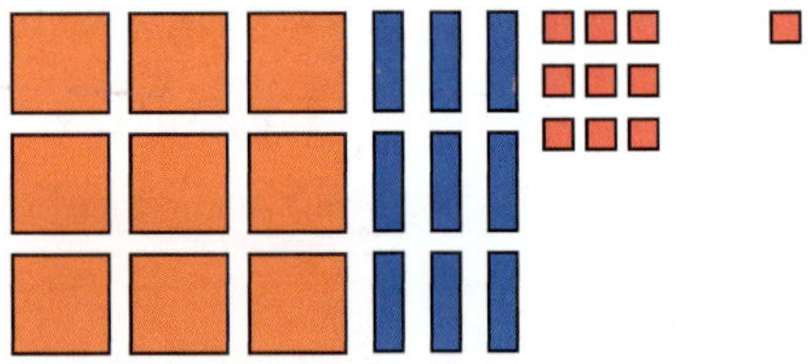

Step 2 Now I subtract 879.

```
 999 + 1
 ~~1000~~
- 879
------
 120 + 1 = 121
```

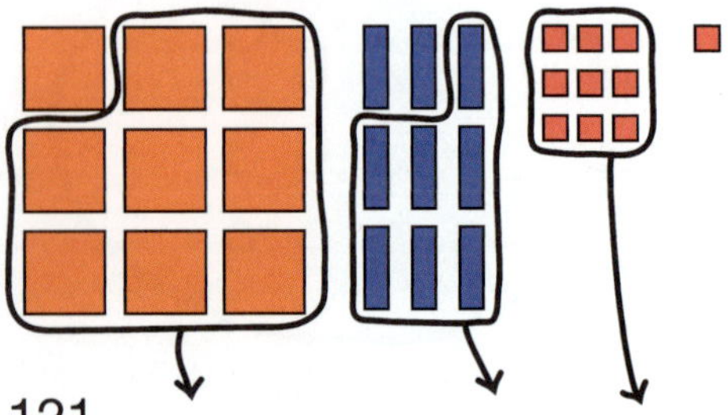

The population of Norman Wells would grow by 121.

My answer seems reasonable because my estimate was between 100 and 200.

Reflecting

1. Why did Paulette add 1 to 120 in Step 2?
2. Would you use Paulette's method to solve these subtractions? Explain your thinking.
   a) 6588 − 699 b) 5000 − 467
3. How is Paulette's method like Jon's method in Lesson 7? How is it different?

Checking

4. The predicted population of a town is 3000 for the year 2050. This year the population of the town is 1598. How much would the population grow for this predicted population?
   Add to check.

Practising

5. The predicted population of a town is 5000 for 2 decades from now. The population of the town today is 929. How much would the population grow for this predicted population?
   Add to check.

6. Estimate and then subtract.
   Use addition to check your answer.

   a) $\begin{array}{r} 1000 \\ -\ 435 \\ \hline \end{array}$ c) $\begin{array}{r} 3000 \\ -\ 278 \\ \hline \end{array}$

   b) $\begin{array}{r} 2000 \\ -\ 849 \\ \hline \end{array}$ d) $\begin{array}{r} 6000 \\ -\ 332 \\ \hline \end{array}$

7. Why don't you have to regroup 4467 to subtract 235?

8. Show 2 different ways to calculate 1000 − 250.
   Which way was easier for you?
   Explain your thinking.

9 Making Change

Goal **Make purchases and change for money amounts.**

Calvin has a $20.00 bill. He plans to buy the skateboard at the yard sale.

? How much change will Calvin get?

Calvin's Solution

My change should be less than $2.00 because $18.00 + $2.00 = $20.00.

I can count on from $18.35 to $20.00 to figure out the actual amount.

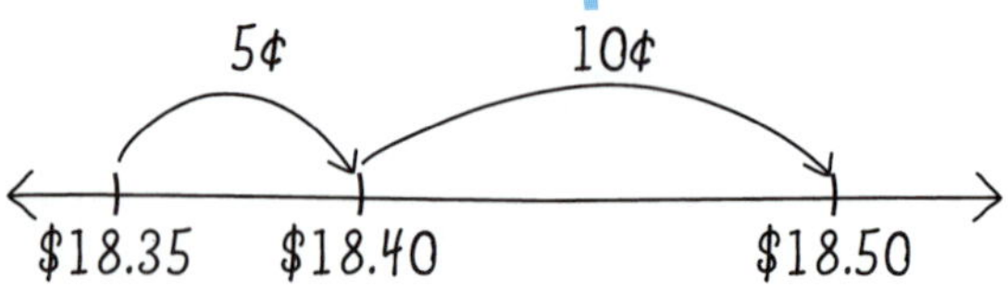

A. Use a number line to show how Calvin might continue counting.

B. Use your number line to find out how much change Calvin will get from $20.00. Explain how you know.

Reflecting

1. Draw a different number line to show how Calvin could count his change. Start with a $1.00 jump.
2. Show another way Calvin could calculate his change. Explain how you could show this using coins and bills.
3. Is counting on addition or subtraction? Explain your thinking.

Checking

4. Tamika wants to use 

to buy 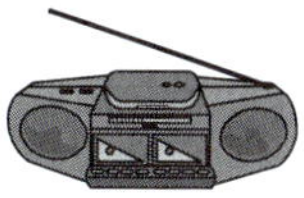 .
   a) Estimate how much change she will get.
   b) Calculate the amount of change she will receive.

Practising

5. Estimate and then calculate the change.

a) 

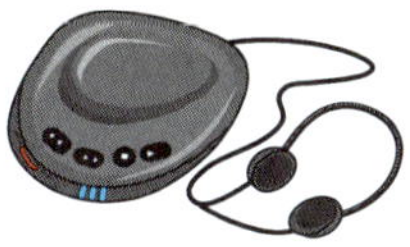

c) 

b) 

d) 

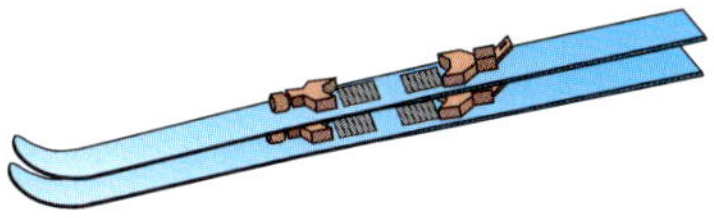

6. You have 

to spend at the yard sale.
Choose an item you could buy.
Which bill or bills would you use to pay for the item?
Show how to calculate your change.

7. Describe 2 different ways to calculate change.
Give an example for each way.

10 Adding and Subtracting Money

Goal **Use different methods to add and subtract money.**

The class raised $50.00 for a retirement gift for the principal. Terry and Shani will buy the gift. Terry wants to buy a plaque that costs $24.99. Shani wants to buy a plant that costs $14.99. They decide to buy both.

? How much change will they get from $50.00?

Terry used mental addition to find the answer. Shani used the pencil and paper method.

Terry's Method

I will add 1¢ to each price to make it easier to add.

$25.00 + $15.00 = $40.00

Then I will take away 2¢ from the answer so the sum will be correct.

$40.00 – 2¢ = $39.98

The total cost of the gift is $39.98.

Shani's Method

I will add $24.99 + $14.99.

```
   11
 $ 24.99
 +14.99
 $ 39.98
```

The answer is in dollars and cents, so I need to write a dollar sign and a decimal point.

The total cost of the gift is $39.98.

A. Describe how Terry used mental math to add.

B. Describe a different mental math strategy Terry could have used.

C. Finish the solution by calculating the change from $50. Show your work.

D. Show how to calculate the change in a different way.

Reflecting

1. a) Give an example of a money addition problem that is easier to solve using mental math than with paper and pencil. Explain why.
   b) Give an example of a money addition problem that you think would be easier to solve using pencil and paper.

Checking

2. You have  to spend at this yard sale.
   a) Choose 2 items you could buy. Calculate the total cost. Show your work.
   b) Calculate the amount of money you will have left. Show your work.

Practising

3. A shirt in Manitoba costs $9.95 plus $1.49 in tax. What is the total cost of the shirt? Show your work.

4. Add or subtract. Show your work.
   a) $10.99 + $5.67
   b) $42.08 − $3.99
   c) $35.00 − $17.03
   d) $23.97 + $11.03 + $6.50

5. Would you calculate each sum using mental math, pencil and paper, or a calculator? Give a reason for each choice.
   a) $17.66 + $4.59
   b) $2.40 + $2.99
   c) $235.97 + $145.88 + $4.56

6. Calculate each sum in Question 5 using your chosen method. Show your work.

Skills Bank

LESSON

1

1. Use mental math.
   a) 73 + 21 c) 71 + 29 e) 54 + 19 g) 43 + 27
   b) 32 + 28 d) 49 + 33 f) 39 + 49 h) 67 + 14

2. Use 20 + 20 = 40 to calculate each sum.
   a) 19 + 21 b) 19 + 19 c) 22 + 21 d) 19 + 23

3. There are 59 students in grade 3.
   There are 36 students in grade 4.
   Use mental math to calculate the total number of students in both grades.

2

4. Estimate. Show your rounded numbers.
   a) 2035 + 915 c) 1392 + 1429 e) 3436+ 2602
   b) 3942 + 495 d) 2688 + 1975 f) 3725 + 4983

5. Estimate the total distance from Vancouver to Montreal. Show your work.

Trip	Flight distance
Vancouver → Regina	1722 km
Regina → Montreal	3320 km

4

6. Estimate. Then calculate the sum.
   a) 1651 + 1237 c) 2627 + 3148 e) 4488 + 2369 g) 6450 + 1762
   b) 2453 + 1832 d) 4255 + 3694 f) 1876 + 3547 h) 4139 + 5367

7. There are 1425 Cubs in the Saskatchewan area.
   There are 1135 Cubs in the Manitoba area.
   What is the total number of Cubs in the Saskatchewan and Manitoba areas? Show your work.

5

8. Use mental math to subtract.

a) 23 − 9
b) 40 − 18
c) 75 − 45
d) 58 − 29
e) 67 − 25
f) 92 − 39
g) 93 − 47
h) 86 − 78

9. Rashad's brother is 22 years old.
Rashad's mother is 51 years old.
Use mental math to find the age difference between Rashad's mother and brother.

6

10. Estimate. Show your rounded numbers.

a) 1597 − 483
b) 2037 − 545
c) 4788 − 826
d) 3615 − 389
e) 2678 − 413
f) 5092 − 478

11. Use the chart to estimate each height difference.
Show your work.

a) between Mount Carleton and Fairweather Mountain
b) between Cypress Hills and Fairweather Mountain
c) between Cypress Hills and Mount Carleton

Mountain	**Province**	**Height or highest point**
Fairweather Mountain	British Columbia	4663 m
Mount Carleton	New Brunswick	817 m
Cypress Hills	Saskatchewan	1468 m

7

12. Estimate. Then calculate each difference.

a) 4376 − 254
b) 1687 − 536
c) 2371 − 640
d) 3695 − 287
e) 7256 − 431
f) 8170 − 962
g) 6223 − 475
h) 5018 − 766

13. A store has 2416 DVDs. 508 of the DVDs are rented.
How many DVDs are left at the store?
Show your work.

8 **14.** Subtract. Use addition to check your answer.

a) 1000 − 145	c) 4000 − 817	e) 6000 − 695	g) 8000 − 999
b) 5000 − 275	d) 2000 − 458	f) 9000 − 918	h) 3000 − 699

15. In a town of 7000 people, 914 of the people are 6 years old or younger.
How many people are older than 6 years old?

9 **16.** Estimate and then calculate the change.

a) A flashlight costs \$7.60.
You pay with a \$10 bill.

b) A backpack costs \$32.69
You pay with a \$50 bill.

17. Pedro paid for a basketball with two \$20 bills.
His change was \$6.65.
How much did he spend?
What coins and bills was he likely to receive as change?

10 **18.** Add or subtract. Show your work.

a) \$12.50 + \$26.25
b) \$17.99 + \$25.99
c) \$16.42 + \$23.39
d) \$36.50 − \$4.30
e) \$42.00 − \$39.98
f) \$38.55 − \$17.95

19. Calculate the total cost. Then find the change.

a)

b)

Problem Bank

LESSON

1 **1.** Keenan had 48¢. Rami had 29¢.
Use mental math to answer each question.
a) How much more money did Keenan have than Rami?
b) How much money did they have altogether?

2. Find the least number that can be used to make this sentence true.

53 + ■ is greater than 100.

Use mental math. Explain your reasoning.

2 **3.** Show 2 different ways to make this number sentence true.

■500 + ●500 = 8000

4. Rey estimated a sum by rounding 2 numbers up to the nearest thousand.
His estimate was 7000.
What could the numbers have been?
Find more than 1 answer.

6 **5.** Find numbers that fit each riddle.
Look for more than 1 answer.
a) There are two 2-digit numbers.
One number is greater than 50 and ends in 0.
When you subtract one number from the other number, the difference is 29.
b) There are 2 numbers with no 0s.
When you subtract the numbers, the difference is close to 3000.

8 **6.** A truck carrying 2 crates has a total **mass** of 8341 kg.
The 1st crate has a mass of 1499 kg.
The truck has a mass of 5443 kg.
What is the mass of the 2nd crate?

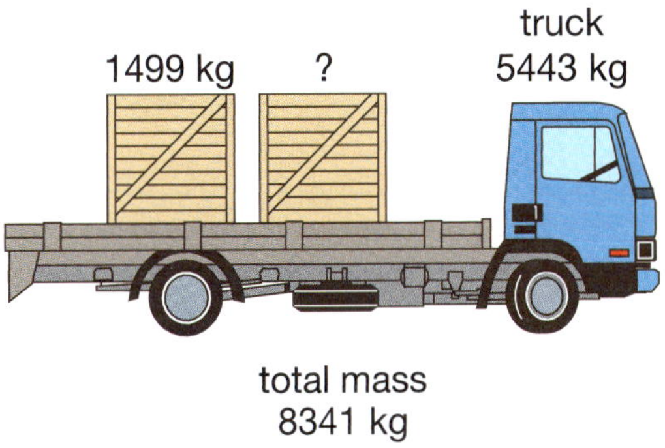

9 **7.** Zola is given a $10 bill.

Zola's Change

Zola can spend $5.50 of the money at a yard sale. She will buy at least one toy for her sister, her brother, and a friend.

a) What can Zola buy?

b) Count to find the amount for Zola's change.

8. Miki's lunch cost $5.56.
She paid with  .

a) How much change did Miki receive?

b) Why might Miki have decided to pay this way instead of just paying  ?

10 **9.** Change 1 digit in each price so the total cost is exactly $50.00. Look for more than 1 answer. Show your work.

10. 2 TV sets cost $1000 altogether.
1 TV set cost $280 more than the other TV set.
What is the price of each TV set?

LESSON

Chapter Review

1 **1.** Find all pairs of these numbers that add to 100.

64, 72, 18, 9, 91, 36, 51, 82, 28

3 **2.** Mandy added 6524 + 1208 and got 7732.
Explain how you can tell if her answer is reasonable.

6 **3.** Estimate each sum or difference. Which answers are greater than 3000? Show your work.

A. $1868 + 2487$ **B.** $1276 + 1367$ **C.** $5000 - 2675$ **D.** $4000 - 264$

4. Estimate to see which answer is closest to 6000.
Show your work.

A. $1756 + 2855$ **B.** $5200 - 857$ **C.** $7000 - 845$ **D.** $3599 + 2001$

7 **5.** Explain why you can use addition to check subtraction answers. Give an example.

8 **6.** An online store sold 3435 books on Saturday and 3675 books on Sunday.

a) Explain how you know the store sold more than 7000 books in total.

b) Exactly how many books were sold?
Show your work.

c) The store also sold 745 DVDs on Saturday.
How many more books than DVDs were sold on Saturday?

10 **7.** You have .

a) What 2 items could you buy at the sale?

b) What coins and bills would you use to pay for the 2 items?

c) How much change would you receive?

Chapter Task

On the Move

Every 3 weeks, about 7500 people move into Ontario.

People Who Moved in During Winter

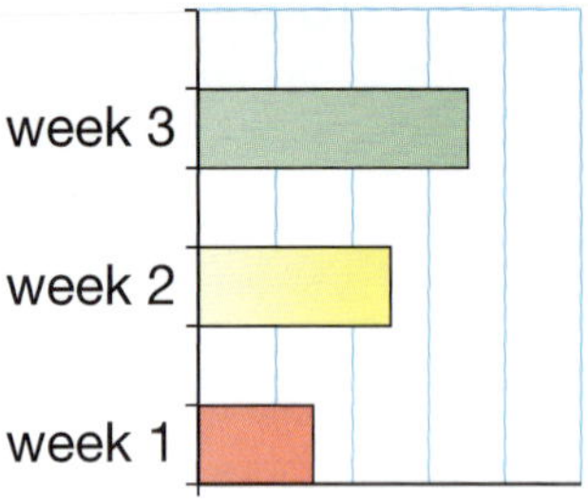

People Who Moved in During Spring

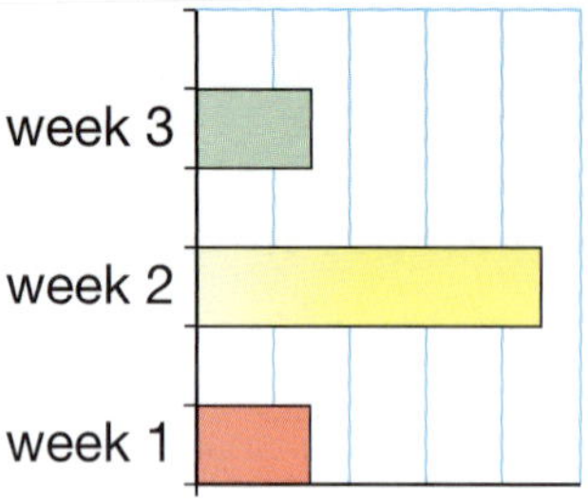

? What are some possible numbers of people who moved in each week?

Part 1

A. For each graph, determine some possible numbers for weeks 1, 2, and 3. Show your work. Explain your thinking.

Part 2

B. Draw another bar graph for 3 weeks in a different season. Show the possible numbers of people that move into Ontario each week.

C. Explain how you created your graph.

Task Checklist

- ✓ Did you add and subtract correctly?
- ✓ Did you verify that your numbers fit with the given information?
- ✓ Did you show all your steps?
- ✓ Did you explain your thinking?

CHAPTER 5

Measuring Length and Time

Goals

You will be able to

- use a variety of length units and describe how they are related
- use a variety of time units and describe how they are related
- estimate, measure, and compare perimeters and circumferences
- measure length and time with appropriate precision
- solve length problems using diagrams

Measuring Zola's height

Getting Started

You will need
- a ruler
- string

Surrounding Shapes

Pedro is making a tray by gluing together pieces of coloured glass.

? How can you draw a shape with the same perimeter as 1 of Pedro's triangles?

A. Estimate the length of the strip of glue around 1 triangle of glass. Explain how you estimated.

B. Why do you think this length is called the **perimeter** of the triangle?

C. Measure to find the perimeter of the triangle.

D. Use string to make a different triangle with the same perimeter as the original triangle.
Draw or describe your new triangle.

E. Create a 4-sided shape that has the same perimeter as the original triangle.
Draw or describe your new shape.

F. Do you think every student created the same shapes for parts D and E? Explain your thinking.

Do You Remember?

1. Arrange these length units in order from shortest to longest.
kilometre metre centimetre

2. Match each unit with its symbol.

a) 1 centimetre **X.** 1 m
b) 1 metre **Y.** 1 km
c) 1 kilometre **Z.** 1 cm

3. a) Write each time in words.

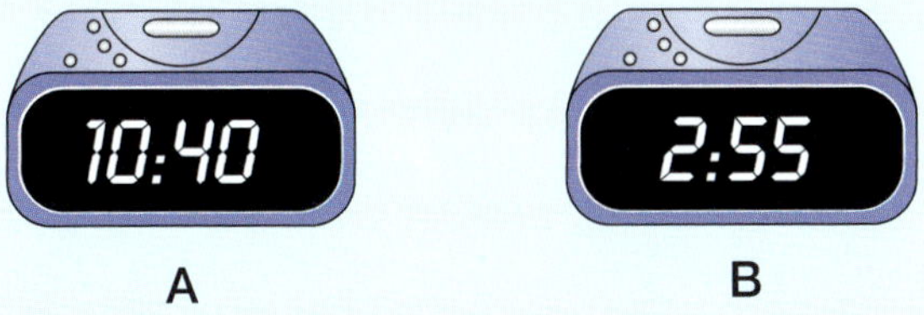

b) How many minutes are in 1 hour?
c) How many hours are in 1 day?
d) How many days are in 1 week?

4. a) What year will it be 5 years from now?
Show your work.
b) What year was it 5 years ago?
Show your work.

CHAPTER 5

1 Measuring with Decimetres

Goal **Measure with decimetres and relate decimetres to centimetres and metres.**

You will need

- a ruler

- a metre stick

- string

- scissors

Human hair grows about 10 cm each year.
10 cm is called 1 **decimetre**, or 1 dm.

? How long will your hair be in 5 years, if you don't cut it?

A. Look at a metre stick.
How many decimetres are in a metre?

B. Cut a piece of string to a length of 1 dm.

C. Use a 30 cm ruler and your 1 dm string.
- **a)** How many centimetres are in a decimetre?
- **b)** How many decimetres are in a 30 cm ruler?

D. **a)** How long is your hair in centimetres?
- **b)** Is it possible to measure the exact length of your hair in decimetres? Explain why or why not.
- **c)** What is the length of your hair to the nearest whole decimetre?

E. Cut a piece of string to model the length of your hair in 5 years. How long is the string?

decimetre
A unit of length that is longer than a centimetre and shorter than a metre

1 dm = 10 cm
10 dm = 1 m

Reflecting

1. What does it mean to measure something to the nearest whole decimetre?
2. a) When might you use decimetres rather than metres?
   b) When might you use decimetres rather than centimetres?

Checking

3. Find an object of each length or height in your classroom. Then measure each object in centimetres.
   a) shorter than 1 dm
   b) between 1 dm and 2 dm
   c) longer than 2 dm

Practising

4. Find an object of each length or height. Then measure each object to the nearest whole decimetre.
   a) about 3 dm
   b) longer than 5 dm
   c) between 10 dm and 20 dm
5. Order the measurements in Question 4 from shortest to longest.
6. Draw a line that is 2 dm long.
   a) How long is the line in centimetres? How do you know?
   b) Which number is greater, the decimetre measurement or centimetre measurement? Why?
7. About how many years would it take you to grow your hair to each length? How do you know?
   a) 2 m   b) 25 cm

2 Measuring with Millimetres

You will need
- printer paper
- construction paper
- a ruler

Goal **Use millimetres to measure and compare with precision, and relate units.**

Chantal wanted to know whether printer paper or construction paper is thicker.

Chantal's Measurement

I made a stack of 50 sheets of white printer paper.

I made another stack of 50 sheets of construction paper.

I measured each stack.

The white paper is about 1 cm thick.

The construction paper is between 1 cm and 2 cm thick.

? How can Chantal make a more precise measurement to compare the stacks?

A. Use a ruler marked with **millimetres**.

- **a)** How many millimetres fit between 50 mm and 70 mm?
- **b)** How many centimetres fit between 50 mm and 70 mm?
- **c)** The length of an object is about halfway between 2 cm and 3 cm. How long could the object be in millimetres?
- **d)** How many millimetres are in 1 dm? Explain how you know.

B. Make 2 stacks of paper like Chantal's.

millimetre

A unit of length that is shorter than a centimetre

10 mm = 1 cm

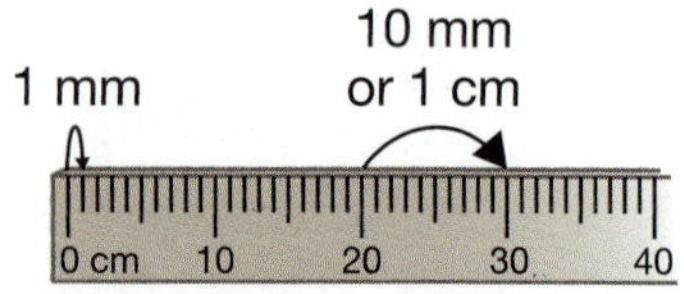

The space between the small markings is 1 mm.

The space from 20 mm to 30 mm is 10 mm, or 1 cm.

C. Measure and record the thickness of each stack of paper in millimetres.

D. How much thicker is the stack of construction paper?

Reflecting

1. How many millimetres are in 1 m? Explain how you know.
2. Why are long distances not measured in millimetres?
3. Why is it possible to be more precise when measuring in millimetres rather than centimetres?

Checking

4. Compare the thickness of 25 sheets of construction paper with the thickness of 25 sheets of white paper. How much thicker is the stack of construction paper?

Practising

5. Estimate in millimetres. Measure to check.
   a) the thickness of a CD case
   b) the length of a paper clip
   c) the width of your thumb

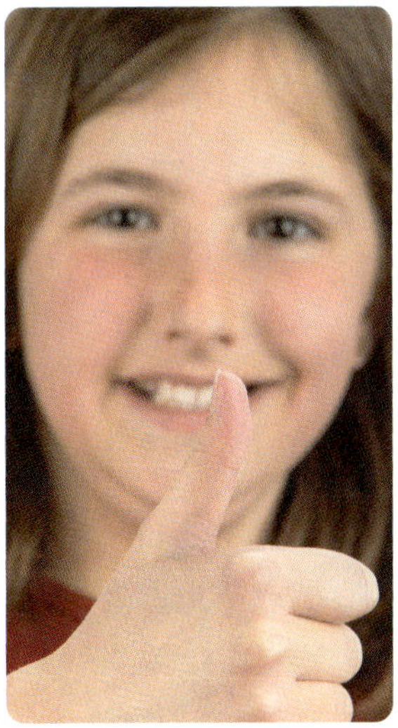

6. Order the measurements in Question 5 from shortest to longest.
7. a) Estimate. Is this line longer or shorter than 100 mm?

   b) Estimate how many millimetres you would add or subtract to make the line 100 mm long.
   c) Measure to check your estimate.
8. Name 3 things you might measure in millimetres.

CHAPTER 5

3 Record Measures Using Multiple Units

Goal **Measure and record using a combination of units.**

You will need

- a metre stick

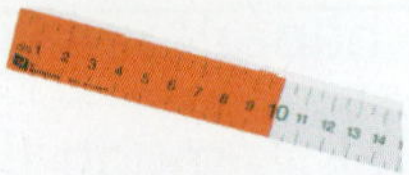

- a ruler

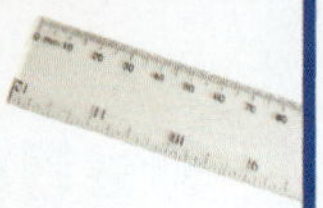

- string

- scissors

Pedro and Manitok competed in the long jump.
Both boys jumped a distance between 3 m and 4 m.
Manitok won by 3 cm.

Miki and Allison raced their snails.
Both snails travelled about 10 cm.
Allison's snail won by 3 mm.

? How could you record each winner's results?

A. Cut 2 lengths of string to represent 2 distances that Pedro and Manitok could have jumped. Each string should be between 3 m and 4 m long. 1 of the strings should be 3 cm longer.

B. Measure the string for the winning long-jump distance.
- **a)** Use both metres and centimetres: ■ m ■ cm
- **b)** Use only centimetres: ■ cm

C. Draw 2 lines to represent 2 distances that Miki's and Allison's snails could have travelled. Each line should be about 10 cm long. 1 of the lines should be 3 mm longer.

D. Measure the line for the winning snail's distance.
- **a)** Use both centimetres and millimetres: ■ cm ■ mm
- **b)** Use only millimetres: ■ mm

Reflecting

1. Why might someone record a measurement like 304 cm as 3 m 4 cm?
2. How could you make a measurement like 3 m 4 cm more precise?

Checking

3. Which measurement is more precise?
   a) 3 m or 2 m 25 cm
   b) 121 mm or 12 cm

4. Complete.
   a) 316 cm = 3 m ■ cm
   b) 175 mm = 17 cm ■ mm
   c) 3 m 15 cm = ■ cm
   d) 14 cm 8 mm = ■ mm

Practising

5. a) Measure the total length of this line using both centimetres and millimetres: ■ cm ■ mm
   b) Measure the line using only millimetres: ■ mm
   c) Draw a straight line that is the same length.

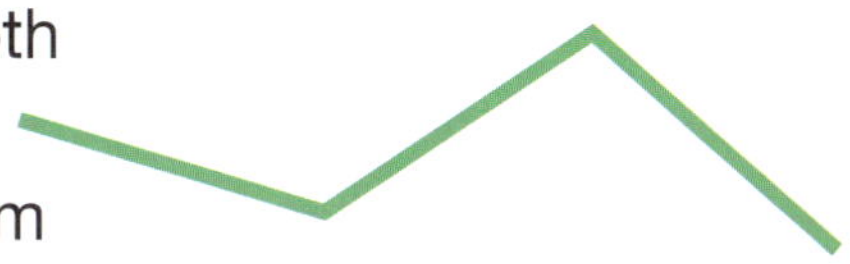

6. Complete.
   a) 37 mm = ■ cm 7 mm
   b) 4 m 8 cm = ■ cm
   c) 217 cm = ■ m ■ cm
   d) 15 cm 6 mm = ■ mm

7. Estimate each length. Work with a partner to measure the length as precisely as you can.
   a) your arm length
   b) your hand span

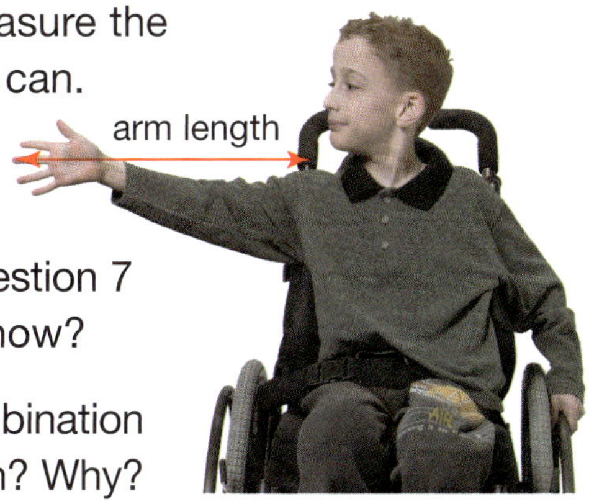

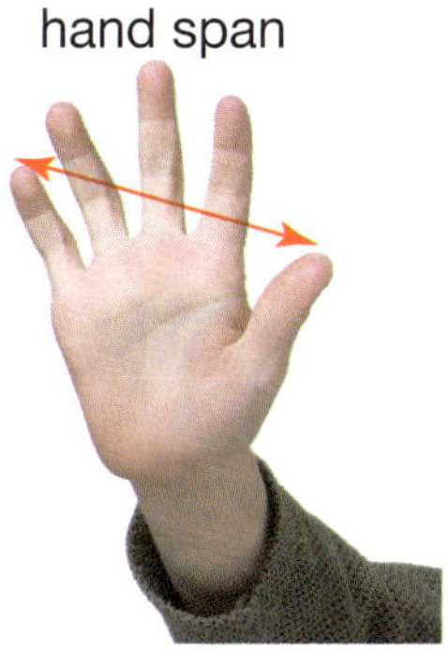

8. Which measurement in Question 7 is longest? How do you know?

9. When might you use a combination of units to measure a length? Why?

4 Using Kilometres

Goal **Use kilometres for measurements, and relate kilometres to other units.**

You will need

- a trundle wheel or a metre stick

The circus is coming to town. The tent is 2 **kilometres**, or 2 km, from Josef's school. Josef's class is deciding if 2 km is a reasonable distance to walk.

kilometre
A unit of length that is longer than a metre

1 km = 1000 m

? How far is 2 km?

Josef's Solution

I used a trundle wheel. I measured the distance around my school and the time it took to walk it. I recorded the information in a chart.

I figured it would take the same length of time to walk around the school again and again.

I used my chart to estimate the length of time it would take my class to walk 2 km.

Number of times around my school	Total time taken	Total distance travelled
1	7 minutes	450 m
2	14 minutes	900 m

A. Complete Josef's chart for 3 more rows.

B. How many metres would Josef walk if he went around the school 3 times?
Write this distance in kilometres and metres: ■ km ■ m

C. About how long would it take Josef to walk 2 km? Explain your thinking.

D. Is 2 km a reasonable distance for a grade 4 class to walk? Why do you think this?

Reflecting

1. Why do you think Josef measured the distance in metres to find out about kilometres?
2. What problem-solving strategies did Josef use?
3. Describe the distance 2 km in different ways. Explain your thinking for each description.

Checking

4. Complete. Explain how you know.
   a) 1644 m = ■ km 644 m b) 3208 m = ■ km ■ m
5. Complete by using >, =, or <. Explain your thinking.
   a) 720 km ■ 702 km b) 5 km ■ 5200 m

Practising

6. Use the data in Josef's chart.
   a) Estimate how long it would take Josef to walk 4 km. Explain how you estimated.
   b) About how far would Josef walk in 40 minutes? Explain how you estimated.
7. Complete.
   a) 3530 m = ■ km ■ m b) 1 km 6 m = ■ m
8. Complete by using >, =, or <. Explain your thinking.
   a) 3 km 7 m ■ 3070 km b) 6 km 410 m ■ 6410 m
9. How much longer or shorter than 2 km is each distance?
   a) 1500 m b) 2023 m c) 912 m
10. Measure a distance at your school. Use your measurement to describe a distance of 1 km.
11. Name 3 distances you might measure in kilometres.

5 Solve Problems by Drawing Diagrams

You will need
- grid paper

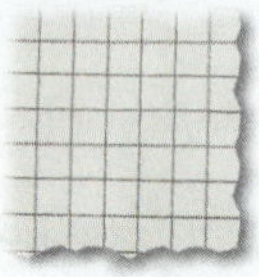

Goal Use diagrams to solve problems.

Sarah's team is playing in a soccer tournament near her home. Her home is also near this sign. Sarah and her mom must pick up players in Penticton and Peachland to bring them to the tournament.

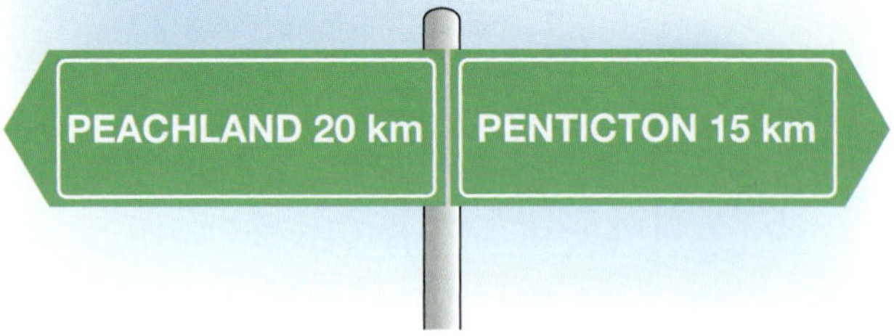

? How far will Sarah travel?

Sarah's Solution

Understand

My trip has 4 parts.

(1) from home to Penticton
(2) from Penticton to home
(3) from home to Peachland
(4) from Peachland to the tournament

Make a Plan

I will draw a diagram to show the parts of my trip.
Then I will add the distances to get the total distance that I will travel.

Carry Out the Plan

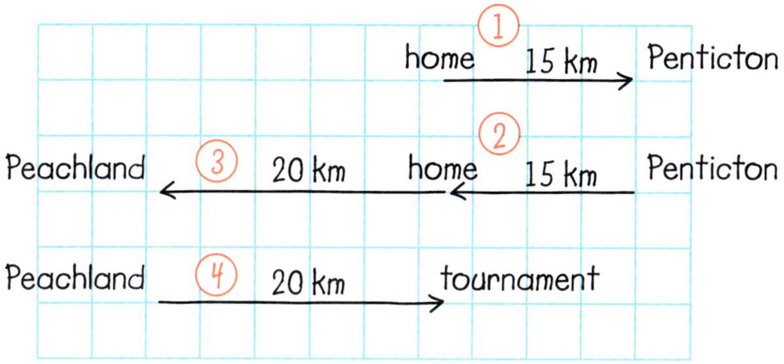

15 km + 15 km + 20 km + 20 km = 70 km

I will travel 70 km altogether.

Reflecting

1. How did Sarah's diagram help her to describe the problem?
2. Was drawing a diagram the only way to solve the problem? Explain your thinking.

Checking

3. Bobby sees this sign in a parking lot. How far will he have to hike to get drinking water, see the waterfall, and then return to the parking lot?

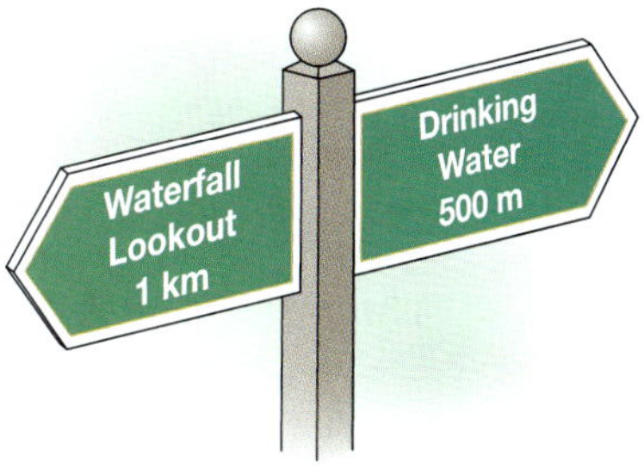

Practising

4. Maria lives 27 km east of Jeff. Zack lives 18 km east of Jeff. How far apart do Maria and Zack live?
5. Write a problem that can be solved using a diagram. Show your solution.

LESSON

Mid-Chapter Review

2 **1.** **a)** The length of a piece of paper is 3 dm. What is the length of the paper in centimetres?

b) The width of a recipe card is 8 cm. What is the width of the card in millimetres?

3 **2.** Five 9-year-olds measured and recorded their heights. Which of these height measurements are probably wrong? Explain your thinking.

Lisa	123 m
Michael	1 m 3 cm
Yuka	132 cm
Angie	182 cm
Carlos	1 m 28 mm

3. Measure each item as precisely as you can.

a) a pen **b)** a pencil case

4 **4.** What unit would you use for each measurement? Explain your thinking.

a) the width of a tack

b) the length of a finger

c) the distance between cities

5. Complete.

a) A room that is 487 cm long is ■ m ■ cm long.

b) A window that is 39 cm high is ■ dm ■ cm high.

c) A table that is 503 mm wide is ■ cm ■ mm wide.

d) A park that is 5480 m away is ■ km ■ m away.

6. Which measurement is more precise? Explain how you know.

a) 7 km or 7 km 540 m **b)** 1200 m or 1 km

5 **7.** Emma has a purse that is 15 cm long and 10 cm wide.

a) Can a pencil that is 120 mm long lie flat inside her purse? Explain your thinking.

b) Can a bookmark that is 1 dm 90 mm long lie flat inside her purse? Explain your thinking.

Cutting and Measuring

You will need

- ribbon

- scissors

1

A ribbon is 100 cm, or 1 m, long. If you fold the ribbon in half and make 1 cut at the fold, you get 2 pieces. Each piece is 50 cm long.

If you fold the metre of ribbon in half and make 1 cut through the middle, you get 3 pieces. 1 piece is 50 cm long and 2 pieces are 25 cm long.

How many pieces could you get if you folded the metre of ribbon twice and then made 1 straight cut? How long would each piece be? Explain your thinking.

2

Some units are even smaller than millimetres. Do some research to find out about nanometres. What types of things are measured in nanometres?

3

You may have heard length measurements given in units of feet. 3 feet is about 1 m. About how many metres is each length?

a) 6 feet
b) 15 feet
c) 30 feet
d) 10 feet

CHAPTER 5

6 Perimeter

Goal **Estimate, measure, and compare perimeters.**

You will need

- a ruler 
- a calculator 
- JILL name sign

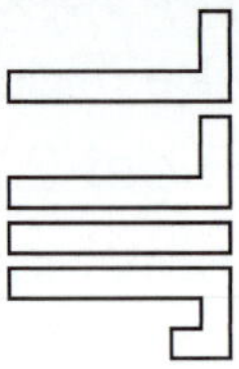

Miki is making a name sign for her friend Jill. Miki wants to cut out the letters and use ribbon to outline each letter. Miki has 1 m 50 cm of ribbon.

? Does Miki have enough ribbon to outline each letter in the JILL sign?

Miki's Measurement

I am going to measure around each letter.

A. Look at the JILL name sign. Which letter do you estimate has the least **perimeter**? Explain your answer.

B. Measure the perimeter of each letter. Which letter has the least perimeter?

C. Does Miki have enough ribbon to outline each letter in JILL? Explain how you know.

Reflecting

1. Why was it easier to find the perimeter of the letter I than the letter J or L?
2. Why would you measure the perimeter of each letter in centimetres when the length of Miki's ribbon is in metres and centimetres?

Checking

3. Estimate each perimeter. Measure to check.

a)

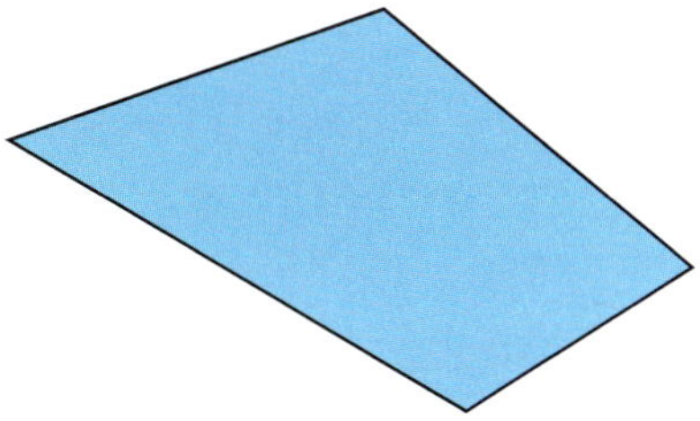

b)

Practising

4. Estimate each perimeter. Measure to check.

a)

b)

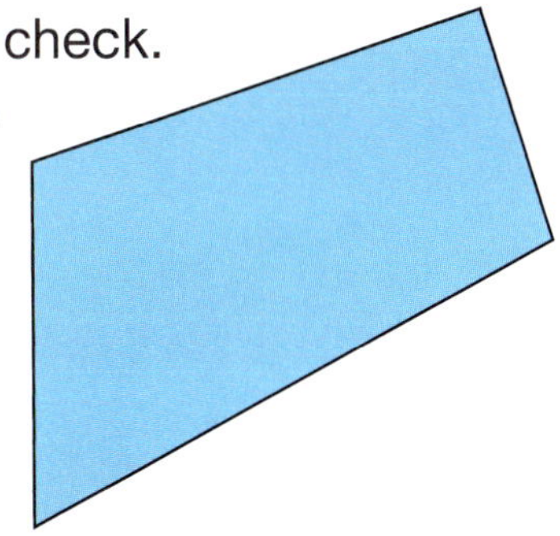

5. Find an object in your classroom with a perimeter that is close to each measurement.
   a) 15 cm b) 40 cm c) 1 m

6. Find 3 objects that you estimate have a perimeter that is greater than 3 m.
   a) Measure each perimeter. Record each perimeter using multiple units: ■ m ■ cm
   b) Order the objects from the greatest to least perimeter.

7. Use a ruler.
   a) Draw a quadrilateral with a perimeter of 32 cm.
   b) Draw a different quadrilateral with the same perimeter.
   c) Explain why different quadrilaterals can have the same perimeter.

8. The perimeter of a triangle is 45 cm.
   The lengths of 2 sides are 15 cm and 20 cm.
   What is the length of the other side?
   Show your work.

CHAPTER 5

7 Circumference

Goal **Estimate, measure, and compare circumferences.**

You will need

- cans

- a measuring tape

Jon is curious about which is greater, the height or the **circumference** of a can.

circumference
The distance around a circle or circular object. Circumference means the perimeter of a circle.

The circumference of this circle is shown in red.

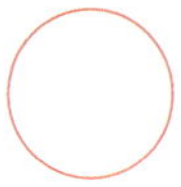

Jon's Question

I'm going to use a measuring tape to find out the height and the circumference of some cans.

Then I'll compare the measurements.

? Which is greater, the height or the circumference of a can?

A. Choose 5 different cans.
For each can, estimate which is greater, the height or the circumference.

B. Estimate the height of 1 of the cans in centimetres.
Record the estimate in a chart.

	Height		Circumference	
Can	Estimate (cm)	Measurement (cm)	Estimate (cm)	Measurement (cm)
Soup				

C. Measure the height of the can to the nearest centimetre.

Record the measurement in the chart.

D. Estimate the circumference of the can in centimetres.

Record the estimate in the chart.

E. Measure the circumference of the can to the nearest centimetre.

Record the measurement in the chart.

F. Repeat parts B to E for each of the other cans.

G. For each can, which is greater, the height or the circumference?

How does this compare with your answer for part A?

Reflecting

1. Suppose you did not have a measuring tape. How could you use string to measure the circumference of a can?
2. For the cans you used, do you think that the greater the height of a can, the greater its circumference?

   Order your cans from least to greatest height and then from least to greatest circumference to check your answer.
3. How are circumference and perimeter the same? How are they different?

8 Decades, Centuries, and Millenniums

You will need

- base ten blocks

- metre sticks

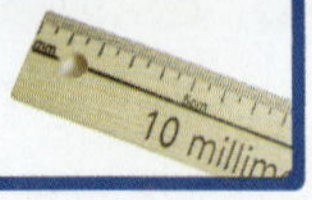

Goal Relate years, decades, centuries, and millenniums.

? How can you use length to make a model of years, decades, centuries, and millenniums?

A. What is the current year?
Use base ten blocks to model the number.

B. Draw a 10 cm long line to represent the age of a 10-year-old person.
This length represents 1 **decade**.
How does 1 decade compare to your age?

C. The Internet became popular about 1 decade ago.
What year was it 1 decade ago?

D. Use a metre stick to represent the age of a 100-year-old person.
This length represents 1 **century**.
How many decades are in 1 century?

E. The ice cream cone was invented about 1 century ago.
What year was it 1 century ago?

F. Use a line of 10 metre sticks to represent 1000 years.
This length represents 1 **millennium**.
How many centuries are in 1 millennium?
How many decades is that?

G. Playing cards were invented about 1 millennium ago.
What year was it 1 millennium ago?

decade
10 years

century
100 years

millennium
1000 years

H. Use a time line like the one below.
Estimate how long ago each event happened.
Write the letter of each event where it belongs on the time line.

a) Vikings settle in Newfoundland
b) first salmon cannery in B.C.
c) snowboarding becomes popular
d) first glass windows appear
e) the zipper becomes popular
f) wearing seat belts becomes law

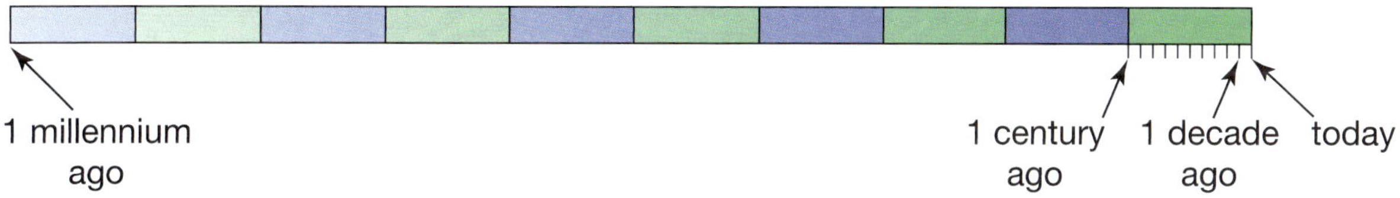

Reflecting

1. a) What things were different a decade ago? What things were the same?
   b) What things were different a century ago? What things were the same?
   c) What things were different a millennium ago? What things were the same?

Mental Imagery

Estimating Length

You will need

- a ruler

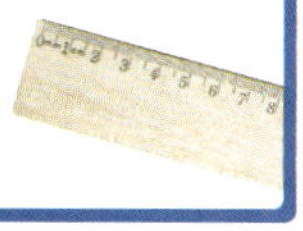

This line is 40 mm long and 1 mm thick.

Try These

1. Estimate the length and thickness of each line. Test your estimates by measuring.
   a)
   b)
   c)
   d)
   e)

9 Telling Time

Goal **Tell and write time to the nearest 5 minutes.**

Paulette arrived at her evening dance class before 7 o'clock.

? What time did Paulette arrive at her dance class?

Paulette's Clock

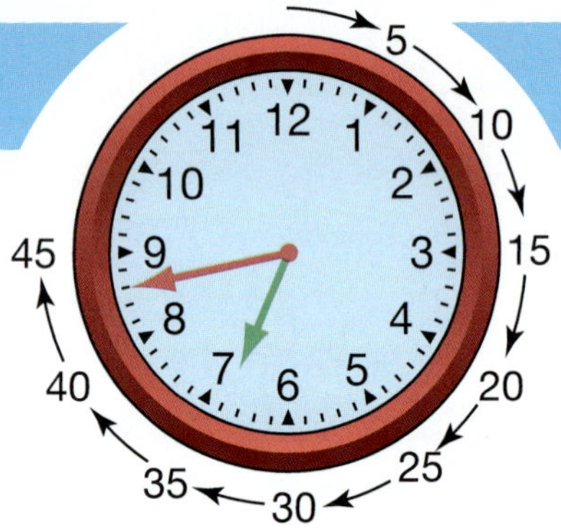

The **hour hand** shows that it is after 6 o'clock and before 7 o'clock.

The **minute hand** is not pointing exactly to the 9. It is pointing close to the 9.

I can skip count by 5s to tell the minutes past 6 o'clock.
5, 10, 15, 20, 25, 30, 35, 40, 45

I can say about forty-five minutes after six or about six forty-five.

Time between midnight and noon is called **a.m.**

Time between noon and midnight is called **p.m.**

I can write about 6:45 p.m.

I can skip count by 5s to tell the minutes until 7 o'clock.
5, 10, 15

Another way to say the time is about 15 minutes to seven or about a quarter to 7.

Reflecting

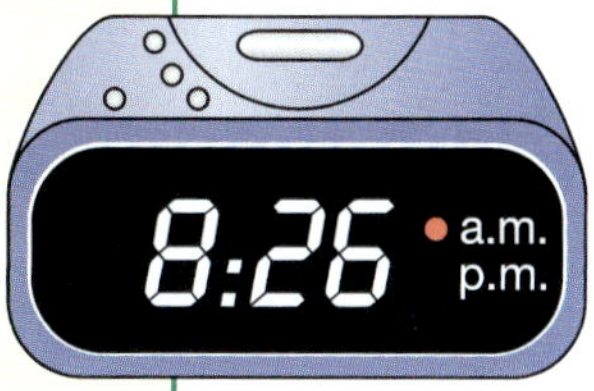

1. The clock shows the time Paulette left for school.
   a) Why is the time to the nearest 5 minutes 8:25 a.m. not 8:30 a.m.?
   b) Why is it 8:25 a.m. not 8:25 p.m.?
2. How does the hour hand show that the time Paulette arrived at her dance class is closer to 7:00 p.m. than to 6:00 p.m.?

Checking

3. Suppose this clock shows the time Paulette's dance class ended.
   a) Tell how to say the time to the nearest 5 minutes.
   b) Write the time to the nearest 5 minutes. Would you use a.m. or p.m.? Why?
   c) Suppose this clock shows the time the sun rose. Write the time to the nearest 5 minutes.

Practising

4. These clocks show times Paulette was at school. Write each time to the nearest 5 minutes. Use a.m. or p.m.

a) 
b) 
c) 
d) 

5. How could you say each time in Question 4?
6. Write each time to the nearest 5 minutes. Use a.m. or p.m.
   a) Last June 15, sunrise was at 4:29 and sunset was at 11:34 in Whitehorse.
   b) Last December 13, sunrise was at 10:01 and sunset was at 3:48 in Whitehorse.
7. Last night, Paulette's plane arrived in Winnipeg at 10:51. Write the time to the nearest 5 minutes. Use a.m. or p.m.

LESSON

Skills Bank

1

1. How long is each measurement in centimetres?

a) 3 dm **b)** 5 dm **c)** 7 dm **d)** 10 dm **e)** 12 dm **f)** 15 dm **g)** 20 dm **h)** 25 dm

2. You hair grows about 1 dm in 1 year. About how many years would it take to grow your hair to each length?

a) 20 cm **b)** 30 cm **c)** 1 m **d)** 60 cm **e)** 3 m **f)** 4 m **g)** 35 cm **h)** 45 cm

2

3. Give an estimate in millimetres. Then measure to check.

a) the thickness of a calculator
b) the thickness of an empty pencil case
c) the thickness of an eraser
d) the length of a CD case

4. A line is 63 mm long. How many millimetres would you need to add to the line to make a line of each length?

a) 7 cm **b)** 1 dm **c)** 12 cm

3

5. Measure each total length using centimetres and millimetres.

a) 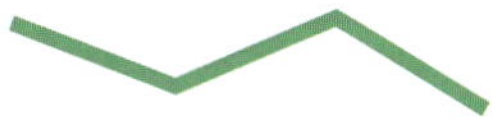

■ cm ■ mm

b) 

■ cm ■ mm

c) 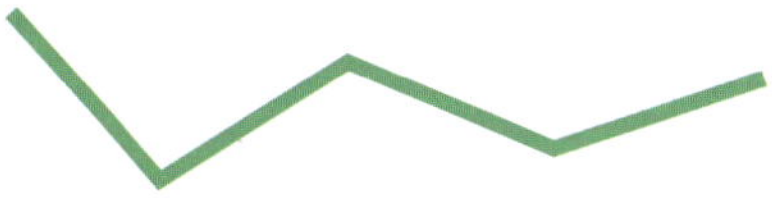

■ cm ■ mm

d) 

■ cm ■ mm

6. Complete.

a) 32 mm = ■ cm 2 mm
b) 47 mm = ■ cm 7 mm
c) 3 m 5 cm = ■ cm
d) 4 m 38 cm = ■ cm
e) 318 cm = ■ m ■ cm
f) 167 cm = ■ m ■ cm
g) 16 cm 15 mm = ■ mm
h) 21 cm 12 mm = ■ mm

4

7. Calgary is 1358 km from Winnipeg.
Regina is 587 km from Winnipeg.
Edmonton is 1308 km from Winnipeg.
Which city is farthest from Winnipeg? Which city is closest?

6 **8.** Find an object that you estimate has a perimeter that is about 2 m.
Measure the perimeter. Record it using multiple units: ■ m ■ cm

9. Measure each perimeter.

a)

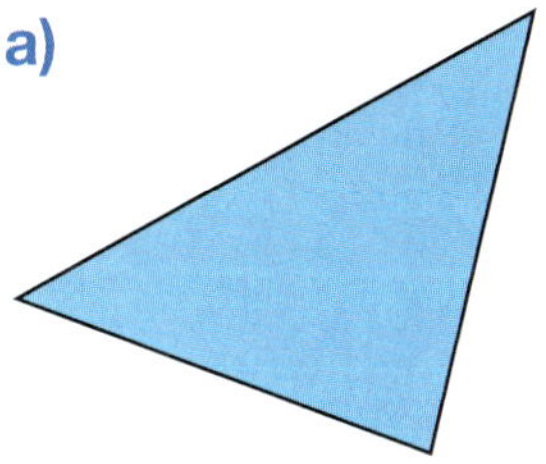

b)

7 **10.** a) Measure the circumference of 3 objects.
b) Order the objects from least to greatest circumference.

11. Name 4 objects that have a circumference you might measure in metres.

8 **12.** For each year, write the year 1 decade earlier.

a) 2050 b) 2000 c) 1995 d) 1967

13. For each year in Question 12, write the year 1 century later.

14. Complete.

a) 7 decades = ■ years
b) 2 millenniums = ■ years
c) 3 centuries = ■ years
d) 20 decades = ■ centuries

9 **15.** Doug's train arrived in Edmonton at 8:03, about an hour after sunrise.
Write the time to the nearest 5 minutes. Use a.m. or p.m.

16. Write each time to the nearest 5 minutes. Use a.m. or p.m.

a) getting ready for school

b) playing soccer

LESSON

Problem Bank

1 **1.** A plant grows about 1 dm every 2 years.
About how long does the plant take to grow 32 cm?

2 **2.** Your hair grows about 1 dm each year.
a) About how long does it take your hair to grow 1 cm?
b) About how long does it take your hair to grow 1 mm?

3 **3.** A regular hexagon has 6 sides of equal length.
Each side of this hexagon is 2 m 6 cm long.
Suppose you rounded each side length to 2 m to find its perimeter. How far off would your estimate be?

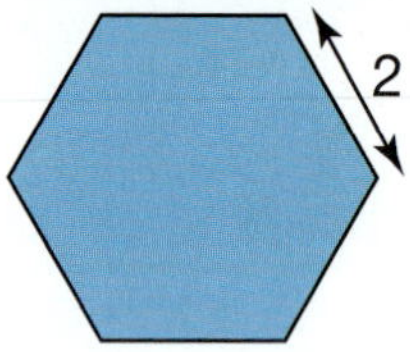

7 **4.** Suppose the circumference of the **red** circle is 7 dm.
Estimate the length of each side of the **blue** square.
Explain how you estimated.

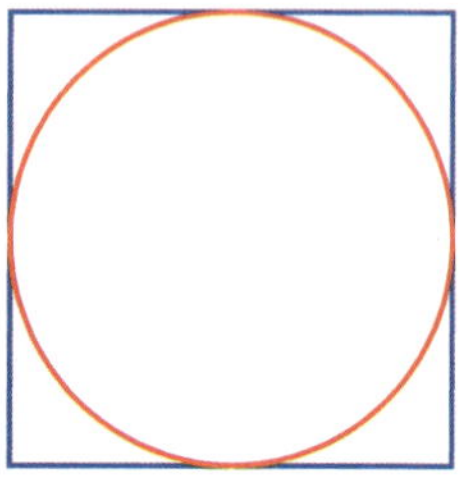

8 **5.** Leon's great grandfather was born in the 1st decade of the 20th century. What year could he have been born?
Why is more than 1 answer possible?

6. The 3rd millennium began in the year 2001.
Will a maple tree planted today be growing by the 1st year of the 4th millennium?
Explain how you know.

9 **7.** Ravi takes 25 minutes to read 17 pages of a book.
About how long will it take him to read 100 pages?

8. Suppose we measured time in decidays and centidays instead of in hours and minutes.
There would be 10 decidays in a day and 100 centidays in a day.
a) About how many hours would a deciday be?
b) About how many minutes would a centiday be?

Chapter Review

LESSON

2 **1.** Name an object that might be about each length.
a) 50 cm **b)** 12 mm **c)** 3 m

3 **2.** **a)** Draw a line that is 2 dm 1 cm long.
b) Draw a line that is 47 mm long.

3. Complete.
a) 165 mm = ■ cm ■ mm
b) 4 m 82 cm = ■ cm

4 **4.** Lee walked to the store. She said she walked 3 km. Do you think this was an exact measurement or an estimate? Explain your answer.

5. Which distance is greater? 10 000 m or 1 km
Explain how you know.

6 **6.** Estimate whether the perimeter of your desk or table is shorter or longer than 20 dm. Measure to check your estimate.

7. **a)** All the sides of this shape are equal in length. Estimate the perimeter of this shape.
b) Draw and label another shape with the same perimeter.

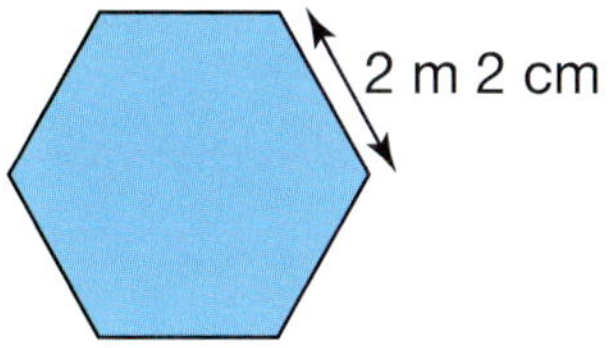

7 **8.** Find 3 objects that you estimate have a circumference greater than 2 dm. Measure each circumference in centimetres.

8 **9.** How many more years are in 2 millenniums than in 2 centuries?

9 **10.** Last July 1, sunrise was at 4:11 and sunset was at 10:28 in Churchill. Write each time to the nearest 5 minutes. Explain how you decided whether to write a.m. or p.m.

Chapter Task

Suncatchers

Beth wants to glue ribbon all around the edges of these 2 suncatchers. She also wants to hang each suncatcher by a loop that is 10 cm long. She has a piece of red ribbon that is 80 cm long.

? Does Beth have enough ribbon for both suncatchers?

Understand

A. What do you need to work out to solve the problem?

Make a Plan

B. What strategy will you use to solve the problem?

Carry Out the Plan

C. Measure as precisely as you can. Record your measurements.

D. Make the necessary calculations.

Look Back

E. Check that your solution fits the original problem.

Communicate

F. Explain why you need to measure precisely.

Task Checklist
✓ Did you check your measurements carefully?
✓ Did you check your calculations?
✓ Did you show all the steps?
✓ Did you use math language?

CHAPTER 6

Multiplication and Division Facts

Goals

You will be able to

- use number strategies to multiply and divide
- explain the relationship between multiplication and division
- apply the meanings of multiplication and division
- solve problems using models

Reflection in an office building

Getting Started

You will need

- a multiplication table

×	1	2	3
1	1	2	3
2	2	4	6
3	3	6	9

Patterns in a Multiplication Table

? What patterns can you find in a multiplication table?

diagonal → ×, column → 2, row → 1

×	1	2	3	4	5	6	7
1	1	2	3	4	5	6	7
2	2	4		8		12	14
3	3	6	9		15	18	
4	4	8	12	16		24	28
5	5	10		20	25		35
6		12	18	24		36	42
7	7	14			35		49

A multiplication table includes many different kinds of patterns. Some of these are

- adding patterns
- subtracting patterns
- digit patterns

Miki's Pattern

I found an adding pattern.

Each number in row 3 is the sum of the 2 numbers above it.

A. Look for patterns and use them to complete the multiplication table.

B. What kinds of patterns did you use? Make a list with examples.

C. Compare your patterns with someone else's. What patterns can you add to your list?

D. Write 5 multiplication facts from the multiplication table.

Do You Remember?

1. Count on.

a) 2, 4, 6, 8, ■, ■, ■

b) 10, 20, 30, 40, ■, ■, ■

c) 5, 10, 15, 20, ■, ■, ■

d) 20, 18, 16, 14, ■, ■, ■

2. Multiply or divide.

a) 5×2

b) $10 \div 5$

c) 5×4

d) $20 \div 5$

e) 6×5

f) $30 \div 5$

g) 5×5

h) $25 \div 5$

3. Explain how factors and products are related.

$4 \times 6 = 24$

factor factor product

1 Using Doubling to Multiply

Goal **Use repeated addition and doubling to multiply.**

Carmen is using green and yellow triangles to make 9 squares for a quilt for her baby sister.

Carmen's Quilt Squares

I want to calculate the number of green triangles I need.

There are 9 squares in my quilt.

There are 2 green triangles in each square.

I need 9×2 green triangles to make my quilt.

I add or skip count to find the number of triangles.

9 x 2 = 2 + 2 + 2 + 2 + 2 + 2 + 2 + 2 + 2

2, 4, 6, 8, 10, 12, 14, 16, 18

? How many green and yellow triangles does Carmen use for her quilt?

A. Write a multiplication sentence for the number of green and yellow triangles.

B. What is the product in the multiplication sentence in part A?

Reflecting

1. How does knowing 9 × 2 help you find 9 × 4?
2. Explain the meaning of **multiples**. Use the word multiplication in your explanation.
3. Why do you think the strategy Carmen used is called the doubling strategy?

multiples
The products when a whole number is multiplied by any whole number except 0

The multiples of 2 are 2, 4, 6, 8, 10, 12, ...

Checking

4. Carmen's new design is a square made of rectangles.
   a) How many yellow rectangles will Carmen use for 4 squares?
   b) Use doubling to find 8 × 3.

Practising

5. a) Count by 4s. 4, 8, 12, ■, ■, ■, ■
   b) How many 4s did you count?
   c) How much is 7 × 4?
   d) How much is 7 × 8?

6. Complete the Venn diagram to sort the numbers from 1 to 21.

Sorting Numbers

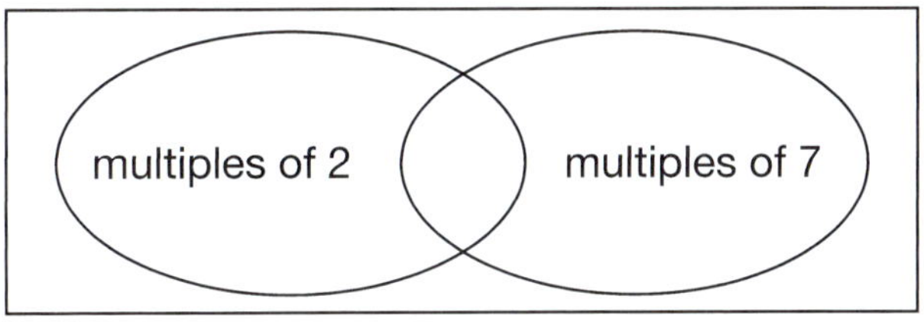

7. Multiply.
   a) 2 × 7　　b) 4 × 7　　c) 8 × 7

8. Find each product. Explain your strategy.
   a) 8 × 4　　b) 8 × 6　　c) 5 × 8

9. a) Design a quilt square with 3 shapes.
   b) Decide how many times your design will repeat in your quilt.
   Write multiplication sentences that describe the number of shapes in your quilt.

2 Sharing and Grouping

You will need

- counters

Goal **Use 2 meanings for division to solve problems.**

Here are 2 meanings of division.

Sharing

You know the number **of groups**.
You are making 6 tarts. How many raisins should be in each tart?

30 ÷ 6 = 5 raisins

(6: number of groups)

Grouping

You know the number **in each group**.
You put 5 raisins in each tart.
How many tarts can you make?

30 ÷ 5 = 6 tarts

(5: number in each group)

Rami is making 6 butter tarts with 30 raisins.
He wants to put the same number of raisins in each tart.

? How many raisins should be in each tart?

Rami's Tarts

I'll put 1 raisin in each tart until there are no raisins left.

There are 6 groups of 5.

I can put 5 raisins in each tart.

Reflecting

1. Which meaning of division did Rami use? Explain your thinking.
2. a) When you divide by the number of groups, what does the **quotient** tell you?
   b) When you divide by the number in each group, what does the quotient tell you?

quotient
The result when you divide

$30 \div 6 = 5$
↑
quotient

Checking

3. a) Write a division sentence for making 8 tarts with 32 raisins. How many raisins should be in each tart?
   b) Write a division sentence for putting 4 raisins in each tart.

Practising

4. Draw pictures or use counters to show each problem. Write a division sentence for each situation.
   a) There are 28 raisins. You put 4 raisins in each tart. How many tarts can you make?
   b) There are 42 raisins. You want to make 6 tarts. How many raisins should be in each tart?
5. Complete each division sentence.
   a) $12 \div 6 = ■$
   b) $24 \div 3 = ■$
   c) $18 \div 2 = ■$
   d) $18 \div 3 = ■$
6. Rami has 24 raisins.
   He wants every tart to have more than 1 raisin.
   What is the greatest number of tarts he can make?
   Show your work.
7. Use the division sentence $42 \div 7 = ■$ to write 2 problems.
   Make 1 problem about grouping.
   Make the other problem about sharing.
   Solve the problems.

3 Division and Multiplication

You will need

- number lines or play money

Goal **Relate multiplication to division.**

Sarah and her friends were paid $20 to clean up a neighbour's yard. They each got $5.

? How many friends helped Sarah?

Rey's Solution

I need to divide $20 by $5.

I subtract 5s from 20 until I am out of money.

This is like working backward on a number line.

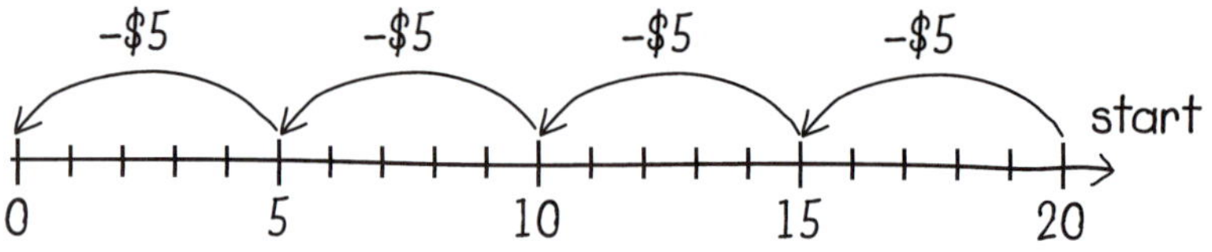

$$20 - 5 - 5 - 5 - 5 = 0$$

I take 4 groups of 5 from 20.

$$20 \div 5 = 4$$

This means 4 people got $5 each.

Since Sarah got $5, she must have had 3 helpers.

I can check by multiplying.

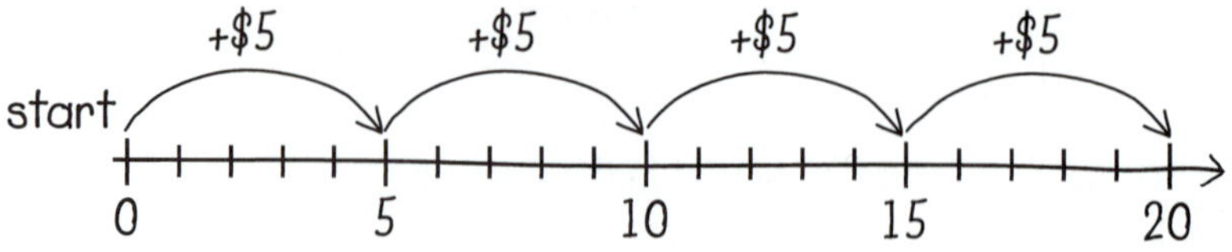

$$5 + 5 + 5 + 5 = 20$$

$$4 \times 5 = 20$$

Reflecting

1. Explain how you can divide by counting multiples.
2. How can you use multiplication to check division? Use an example to help you explain.

Checking

3. A clean-up team was paid \$16. Each person got \$4. How many people were on the team?

4. a) Write a division sentence for this number line.

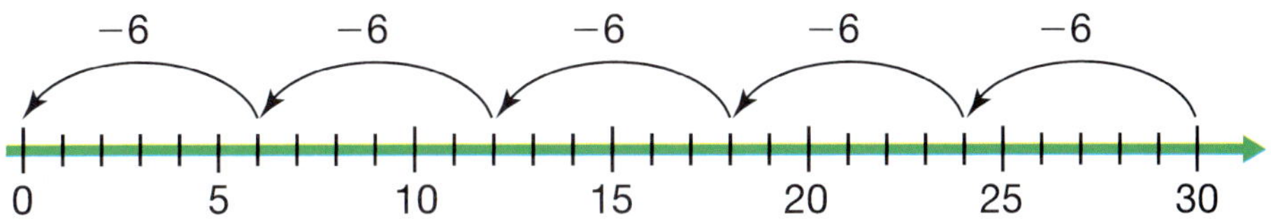

   b) Write a related multiplication sentence.

5. Divide. Check by multiplying.
   a) $18 \div 3$ b) $18 \div 6$ c) $28 \div 7$

Practising

6. Sarah and her 3 friends shared \$20 equally. Rey and his 2 friends shared \$18 equally. Which group would you choose to be in? Explain.

7. a) Write a division sentence for this number line.

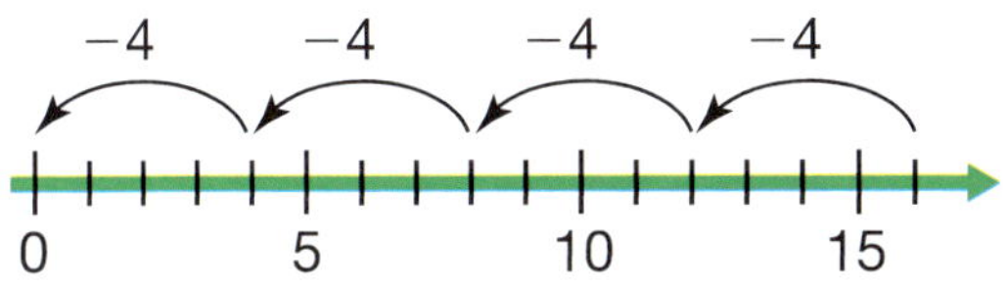

   b) Write a related multiplication sentence.

8. a) Write a division sentence for this subtraction.
   $24 - 6 - 6 - 6 - 6 = 0$
   b) Write a related multiplication sentence.

4 Arrays for Fact Families

You will need
- cards

Goal **Describe arrays using fact families.**

In the game Concentration, cards are used to make an **array**.

array
Objects arranged in the shape of a rectangle in rows and columns

A. How many rows does this card array have?

B. How many cards are in each row?

C. How many cards are in the whole array?

? How can you use multiplication and division to describe this array?

Columns go up and down.

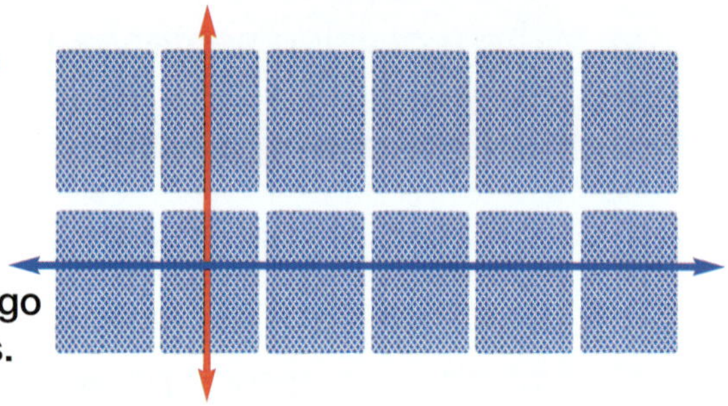

Rows go across.

D. Complete the multiplication sentences for the card array. Replace each symbol with the correct number.

a) Read: ■ rows with ● cards in each row = ◆ cards in all
Write: ■ × ● = ◆

b) Read:
● columns with ■ cards in each column = ◆ cards in all
Write: ● × ■ = ◆

E. Complete the division sentences for the card array.

a) Read: ◆ cards in ■ rows = ● cards in each row
Write: ◆ ÷ ■ = ●

b) Read:
◆ cards with ● columns = ■ cards in each column
Write: ◆ ÷ ● = ■

fact family
Number facts that describe an array. Each fact uses the same numbers.

3 rows and 4 columns:
$3 \times 4 = 12$ $\quad 12 \div 3 = 4$
$4 \times 3 = 12$ $\quad 12 \div 4 = 3$

F. Use cards to make your own array.
Write all the number sentences that describe the array.
These number sentences form the **fact family** for the array.

Reflecting

1. How does knowing $6 \times 2 = 12$ help you calculate $2 \times 6 = \blacksquare$?
2. a) How does knowing $6 \times 2 = 12$ help you check the quotient for $12 \div 6$?
   b) How does knowing $12 \div 6 = 2$ help you check the product for 6×2?
3. Can a fact family have only 2 number sentences? Explain your answer.

Checking

4. Complete the fact family.

   $2 \times \blacksquare = \blacklozenge$ $3 \times \bullet = \blacklozenge$

   $6 \div \bullet = \blacksquare$ $6 \div \blacksquare = \bullet$

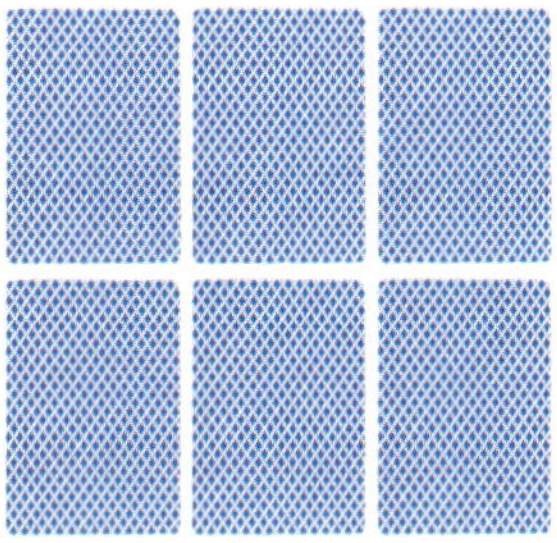

Practising

5. Use the numbers 6, 7, and 42 to create a fact family.
6. Write the fact family that goes with the array on the right. Explain how each fact describes the array.

7. a) Draw a different card array.
   b) Write the fact family for your cards.
8. Use the given fact to write the whole fact family.
   a) $8 \div 2 = 4$
   b) $3 \times 8 = 24$
   c) $45 \div 9 = 5$

5 Using Facts to Multiply Larger Numbers

You will need

- a calculator

Goal Use basic facts, patterns, and mental math to multiply.

Vinh sleeps about 8 hours every day.

? About how much time will Vinh spend sleeping in the month of April?

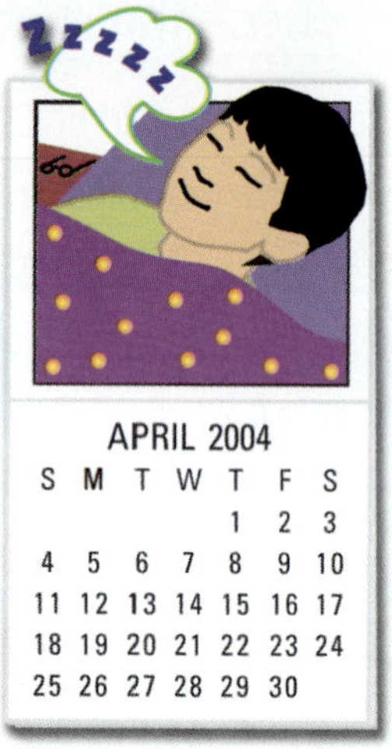

Vinh's Solution

Understand

There are 30 days in April.

I sleep about 8 hours each day.

This is a multiplication problem because there are 30 groups of 8.

Make a Plan

30 groups of 8 is 30×8.
I know that $30 \times 8 = 8 \times 30$ because they are in the same fact family.

I will multiply 8×30 because it is easier.

Carry Out the Plan

I know that $30 = 3$ tens.

I imagine 8 groups of 3 tens.

There are 24 tens.

I know that 24 tens = 240.

So $8 \times 30 = 240$.

8 x 3 tens = 24 tens

I will sleep about 240 hours in the month of April.

Reflecting

1. Explain how knowing $8 \times 3 = 24$ helps Vinh calculate 8×30.
2. Explain how Vinh could use his answer to find the number of hours he will sleep in May.
3. Explain how Vinh could calculate 8×300.

Checking

4. Show how each fact can help you multiply the others.

a)	b)	c)
3×5	7×7	5×4
3×50	7×70	5×40
3×500	7×700	5×400

Practising

5. Find the products. Use a calculator to check.

a) 6×40	d) 2×60	g) 6×50
b) 6×400	e) 2×600	h) 6×500
c) 6×4000	f) 2×6000	i) 6×5000

6. Show why these products are equal.
   a) $4 \times 50 = 5 \times 40$ b) $3 \times 500 = 5 \times 300$

7. a) How many tissues are in 7 boxes? Show your work.
   b) How many tablets are in 5 bottles? Show your work.

8. To estimate the **perimeter** of this regular **hexagon**, you can **round** 29 cm to 30 cm. Then multiply by 6. The perimeter is about 6×30 cm $= 180$ cm. Estimate the perimeter of regular hexagons with these side lengths.
   a) 68 cm   b) 42 m   c) 196 cm

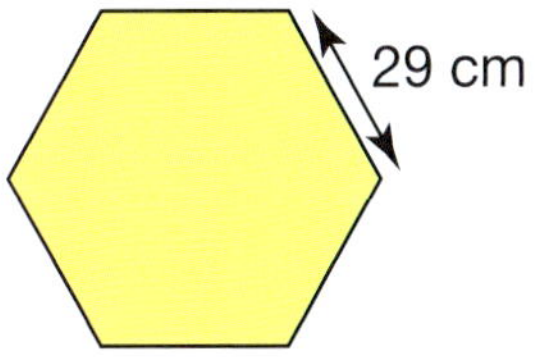

hexagon

6 Solve Problems by Making Models

You will need

- linking cubes

Goal Make models to solve problems.

Terry and his mother made 8 jars of jam.
Terry will put the jars on a shelf.

? How can Terry arrange the jam jars in arrays?

Terry's Arrays

Understand

I have to make different arrays of 8 jars.

Make a Plan

I can use linking cubes to make models.

Carry Out the Plan

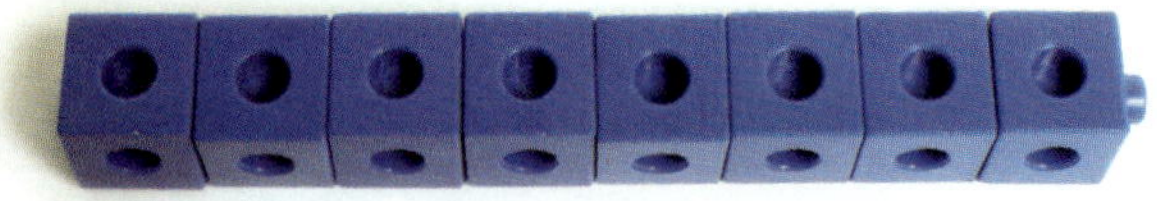

1 by 8
$1 \times 8 = 8$

2 by 4
$2 \times 4 = 8$

I can arrange the jars in 1 row of 8 jars, or 2 rows of 4 jars.

Reflecting

1. How did making a model help Terry solve the problem?
2. Has Terry found all the ways to arrange the jars in an array? Explain how you know.

Checking

3. Show as many ways as you can to arrange 18 jars of jam in arrays.

Practising

4. Katie has 15 apples. How many more apples does she need to make 5 rows of 4 apples?
   Show your work.
   Explain why your strategy is reasonable.

5. Justin packs apples. After he packs every 5 apples, he stamps a  on the box.
   Show the number of stamps for 35 apples.

6. 12 fruit trees are planted in equal rows.
   Show all possible arrangements.

7. A 2-by-2 array of blocks has a square top. It uses 4 blocks.
   If you have 30 blocks, what is the biggest square you can make?
   Show your work.

8. A number of jars of jam can be shared equally by 4 people. The same number can also be shared equally by 3 people.
   How many jars might there be?
   Show your work.

9. 2 out of every 3 apples grown in Canada are sold fresh.
   1 out of every 3 apples is processed to make a product like juice or sauce.
   Out of 15 apples, how many will be sold fresh? Show your work.

10. Write and solve a problem about arranging fruit.
    Ask someone else to solve your problem.
    Check the answer.

Mid-Chapter Review

LESSON

1 **1.** a) Count by 6s. 6, 12, 18, ■, ■, ■, ■
b) How many 6s did you count?
c) How much is 6×4?
d) How much is 6×8?

2 **2.** There are 42 apples in groups of 6.
How many groups are there altogether?
Show your work.

3 **3.** Give an example for each idea.
a) You can use division to find out how many groups there are.
b) You can use division to find out how many things are in each group.
c) Multiplication and division are related.
d) You can count multiples to divide.
e) You can count multiples to multiply.

4. Explain with words, pictures, or numbers.
a) How is multiplication related to addition?
b) How is division related to subtraction?

4 **5.** Use an array to show that $3 \times 4 = 4 \times 3$.

6. Complete each fact family.
a) $3 \times 8 = 24$ b) $28 \div 7 = 4$

5 **7.** Complete.
a) $7 \times 2 =$ ■ c) $7 \times 40 =$ ■ e) $7 \times 800 =$ ■
b) $7 \times 20 =$ ■ d) $7 \times 400 =$ ■ f) $7 \times 8000 =$ ■

6 **8.** Vikram is arranging 6 rows of 3 chairs for a show.
a) How many chairs are there?
b) How many chairs will there be if he adds another row? Show your work.

Curious Math

Multiplying and Dividing with 0

You will need

- pennies 
- a calculator 

1

a) If you give 0 groups of pennies to your friend, how many pennies will your friend get?

b) Explain what happens when you multiply a number by 0.

2

a) If you share 0 pennies with some friends, how many pennies will each person get?

b) Explain what happens when you divide 0 by another number.

3

a) If you have some pennies and you start to make groups of 0 pennies, when will you run out of pennies?

b) Explain why you can't divide by 0.

4

Complete each number sentence. Which ones can't be completed?

a) $5 \times 0 = ■$

b) $0 \div 5 = ■$

c) $0 \times 5 = ■$

d) $5 \div 0 = ■$

e) $0 \times 1 = ■$

f) $1 \div 0 = ■$

g) $0 \times 0 = ■$

h) $2 \div 0 = ■$

5

Enter any number in your calculator. Divide by 0. What does your calculator show?

6

Use an example to show how many facts can be in a fact family with 0.

7 Halving Strategies: Facts with 5 and 10

You will need

- counters

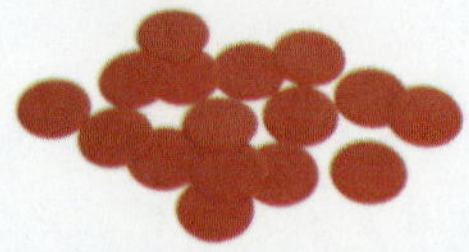

Goal **Find patterns in multiplication and division facts with 5 and 10.**

Natalie has 8 nickels. She wants to know how much money she has, but she can *never* remember 8 × 5.

? What strategy could Natalie use to figure out how much money she has?

Natalie's Strategy

I used 5 counters to represent each nickel.

8 groups of 5 is the same as 4 groups of 10.

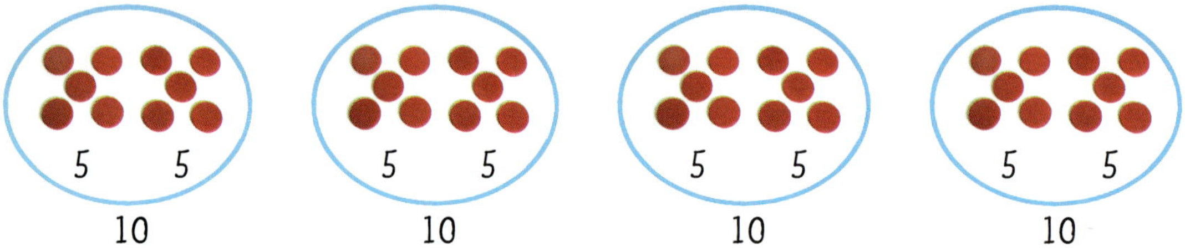

$8 \times 5 = 4 \times 10$

So now I know that 8 × 5 = 40.

Since 4 is half of 8 and 10 is double 5, Natalie's strategy is called **halve and double**.

halve
To divide a number by 2

Halve and Double Table

Multiply an even number by 5	Halve the even number and multiply by 10
2 × 5	1 × 10
4 × 5	2 × 10
6 × 5	3 × 10
8 × 5	4 × 10

Reflecting

1. What do you notice about the products in each row of the Halve and Double table?
2. What is the next row in the Halve and Double table? Explain your answer.
3. How could you use the table to find 7×5?

Checking

4. Show how to extend the Halve and Double table to find the value of 12 nickels.

5. Use the halve and double strategy to multiply.

   a) $8 \times 5 = ■ \times 10$
   $8 \times 5 = ◆$

   b) $16 \times 5 = ■ \times 10$
   $16 \times 5 = ◆$

Practising

6. Show how 4 groups of 5 is the same as 2 groups of 10.

7. a) What do you notice about the quotients in each row of this table?
   b) Explain how to use $40 \div 10$ to find $40 \div 5$.
   c) Explain why your strategy is reasonable.

Divide by 10	Divide by 5
$10 \div 10 = 1$	$10 \div 5 = 2$
$20 \div 10 = 2$	$20 \div 5 = 4$
$30 \div 10 = 3$	$30 \div 5 = 6$

8. Complete.

   a) $6 \times 5 = ■$
   b) $8 \times 5 = ■$
   c) $2 \times 5 = ■$
   d) $5 \times 4 = ■$
   e) $4 \times 5 = ■$
   f) $5 \times 8 = ■$
   g) $7 \times 5 = ■$
   h) $9 \times 5 = ■$
   i) $5 \times 3 = ■$

9. Complete.

   a) $5 \times 4 = ■$
   b) $5 \times 40 = ■$
   c) $5 \times 400 = ■$
   d) $6 \times 5 = ■$
   e) $6 \times 50 = ■$
   f) $6 \times 500 = ■$
   g) $7 \times 50 = ■$
   h) $7 \times 500 = ■$
   i) $7 \times 5000 = ■$

CHAPTER 6

8 Adding On: Facts with 3 and 6

You will need
- counters

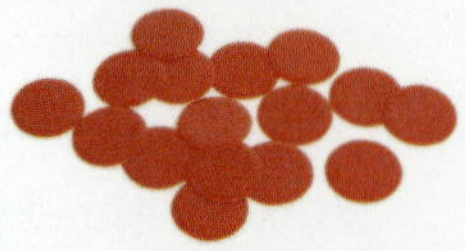

Goal **Use addition strategies to multiply and divide with 3 and 6.**

Paulette uses addition to find new facts from facts she already knows.

Paulette's Adding-On Strategy

I know that five 3s make 15.

$5 \times 3 = 15$

If I want to know 6×3, I can add 1 group of 3.

$6 \times 3 = 18$

So six 3s make 18.

? How can you use the adding-on strategy to complete this multiplication table?

×	3	4	5	6	7	8	9	10
5	15							
6	18							

A. Use counters to make 5 groups of 4. Write $5 \times 4 = ■$.

B. Add 4 counters to make 6 groups of 4. Write $6 \times 4 = ■$.

C. Complete the 4 column in the multiplication table.

D. Repeat parts A, B, and C to complete the remaining columns.

Reflecting

1. How are the rows for 5× and 6× related?

Checking

2. a) Complete a multiplication table for 2 and 3.
   b) How are the rows for 2× and 3× related?

×	3	4	5	10
2	6			
3	9			

Practising

3. Show how knowing 6 × 3 = 18 can help you find 7 × 3.

4. a) Complete a multiplication table for 3 and 4.
   b) How are the rows for 3× and 4× related?

×	1	2	3	10
3				
4				

5. Explain how one fact can help you find the other.
   a) 3 × 3 = 9, so 4 × 3 = ■.
   b) 6 × 7 = 42, so 7 × 7 = ■.
   c) 6 × 5 = 30, so 7 × 5 = ■.

6. a) Describe a pattern in this table.
   b) Explain how to use 30 ÷ 3 to find 30 ÷ 6.

Divide by 3	Divide by 6
6 ÷ 3 = 2	6 ÷ 6 = 1
12 ÷ 3 = 4	12 ÷ 6 = 2
18 ÷ 3 = 6	18 ÷ 6 = 3
24 ÷ 3 = 8	24 ÷ 6 = 4

7. Complete.
   a) 3 × 4 = ■
   b) 3 × 40 = ■
   c) 3 × 400 = ■
   d) 6 × 7 = ■
   e) 6 × 70 = ■
   f) 6 × 700 = ■
   g) 6 × 5 = ■
   h) 6 × 50 = ■
   i) 6 × 500 = ■

9

Subtracting Strategy: Facts with 9

You will need

- a 100 chart

1	2	3	4
11	12	13	14
21	22	23	24
31	32	33	34

Goal **Use counting patterns to multiply and divide with 9.**

9 is a special number for baseball players. There are

- 9 positions on the field
- 9 innings in a game

? Rey played 9 games in a tournament. How many innings did he play?

You can answer this question if you know the 9× facts. Patterns can help you find the 9× facts.

A. Use a 100 chart.

B. Colour every 9th number green and every 10th number blue.

row 1	1	2	3	4	5	6	7	8	9	10
row 2	11	12	13	14	15	16	17	18	19	20
row 3	21	22	23	24	25	26	27	28	29	30

C. Describe the green pattern in the chart.

D. Explain how the numbers in the green and blue squares in each row are related.

E. Describe how to use the relationship to find the number of innings Rey played.

Reflecting

1. Describe how to use a 10× fact to find a 9× fact.

Curious Math

Multiplying with 9

1

a) Complete the table.
b) Describe the patterns in the Tens and Ones columns.
c) Add the tens and ones digits in each product. What do you notice?
d) Describe how the numbers in the 9× column are related to the numbers in the Tens column.
e) Explain how to use patterns to find any 9× fact.

Products Table

9×	Tens	Ones
1	0	9
2	1	8
3		
4		
5		
6		
7		
8		
9		

Mental Math

Adding the Middle

Sometimes 2 numbers can be added by adding the middle number to itself. To add 47 + 53:

50 is the middle number.
50 + 50 = 100
So 47 + 53 = 100.

A. How do you know that 50 is the middle number?
B. Would you use this method to add 37 and 38? Explain.

Try These

1. Use this method to add.
   a) 39 + 41
   b) 28 + 32
   c) 66 + 74
   d) 297 + 303
   e) 153 + 147
   f) 496 + 504

CHAPTER 6

10 Number Neighbours: Facts with 7 and 8

Goal **Use facts you know to multiply and divide with 7 and 8.**

Shani's class is making bugs to decorate the set for a play. Shani is making 7 ladybugs. Each ladybug has 6 legs.

Shani's Ladybugs

I need 7×6 legs. I can't remember 7×6, but I know its neighbour, $6 \times 6 = 36$.

$6 \times 6 = 36$

6 12 18 24 30 36

I need another group of 6.

$7 \times 6 = ?$

$36 + 6 = 42$

So $7 \times 6 = 42$.

? How can you use other facts to show that $7 \times 6 = 42$?

A. Use $8 \times 6 = 48$.

B. Use $5 \times 7 = 35$.

C. Use $7 \times 7 = 49$.

Reflecting

1. List the neighbouring facts you could use to find 7 × 8. Explain your answers.

Checking

2. a) Complete the multiplication table.

×	5	6	7	8	9	10
6	30		42		54	
7		42		56		70
8	40		56		72	
9		54		72		90

b) Explain how the rows for 6× and 7× are related.
c) Explain how the rows for 8× and 9× are related.

Practising

3. Show how to use each fact to find 8 × 8.
a) 9 × 8 = 72 b) 8 × 7 = 56 c) 4 × 8 = 32

4. Matt has enough pipe cleaners for 60 legs. Can he make 8 spiders? Explain your strategy.

5. The Science Fair is in exactly 6 weeks. How many days is that? Show your work.

6. Explain how you can use the multiplication table in Question 2 to check each quotient.
a) 72 ÷ 8 = 9 b) 56 ÷ 7 = 8 c) 90 ÷ 10 = 9

7. Multiply.
a) 9 × 80 b) 800 × 7 c) 8 × 500

8. Write the fact family for each fact.
a) 7 × 8 = ■ b) 7 × 9 = ■ c) 8 × 9 = ■

Circles and Digits

You will need

- counting circles

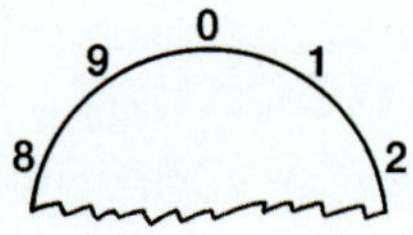

1

If you count by 2s from 0 to 20 and join the ones digits on a circle, the result looks like "Counting by 2s."

If you count by 8s from 0 to 80, the result looks like "Counting by 8s."

a) Make circles like these for all the other facts from 1× to 9×.

b) Which pairs of circles look alike?

c) What happened with 5×?

Counting by 2s (2×)

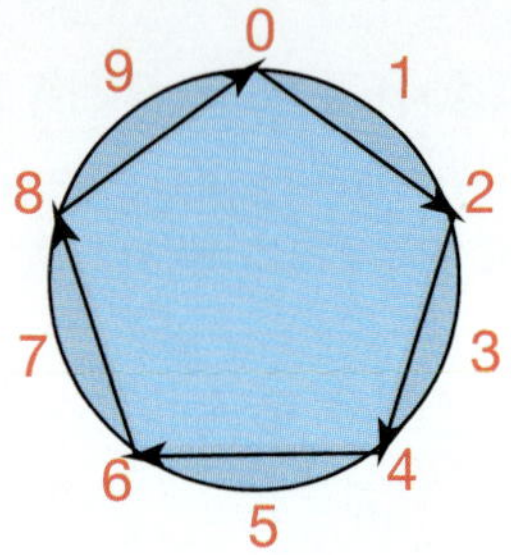

Counting by 8s (8×)

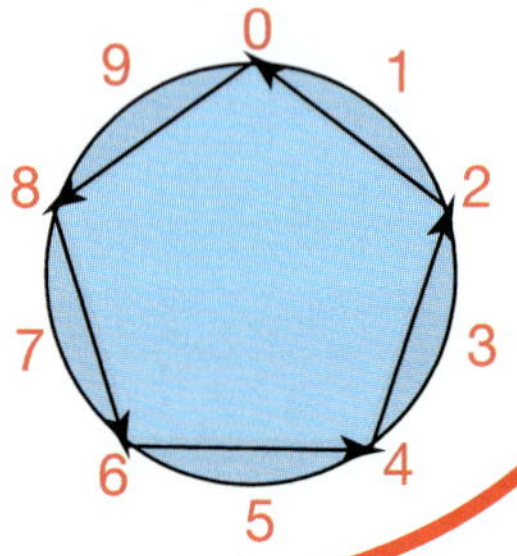

2

Look at fingers touching and not touching. How does it work? Try it with other facts. Explain it to someone else.

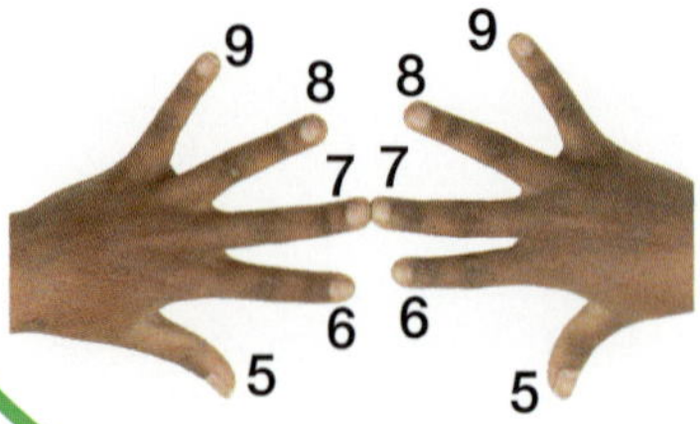

7 × 7 = 49

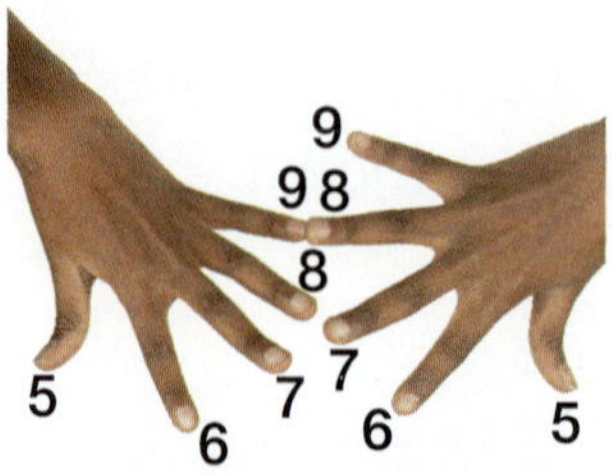

9 × 8 = 72

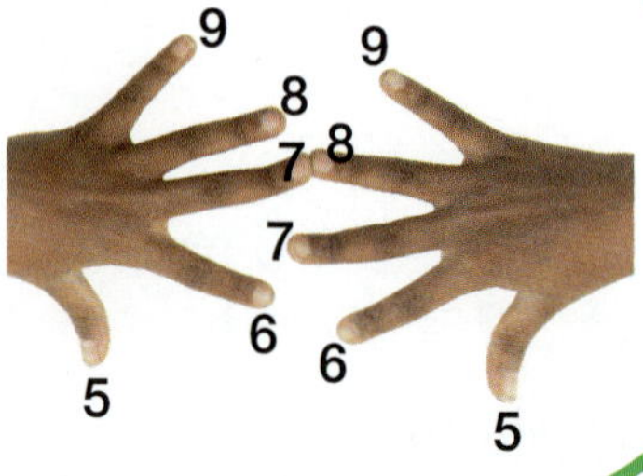

7 × 8 = 56

Math Cat

Number of players: 2
How to play: Add the **quotients** on your cards.

You will need

- division fact cards

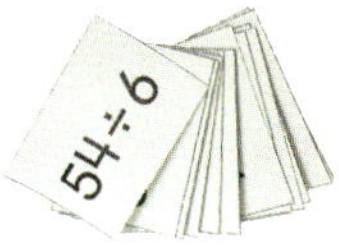

- counters

- Math Cat (optional)

Step 1 Deal 6 cards to each player.

Step 2 Each player keeps 4 cards and chooses 2 cards to give (face down) to Math Cat.

Step 3 Add the quotients for your cards to get your score. If you aren't sure of the quotients, think of patterns and fact families.

Step 4 Record the score for each player, including Math Cat. The player with the highest score gets 1 counter.

Allison	Vinh	Math Cat
28	30	29

Play until a player has 5 counters.

Allison's Turn

My quotients are 7, 8, 5, and 8.
My cards are worth 7 + 8 + 5 + 8, or 28 points.

42÷6	72÷9
25÷5	56÷7

Skills Bank

LESSON

1

1. Count on. Then multiply.

a) 5, 10, 15, ■, ■, ■, ■, ■ $8 \times 5 = $ ■

b) 7, 14, 21, ■, ■, ■ $6 \times 7 = $ ■

2. Multiply.

a) $5 \times 2 = $ ■
$5 \times 4 = $ ■

b) $2 \times 8 = $ ■
$4 \times 8 = $ ■

c) $6 \times 2 = $ ■
$6 \times 4 = $ ■

3. Complete the Carroll diagram to sort as many of the numbers 5, 6, 9, 14, 16, 18, 20, 24, 27, 30, 32, 36, 40, 45, 49, 54, 58, 63, 73, 81 as possible.

Sorting Numbers

	Multiples of 5	Multiples of 9
Even		
Odd		

2

4. Write a division sentence for each.

a) There are 21 cookies.
There are 3 cookies in each package.
How many packages are there altogether?

b) There are 15 cookies.
There are 5 packages altogether.
How many cookies are in each package?

c) There are 42 students.
There are 6 groups altogether.
How many students are in each group?

d) There are 27 students.
There are 3 students in each group.
How many groups are there altogether?

3

5. Write a division and a multiplication sentence for each.

a)

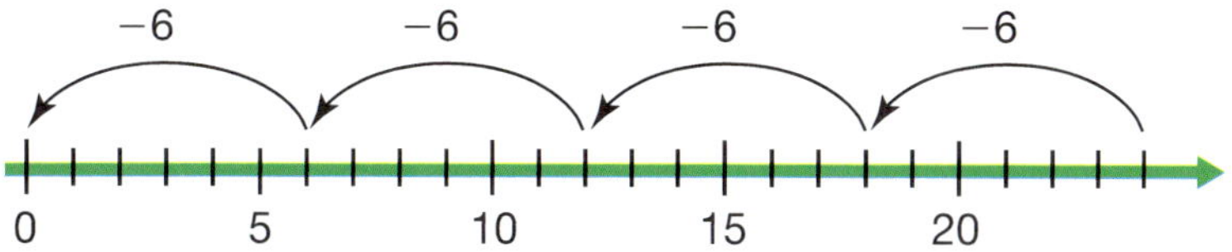

b)

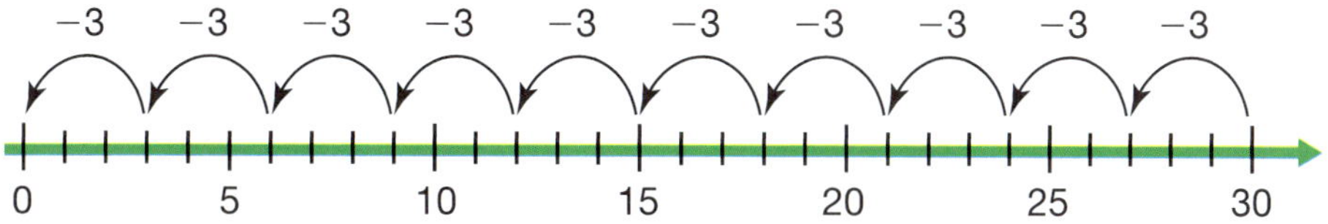

c) $40 - 5 - 5 - 5 - 5 - 5 - 5 - 5 - 5 = 0$

6. Divide. Check by multiplying.

a) $18 \div 2$
b) $18 \div 3$
c) $20 \div 4$
d) $20 \div 5$
e) $24 \div 4$
f) $24 \div 6$

7. Write the missing number for each section.

a) yellow
b) blue
c) red
d) green
e) orange
f) black
g) purple
h) brown
i) pink
j) grey

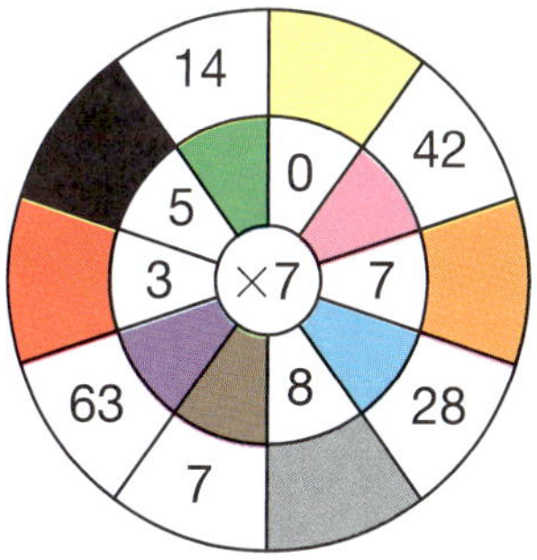

4

8. Write the fact family for each.

a) 7, 3, 21
b) $5 \times 6 = ■$
c)

d) $8 \div 4 = ■$
e) 4, 6, 24
f)

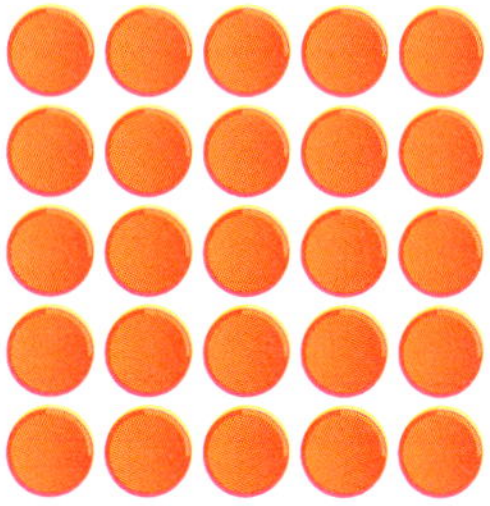

g) $40 \div 5 = ■$
h) $2 \times 4 = 8$
i)

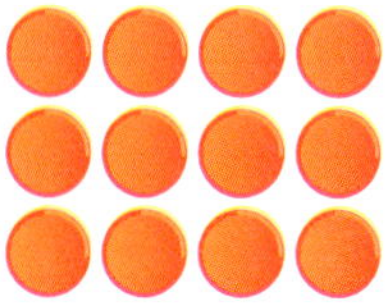

5

9. Find the product.

a) 2×70
b) 7×700
c) 5×700
d) 6×50
e) 7×7000
f) 4×60

5 **10.** A box holds 200 straws.
How many straws are in 4 boxes?

6 **11.** Solve each problem. Show your work.

a) 20 tulips are planted in equal rows. Show all possible arrangements.

b) Kyle has 23 stamps. How many more stamps does he need to make 8 rows of 6 stamps?

c) A number of marbles can be shared equally by 3 people. The same number of marbles can also be shared equally by 5 people. How many marbles might there be?

7 **12.** Complete.

a) $4 \times 5 = ■ \times 10$ **b)** $■ \times 5 = 3 \times 10$ **c)** $8 \times ■ = 4 \times 10$

13. Complete.

a) $4 \times 5 = ■$ **c)** $40 \div 5 = ■$ **e)** $8 \times 50 = ■$

b) $6 \times 5 = ■$ **d)** $7 \times 5 = ■$ **f)** $20 \div 5 = ■$

8 **14.** Use the first fact to find the second fact.

a) $4 \times 3 = ■$, so $5 \times 3 = ■$.

b) $7 \times 3 = ■$, so $8 \times 3 = ■$.

c) $3 \times 6 = ■$, so $4 \times 6 = ■$.

15. Complete.

a) $9 \times 60 = ■$ **b)** $36 \div 6 = ■$ **c)** $21 \div 3 = ■$

9 **16.** Complete.

a) $8 \times 9 = ■$ **c)** $3 \times 9 = ■$ **e)** $7 \times 9 = ■$

b) $9 \times 9 = ■$ **d)** $9 \times 20 = ■$ **f)** $9 \times 300 = ■$

10 **17.** Show how to use each fact to find 7×8.

a) $7 \times 7 = 49$ **b)** $8 \times 8 = 64$ **c)** $7 \times 4 = 28$

18. Complete.

a) $56 \div 7 = ■$ **c)** $7 \times 80 = ■$ **e)** $800 \times 9 = ■$

b) $7 \times 9 = ■$ **d)** $8 \times 9 = ■$ **f)** $64 \div 8 = ■$

LESSON

Problem Bank

2 **1.** Every 4th year is a leap year.
In leap years, there is an extra day on February 29.
Omar was born on February 29.
How many actual birthdays will Omar have had when he is 28 years old?

3 **2.** Estimate the number of times you can multiply 2 by itself before you pass 1000.
Check with a calculator. $2 \times 2 \times 2 \times 2$ ...

3. Use these numbers to make 4 division sentences.
Use each number only once.

9, 6, 28, 3, 7, 30, 24, 4, 2, 5, 8, 18

6 **4.** Some friends shared \$35 evenly.
They received more than \$5 each.
How many friends shared the money?
Show your work.

5. There are 50 boxes with 6 cans in each box.
How many cans are there in total?
Explain your answer.

6. 2 steps turn each of Jack's numbers into one of Jill's numbers.
What are the 2 steps?
Explain using more examples.

Jack		Jill
3	→	11
4	→	15
5	→	19
7	→	27
10	→	39

7. Write a number sentence for each statement.

a) There are ■ sides in 8 hexagons.
b) There are 24 sides in ■ hexagons.

7 **8.** Norah has 3 dimes and 9 nickels.
How much money does she have altogether?
Show your work.

8 **9.** **a)** Which number doesn't belong in the group 12, 16, 24, 30, 42? Explain.
b) Write another number that belongs.

10 **10.** A whole number multiplied by itself makes a square number.

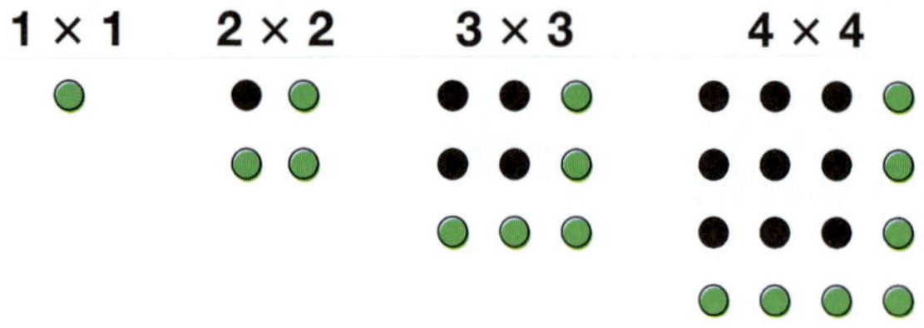

Whole numbers	Green circles	Square numbers
1	1	1
2	3	4
3	5	9
4	7	16
5	9	
6	11	
7	13	
8	15	
9	17	

a) Describe the number pattern for the green circles.
b) What do you add to a square number to get the next square number?
c) Continue the pattern to complete the Square numbers column.
d) What is the relationship between the Whole numbers column and the Green circles column?
e) Describe 2 ways a 4×4 array shows $4 + 4 + 4 + 4$.
f) Does a 3×3 array show the sum of 3 odd numbers? Explain how you know.
g) Does a 6×6 array show the sum of 6 odd numbers? Explain how you know.

Chapter Review

LESSON

2 **1.** 5 friends are sharing 20 jellybeans.
How many jellybeans will each person get?

2. There are 6 eggs in 1 carton.
How many eggs are in 3 cartons?

3 **3.** Which quotient is greater, $32 \div 4$ or $32 \div 8$?
Explain how you can tell without dividing.

4 **4.** Tara made an array with 36 tiles.
- **a)** Show as many different ways to make the array as you can.
- **b)** Write the fact family for each array.

8 **5.** Find each result. Explain what you did.
- **a)** Use $4 \times 6 = 24$ to find $24 \div 6$.
- **b)** Use $6 \times 5 = 30$ to find 3×5.
- **c)** Use $2 \times 5 = 10$ to find 4×5.
- **d)** Use $10 \times 3 = 30$ to find 5×6.
- **e)** Use $5 \times 3 = 15$ to find 10×3.
- **f)** Use $7 \times 8 = 56$ to find 8×8.

6. Maggie says "5×8 is 40, so 6×8 must be $40 + 6$, or 46."
Write a note to explain why her reasoning is wrong.

10 **7.** Are all these answers correct?
Check by multiplying. Correct any errors.
- **a)** $12 \div 2 = 8$
- **b)** $63 \div 7 = 8$
- **c)** $25 \div 5 = 5$
- **d)** $24 \div 4 = 4$
- **e)** $81 \div 9 = 8$
- **f)** $60 \div 10 = 6$

8.
- **a)** Complete the multiplication table.
- **b)** Colour any square that contains a product that is an odd number.
- **c)** How are odd and even products related to their factors?

×	6	7	8	9	10
6					
7					
8					
9					
10					

Chapter Task

Arrays on Parade

You will organize the school band so it can march in this year's community parade.

- The band will have between 60 and 85 people.
- The band will march in rows and columns.
- Each row has the same number of people in it.
- You cannot have rows or columns with more than 9 people.

? How can you arrange the people in the band?

A. Create all the arrays you can to fit between 60 and 85 people in the band.

B. Write the fact family for each of your arrays.

C. Explain how you know you found all the possible arrays in part A.

D. Choose 1 array. Explain how the fact family describes it.

Task Checklist

- ☑ Did you find all the arrangements where rows and columns are not greater than 9?
- ☑ Did you organize your work so it is easy to follow?
- ☑ Did you explain your thinking?
- ☑ Did you use math language?

CHAPTER 7

2-D Geometry

Goals

You will be able to

- identify and describe lines
- compare angles with right angles
- identify and sort 2-dimensional (2-D) shapes
- create and describe symmetrical 2-D shapes
- solve geometry problems by acting them out

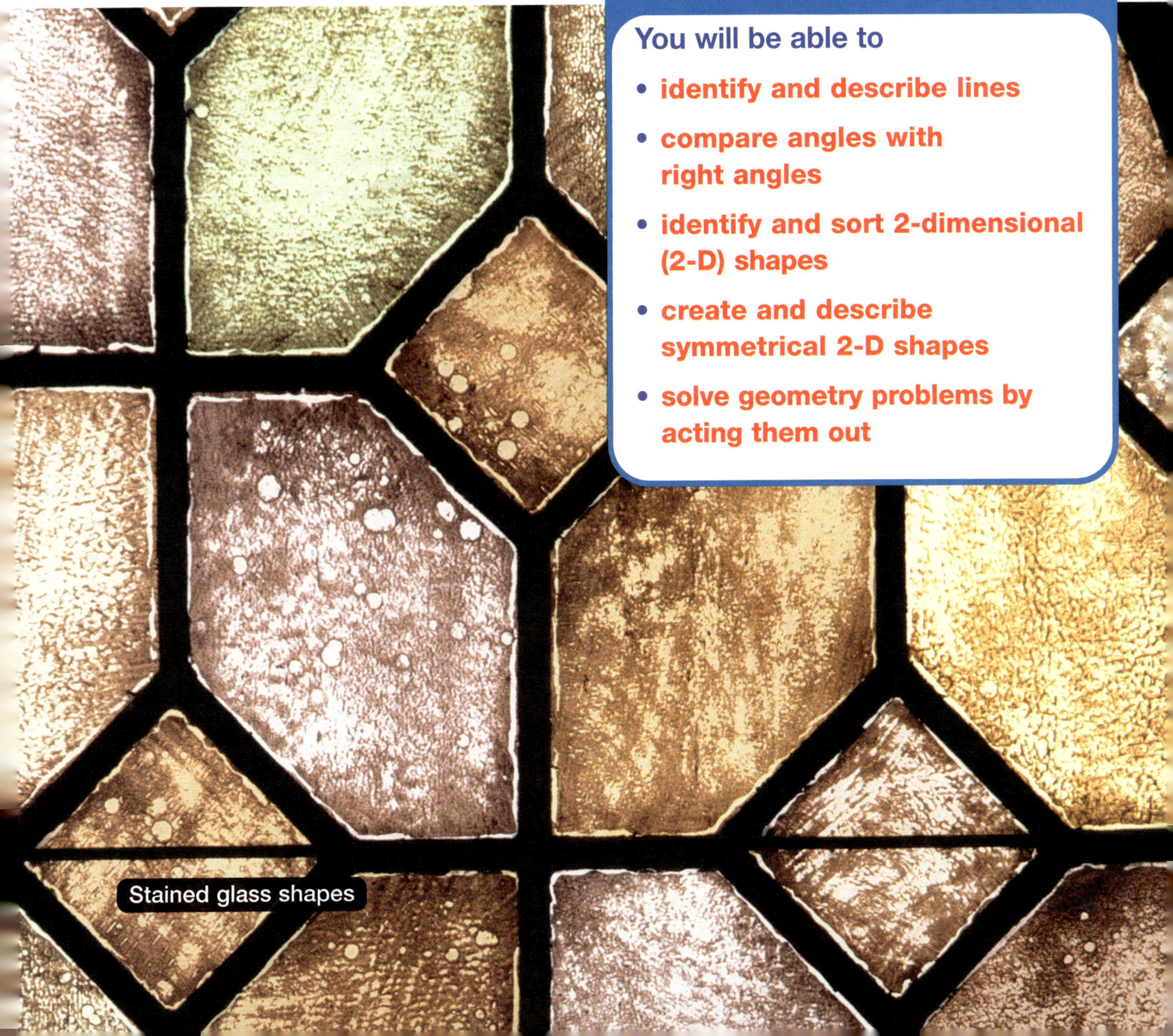

Stained glass shapes

Getting Started

Exploring Geometry with Puzzles

You will need

- pattern blocks

- a rectangle frame

- a ruler

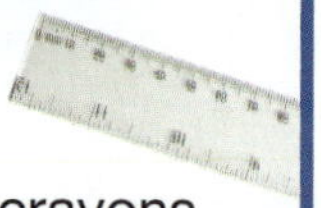

- pencil crayons

The butterfly puzzle is made of eight 2-D pieces.

? How can you create a puzzle that follows these rules?

- The whole puzzle must be a rectangle.
- There are 8 or more pieces.
- At least 2 pieces are **congruent**.
- At least 1 piece is a **square**.
- At least 1 piece is a **rectangle**.
- At least 1 piece is a **triangle**.

A. Does the butterfly puzzle follow the rules?
Explain your answer.

B. Sketch a plan for a puzzle that follows the rules.

C. Use a ruler to draw your puzzle pieces or trace around pattern blocks.

D. Create your puzzle.

E. Some of your pieces have **square corners**.
What is the smallest number of these pieces that could be in your puzzle?
Explain your answer.

F. Show how you could cut pieces in your puzzle in ways that would create more rectangles.

Do You Remember?

1. Which shapes match each name?
   a) rectangle **b)** hexagon **c)** pentagon **d)** triangle

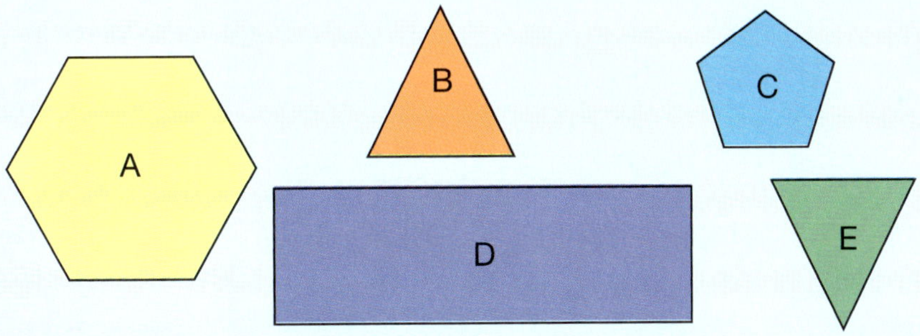

2. Which shapes do not have square corners?
3. Sort the 5 shapes using 2 **attributes**.
   Record and label your sorting.
4. Which 2 shapes are congruent?
   Explain your answer.

1 Classifying Angles

Goal **Describe angles by comparing them to right angles.**

Triangles have 3 sides and 3 **angles**.

? How can you use a 24 cm loop of string to make triangles with different size angles?

Vinh's Triangle

I used a 24 cm loop of string to make a triangle.

I made the sides 6 cm, 8 cm, and 10 cm.

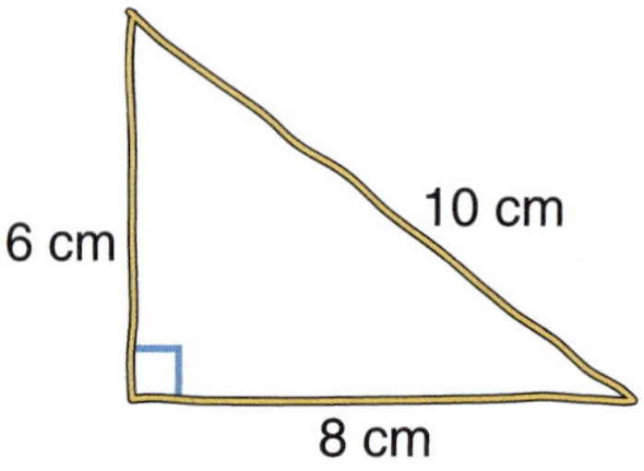

I compared each angle to a square corner.

The angle with the **blue** square is a **right angle** because it's a square corner.

The angle with the **red** arc is less than a right angle because it's smaller than a square corner.

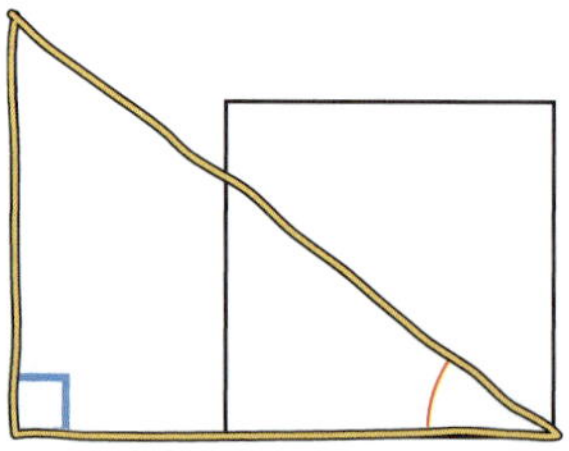

A. Compare the other angle in Vinh's triangle to a square corner.
Is it a right angle, less than a right angle, or greater than a right angle?

B. Work with a partner. Use a 24 cm loop of string and tape to make different triangles. Copy at least 4 triangles on paper.

You will need

- a ruler 
- string 
- tape 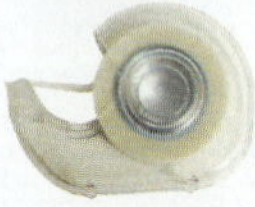
- a square corner 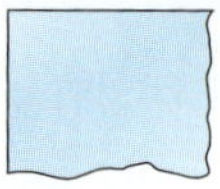

angle
The space between 2 arms that meet at a vertex

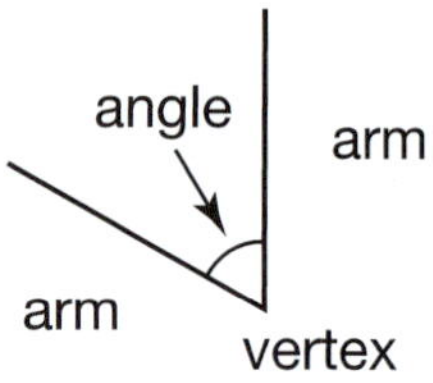

The arc shows the angle.

right angle
An angle that forms a square corner

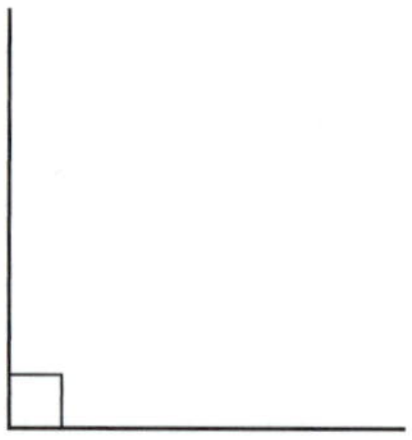

The square symbol shows the angle is a right angle.

C. Look at each triangle you drew.
  a) Mark each right angle with a **blue** square.
  b) Mark each angle that is less than a right angle with a **red** arc.
  c) Mark each angle that is greater than a right angle with a **green** arc.

Reflecting

1. Why do you think a right angle can be called a square corner?
2. Explain each answer. Include a drawing with each explanation.
  a) Can a quadrilateral have all right angles?
  b) Can a triangle have all right angles?

Checking

3. Is each angle a right angle, less than a right angle, or greater than a right angle?

a)
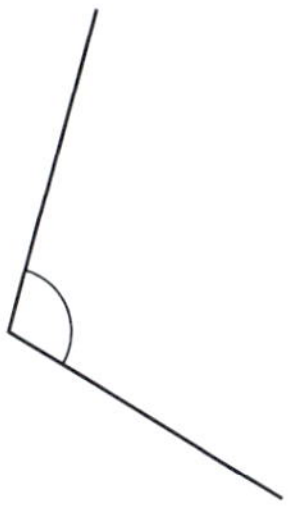

b)

c)
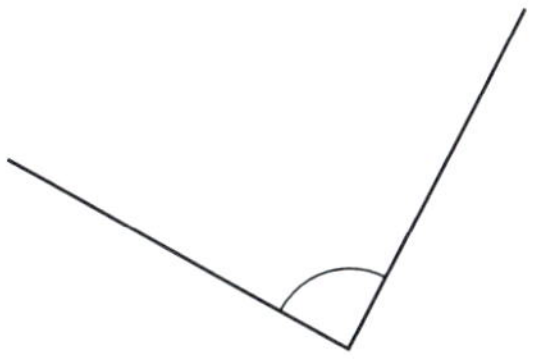

Practising

4. Is each angle a right angle, less than a right angle, or greater than a right angle?

a)
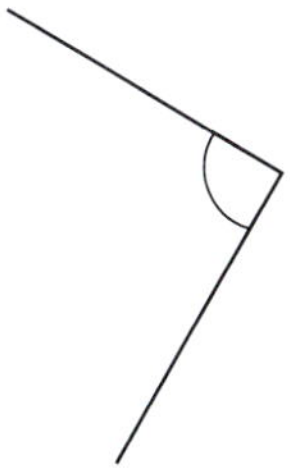

b)
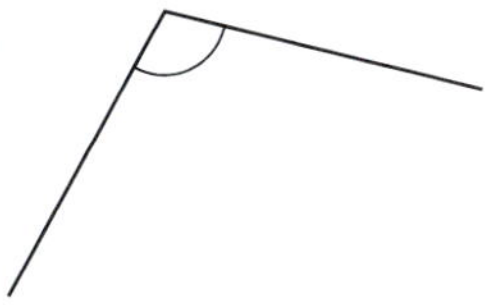

c)
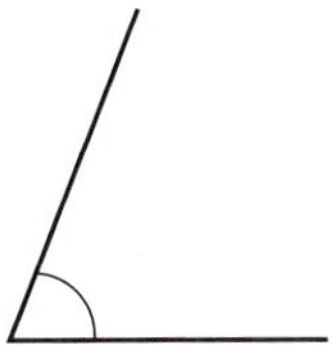

5. Look at each marked angle. Is it a right angle, less than a right angle, or greater than a right angle?

a)
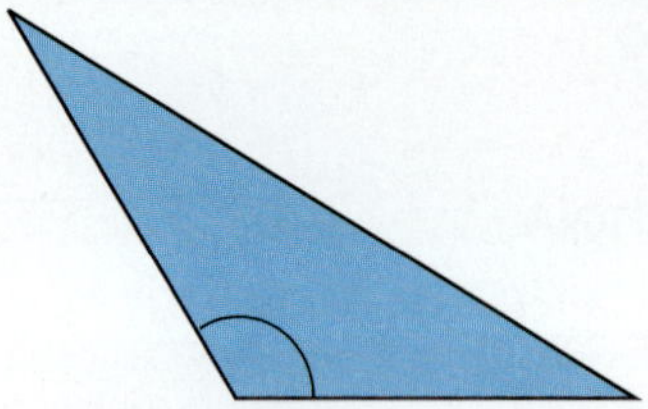

c)
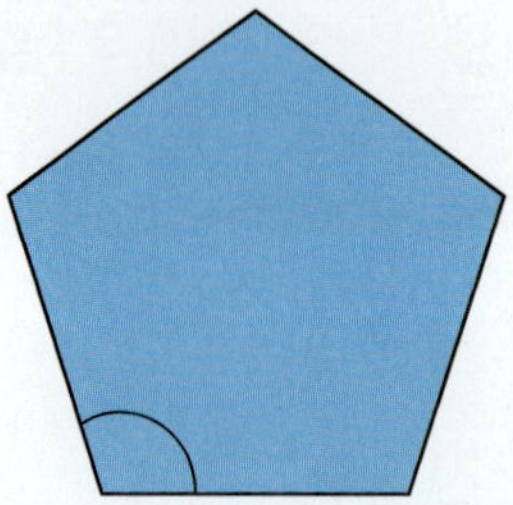

b)
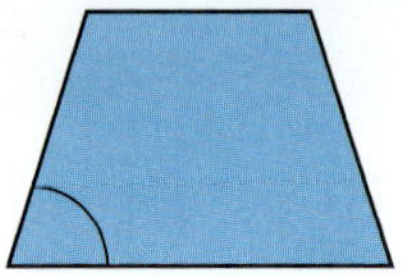

d)
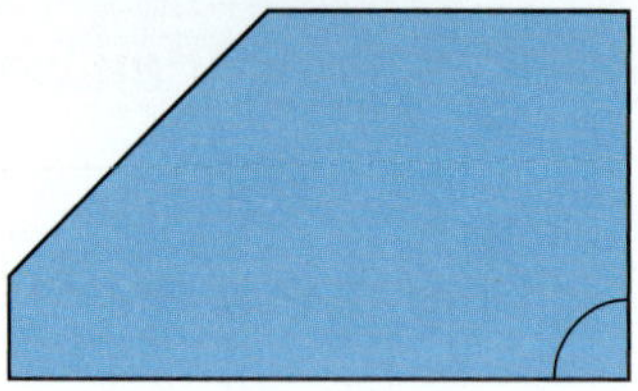

6. Find an object in your classroom with each angle. Sketch the objects.
   - Mark the right angle with a **blue** square.
   - Mark the angle that is less than a right angle with a **red** arc.
   - Mark the angle that is greater than a right angle with a **green** arc.

   a) a right angle
   b) an angle that is less than a right angle
   c) an angle that is greater than a right angle

7. With a partner, use a 24 cm loop of string to make each shape. Sketch each shape.
   - Mark each right angle with a **blue** square.
   - Mark each angle that is less than a right angle with a **red** arc.
   - Mark each angle that is greater than a right angle with a **green** arc.

   a) a triangle with 3 angles less than a right angle
   b) a quadrilateral with 2 right angles
   c) a pentagon with 1 angle greater than a right angle

Semaphore Flags

Sailors use the Semaphore Flag Signalling system to spell words by waving 2 flags. The position of the 2 flags represents a letter. Each flag is usually a red and yellow square.

These flags spell the word BAGS.

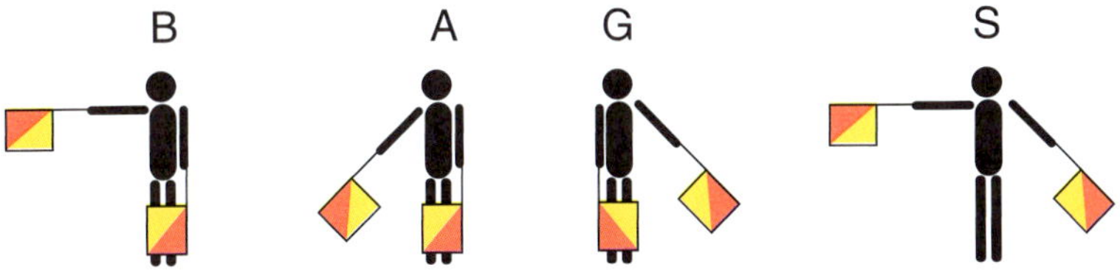

The flag angles for the letters show 1 right angle, 2 angles less than a right angle, and 1 angle greater than a right angle.

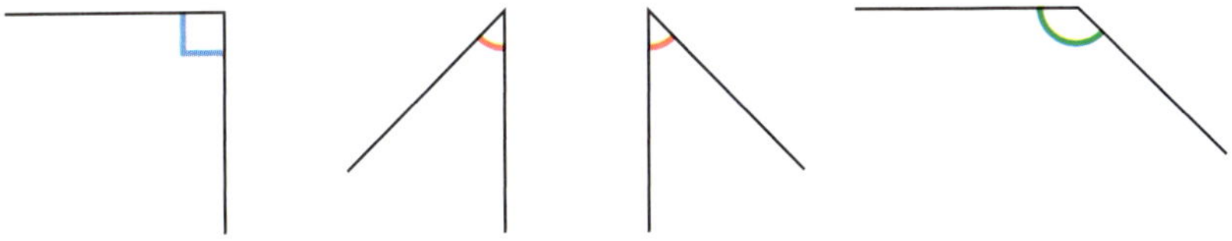

You will need

- semaphore flags

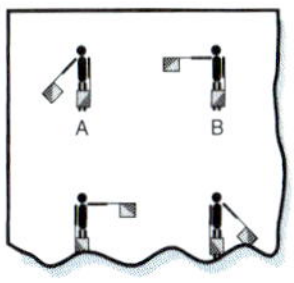

- a square corner

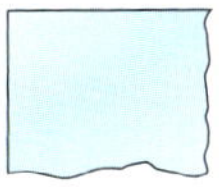

1 Write another word that has 2 letters with flags showing angles less than a right angle. Draw the flag angle for each letter.

2 Write a word that has 2 letters with flags showing right angles. Draw the flag angle for each letter.

a) How many letters have flags showing an angle greater than a right angle?

b) How many letters have flags showing an angle less than a right angle?

2 Lines and Points

Goal Recognize and identify lines and points.

Manitok says that objects with sides that are **parallel lines** often have sides that are **perpendicular lines**.

? Do objects with sides that are parallel lines often have sides that are perpendicular lines?

Find these objects in the picture.
Sketch diagrams with your lists if that helps.

A. **a)** List 3 objects with sides that are parallel lines. Explain how you can tell they are parallel.
b) Do any objects in your list have sides that are perpendicular lines? If so, underline the objects. Explain how you can tell they are perpendicular.

B. **a)** List 3 objects with perpendicular sides. Choose different objects than you used for part A.
b) Underline any objects in your list with parallel sides.

You will need

- a ruler

- a square corner

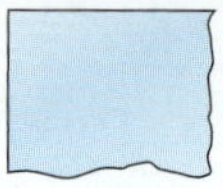

- a square design

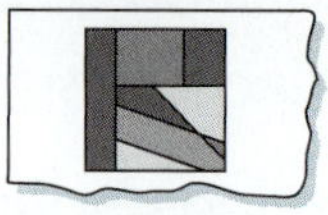

- pencil crayons

parallel lines
Lines that never cross or meet

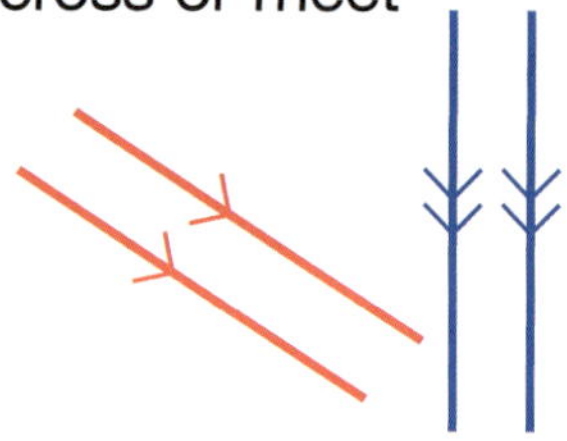

Parallel lines are marked with the same number of small arrows or drawn in the same colour.

perpendicular lines
Lines that cross or meet at right angles

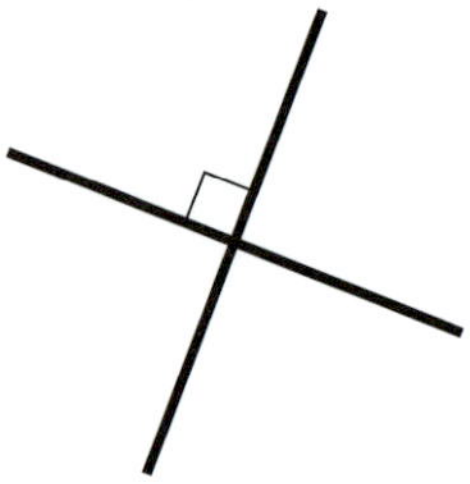

C. a) List 2 objects with sides that are **intersecting lines** but not perpendicular lines.
Explain how you can tell they are not perpendicular.
b) Underline any objects in your list with parallel sides.

D. List 3 objects for each.
a) At least 1 side is a **vertical line**.
b) At least 1 side is a **horizontal line**.

intersecting lines
Lines that cross or meet

vertical line
A line that goes up and down

horizontal line
A line that goes left and right

point
Where intersecting lines cross or meet

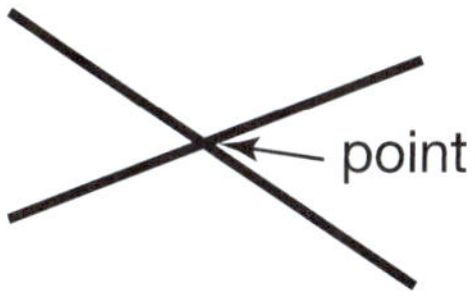

Reflecting

1. Do you agree with Manitok about objects with sides that are parallel lines? Explain reasons for your answer.
2. Explain each answer using a diagram.
   a) Can a vertical line and a horizontal line be parallel?
   b) Can 2 vertical lines meet at a **point**?
3. Manitok says that all perpendicular lines intersect but not all intersecting lines are perpendicular. Draw a diagram to explain what he means.

Checking

4. Match each picture with as many descriptions as possible.
   - point
   - perpendicular
   - parallel
   - intersecting
   - vertical
   - horizontal

a) 

c) 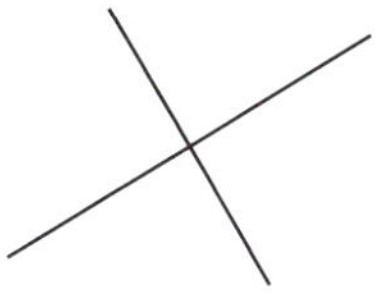

e)

b)

d) 

Practising

5. Use the square design.
   a) Colour a pair of parallel lines **green**.
   b) Colour a pair of perpendicular lines **red**.
   c) Write the letter **V** on a vertical line.
   d) Write the letter **H** on a horizontal line.
   e) Colour 2 lines that intersect, but are not perpendicular **blue**.

6. Create a design that includes these lines.
   - at least 3 parallel lines
   - more than 1 pair of perpendicular lines
   - at least 2 pairs of intersecting lines that are not perpendicular
   - exactly 5 vertical lines
   - exactly 3 horizontal lines

   Show a classmate the lines in your design.

7. Find an example of each of these in your classroom. Name or sketch each example.
   a) parallel lines that are not vertical
   b) perpendicular lines that are not vertical or horizontal
   c) intersecting lines that are not perpendicular

8. The blue line is 2 cm long.
   a) Without measuring, how long do you think the red line is? Why do you think this?
   b) Measure to check your answer.
   c) What does this show about the distance between parallel lines?

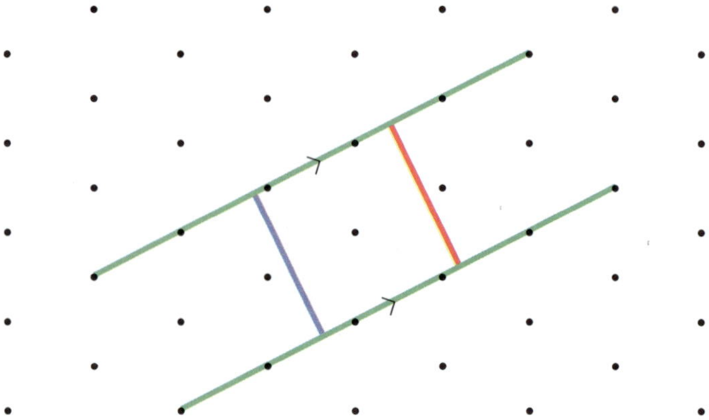

Building Shapes from Triangles

You will need

- pattern block triangles

Shani's Hexagon

I created a hexagon with 6 triangles.

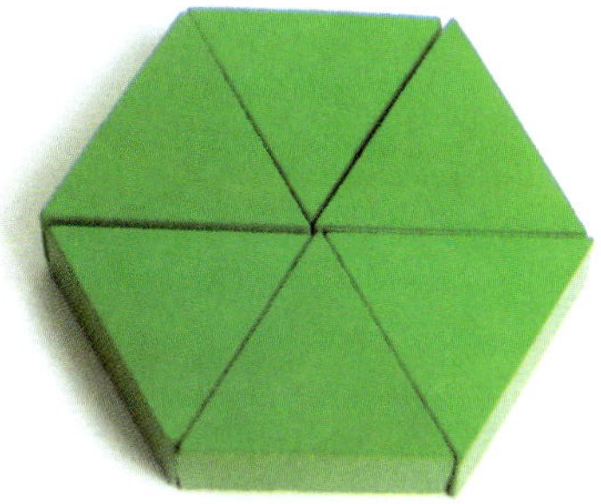

A. Use 6 pattern block triangles to make a hexagon like Shani's.

Try These

1. How many different hexagons can you make with 6 pattern block triangles? Show your work.
2. Make each of these shapes with 6 triangle pattern block triangles.
   a) an octagon (8-sided) b) a quadrilateral
3. Make each of these shapes with 4 triangle pattern blocks.
   a) a triangle b) a hexagon

3 Classifying Quadrilaterals

Goal Identify and sort quadrilaterals.

A furniture store is creating a catalogue of its tables. Tables with these **quadrilateral** tops will be in the catalogue.

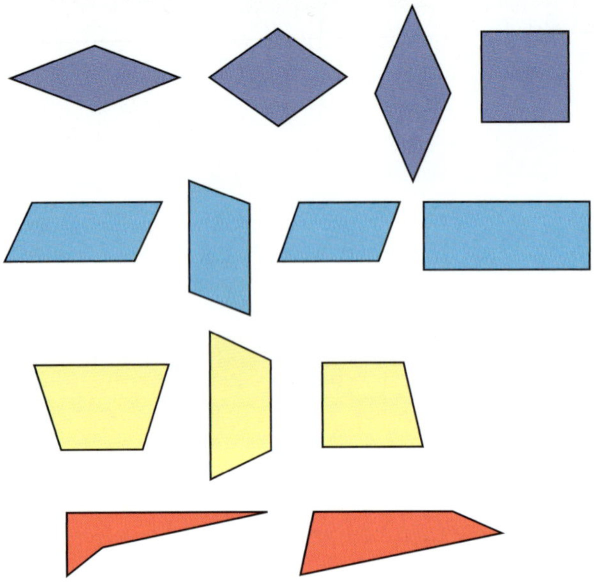

quadrilateral
A shape with 4 straight sides and 4 angles

parallelogram
A quadrilateral with equal and parallel opposite sides

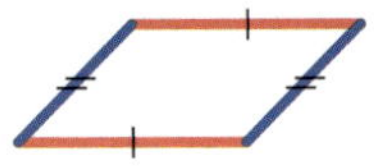

equal lines
Lines that are the same length

Equal lines are marked with the same number of ticks.

? How are the tables organized?

All the tables share the attribute of having 4 straight sides.

A. What other attributes do the purple tables share?

B. A square table is included with the purple tables. Is that a mistake? Explain your answer.

C. What other attributes do the blue tables share?

D. What other attributes do the yellow tables share?

E. Do the red tables share any attributes? Explain your answer.

F. Where does each shape belong in the Venn diagram?

a) a **rectangle**

b) a **rhombus**

c) a **square**

Sorting Quadrilaterals

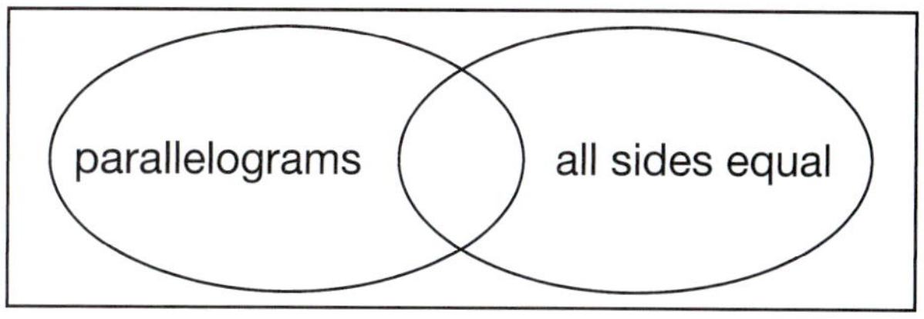

G. What shape name would you give to each group of table tops?

a) purple **b)** blue **c)** yellow

Reflecting

1. In the word *quadrilateral*, what does *quad* mean?
2. Use examples to explain why a quadrilateral can have more than 1 shape name.

Checking

3. List all the possible shape names for each table top.

a)

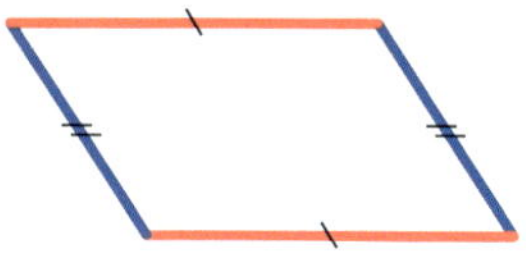

b)

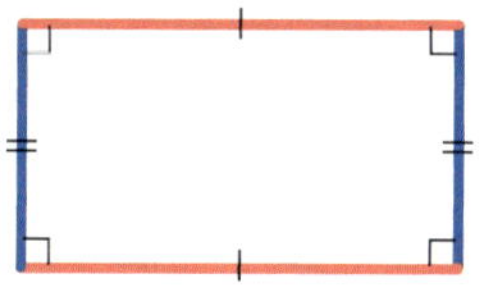

c)

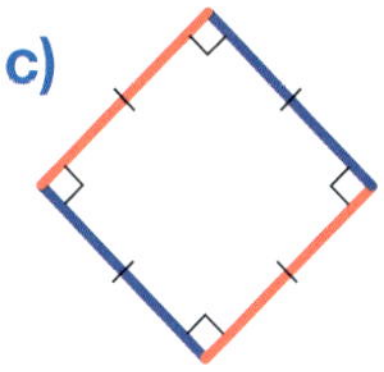

Practising

4. Draw your own table tops to show each shape.

   a) a **trapezoid**

   b) a quadrilateral that is not a **parallelogram** *or* a trapezoid

5. A table top has 4 sides, but only 2 right angles. What shape name can you use to describe it?

6. A table top has 2 pairs of parallel sides. What shape is it?

rectangle
A parallelogram with 4 right angles

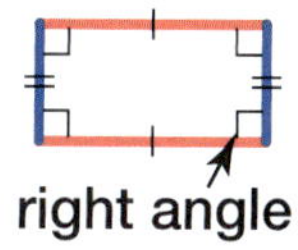

rhombus
A parallelogram with 4 equal sides

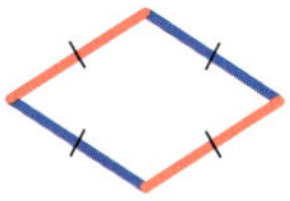

square
A parallelogram with 4 equal sides and 4 right angles

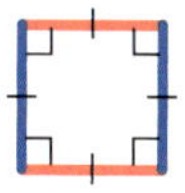

trapezoid
A quadrilateral with only 1 pair of parallel sides

CHAPTER 7

4 Building Quadrilaterals

You will need

- straws cut to 10 cm, 15 cm, and 20 cm lengths

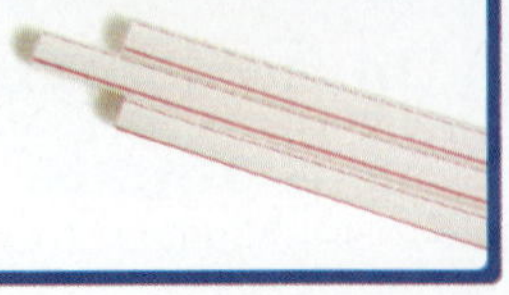

Goal Relate properties of quadrilaterals to their side lengths.

? What quadrilaterals can you model with different sets of straws?

A. What quadrilaterals can you model with each set of straws?
   a) 4 short straws
   b) 2 short straws and 2 medium straws
   c) 2 short straws, 1 medium straw, and 1 long straw

B. Draw a picture of each model.

C. Label each picture with as many quadrilateral names as you can.

D. Use any of the straws to build a quadrilateral that is not a parallelogram or a trapezoid.

E. Draw the quadrilateral from part D, and list the straws you used to make it.

Reflecting

1. Name 2 different shapes that can be made with the same straws you used in part D.
2. Can you form a parallelogram and then a trapezoid with the same set of straws? Explain why or why not.

LESSON

Mid-Chapter Review

1 **1.** Copy each shape.

- Mark each right angle with a **blue** square.
- Mark each angle that is less than a right angle with a **red** arc.
- Mark each angle that is greater than a right angle with a **green** arc.

a)

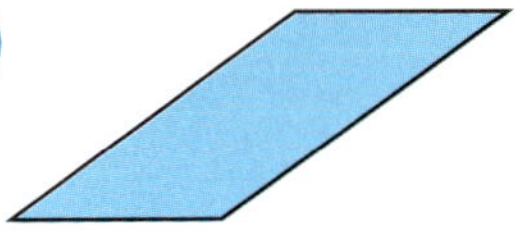

b)

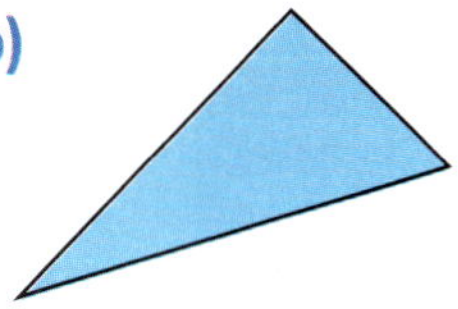

2 **2.** Copy this **red** line on dot paper. Then draw each of these lines.

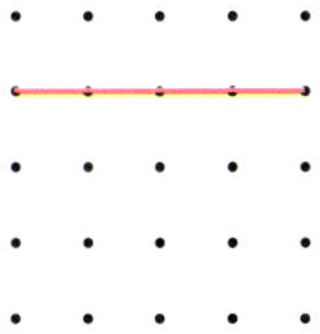

a) a **blue** line parallel to the **red** line
b) a **green** line perpendicular to the **red** line
c) an **orange** vertical line
d) a **black** line that intersects the **red** line but is not perpendicular to it

3. Circle the points where the lines in your diagram for Question 2 meet.

3 **4.** Use each shape name once to make these sentences true.

- parallelogram
- trapezoid
- rhombus
- square
- rectangle

a) A ■ is a special ■.
b) A ■ is never a ■.
c) A ■ always has 4 right angles.

5. What do you know about the lengths of the sides of each quadrilateral? Explain using a diagram.

a) a parallelogram
b) a square
c) a rectangle
d) a rhombus

CHAPTER 7

5 Solve Problems by Acting Them Out

You will need

- a ruler 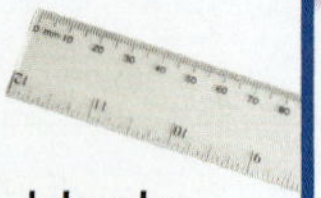
- pattern blocks

Goal **Act out a problem to solve it.**

Sarah is pushing together 20 tables shaped like trapezoids to make a single row.

? How many people can sit at the tables if 1 person sits at each side?

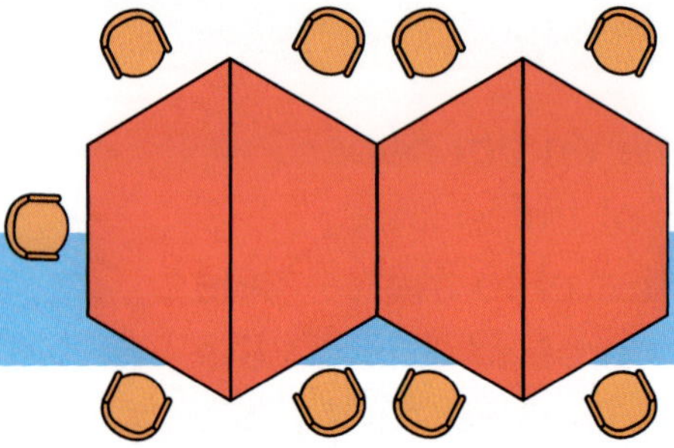

Sarah's Solution

Understand

It isn't possible to sit where 2 tables come together.

That means I have to count only the open sides.

Make a Plan

I will act out the problem using trapezoid blocks for the tables.

Carry Out the Plan

First I count the number of open sides on top.

There are 20 open sides on top.

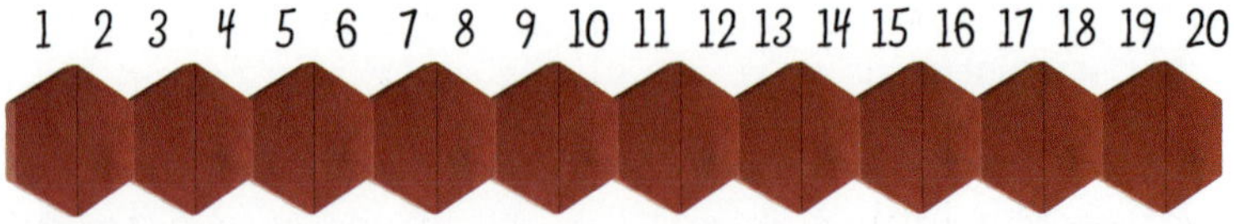

There are 20 more open sides on the bottom and 1 open side at each end.

$20 + 20 + 1 + 1 = 42$ open sides

42 people can sit at the tables.

Reflecting

1. Why didn't Sarah need to count the sides on the bottom?
2. If Sarah had arranged the tables in 2 rows of hexagons, would there have been more or fewer places to sit? Explain your answer.
3. How did acting out the problem help Sarah find a solution?

Checking

4. Rami is pushing together 7 tables shaped like hexagons to make 1 row. How many people can sit at the tables if 1 person sits on each open side? Show your work.

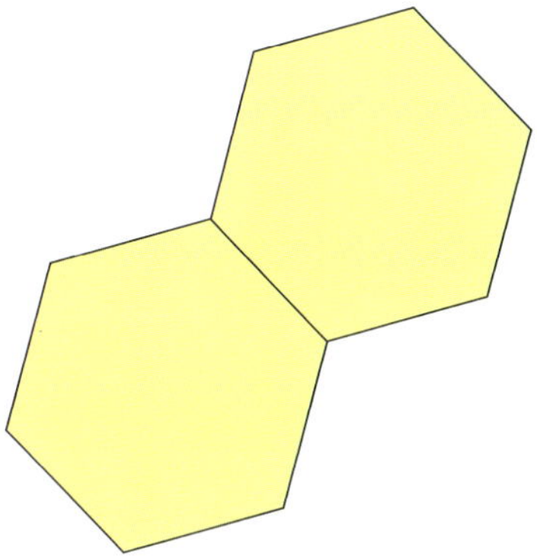

Practising

5. Terry is pushing together 9 tables shaped like rhombuses to make 1 row. How many people can sit at the tables if 1 person sits on each open side?
6. Tamara is making 1 big table by pushing together 6 small tables shaped like equilateral triangles. 1 person sits at each open side.
   a) How many people can sit at the larger table?
   b) Explain why there is more than 1 answer to this problem.

6 Lines of Symmetry

Goal **Draw lines of symmetry.**

Symmetry can be used to see patterns and solve puzzles. These 2 names are half covered. Which name is easier to figure out?

The name on the left is probably easier to figure out because each letter in TOM is symmetrical across the vertical line. The letters in JEN are not.

? If you fold a shape on its line of symmetry, is it easy to visualize the whole shape?

A. Trace a square pattern block or make a square on dot paper.

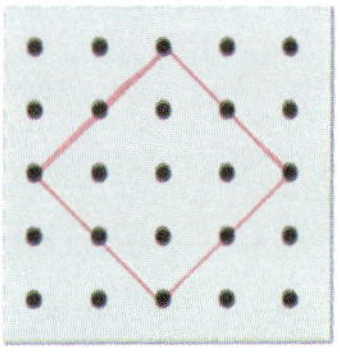

B. Draw all the **lines of symmetry** of the square.
Try each of these methods.

Use a transparent mirror.

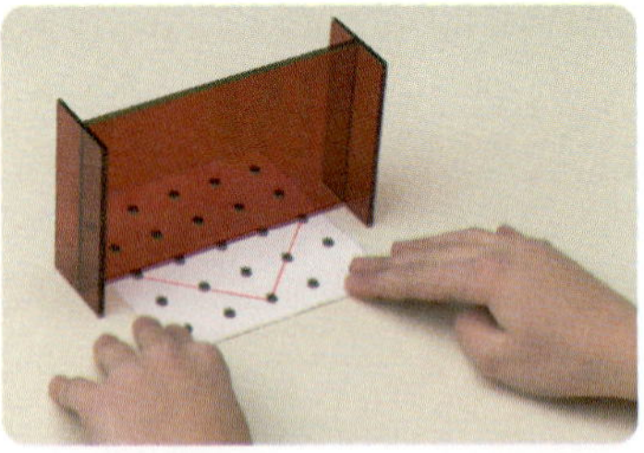

Measure with a ruler.

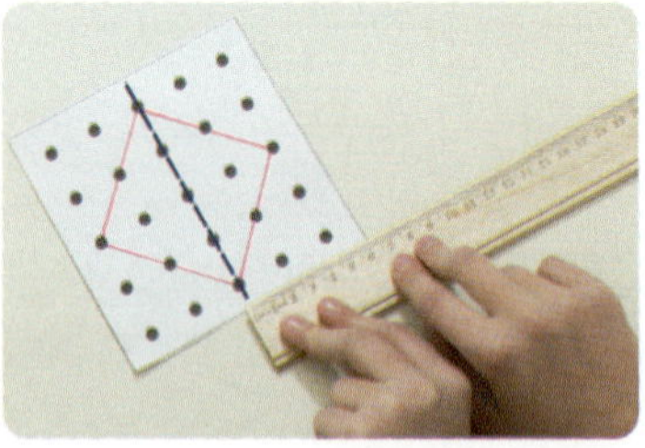

Cut it out and fold.

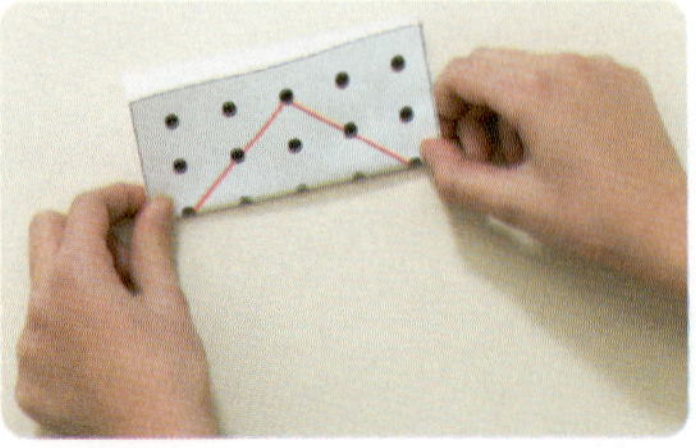

Use the dots on the grid.

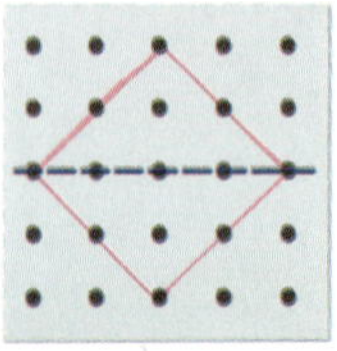

You will need

- a transparent mirror

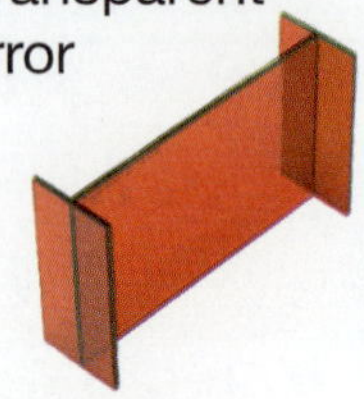

- pattern blocks

- scissors

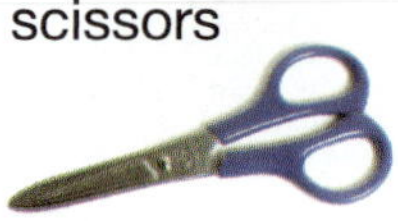

- dot paper

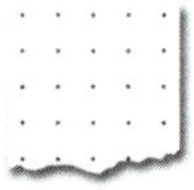

line of symmetry
A line that divides a shape in half so that if you fold the shape on the line, the halves match

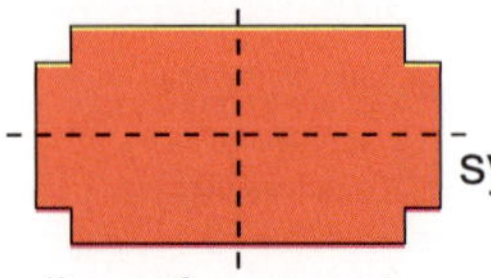

line of symmetry

line of symmetry

This shape has 2 lines of symmetry.

C. Repeat parts A and B with a triangle, a hexagon, and a rhombus.

D. Trace a trapezoid pattern block. Draw any lines of symmetry. Which method did you use? Why?

E. Fold each of the shapes in parts C and D on 1 of its lines of symmetry. Hide half of the shape. Is it easy to visualize the whole shape from the half? Why or why not?

F. Draw these 3 quadrilaterals on dot paper. For each quadrilateral, find all the lines of symmetry, if there are any.

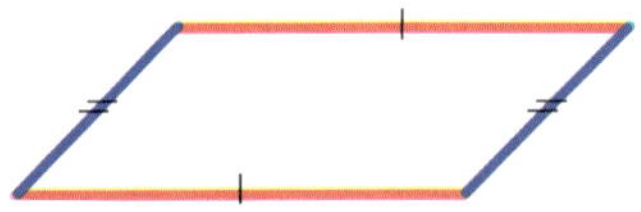

parallelogram that is not a rhombus

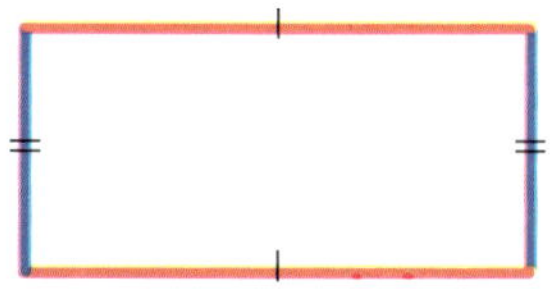

rectangle that is not a square

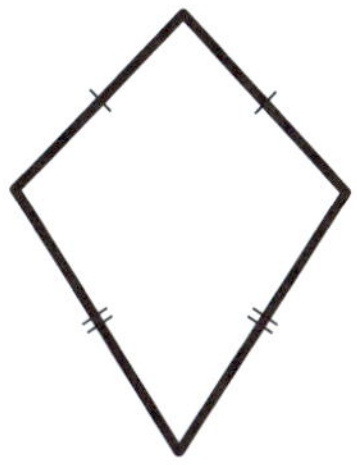

kite

kite
A quadrilateral with 2 pairs of equal sides and no sides parallel

Reflecting

1. What do you notice about the shapes that have more than 1 line of symmetry?
2. How does dot paper help you to find lines of symmetry?
3. What is an easier tool for drawing lines of symmetry, a transparent mirror or a ruler? Explain your thinking.

7

Symmetrical Designs

Goal **Create symmetrical designs.**

You will need

- a transparent mirror

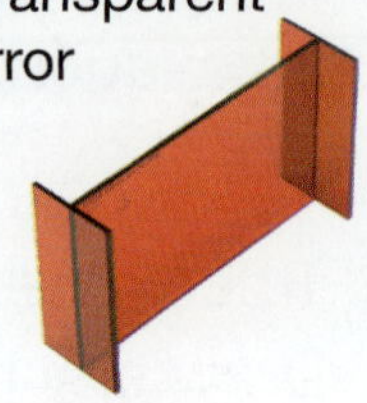

- dot paper

- grid paper

Mandy wants to create a symmetrical design for the cover of her project. She began by drawing 2 lines of symmetry. Then she drew $\frac{1}{4}$ of the design.

Mandy's Design

I'll finish my design using a transparent mirror.

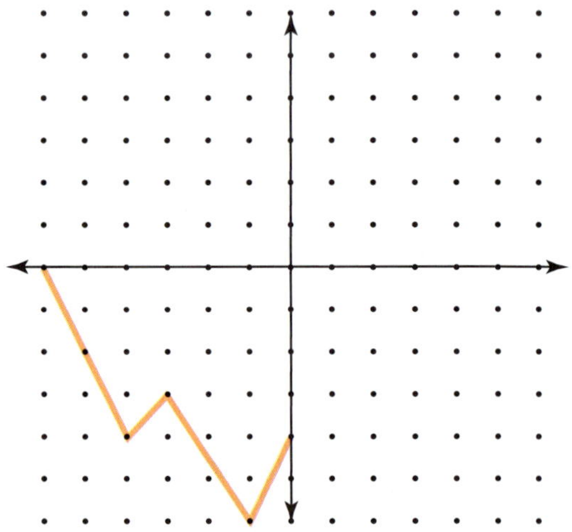

? If $\frac{1}{4}$ of a symmetrical design has 4 sides, how many sides will the design have?

A. Predict the number of sides Mandy's design will have. Explain your prediction.

B. Complete Mandy's design.

a) Copy Mandy's partial design.

b) Place a transparent mirror along 1 line of symmetry. Then finish half of the design.

c) Place the mirror along the other line of symmetry. Now complete the design.

C. How many sides does Mandy's design have? Is this what you predicted?

D. Use Mandy's method. Create a symmetrical design with a different number of sides. Create at least 3 more designs.

Reflecting

1. How can you check to make sure your designs are symmetrical?
2. Suppose Mandy did not have a mirror. How could she have completed her design?
3. a) If a design has 2 lines of symmetry, why must the number of sides be at least 4?
   b) How can you predict the number of sides in a design?

Checking

4. Copy each partial design and then complete it.

a) 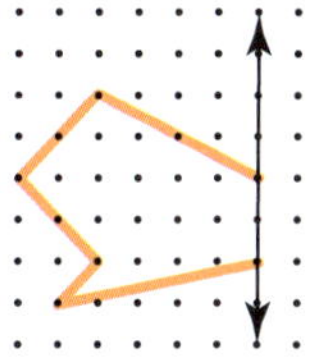

b) 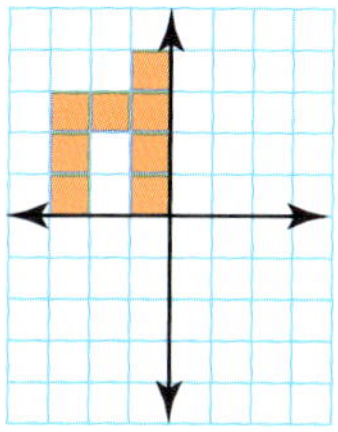

Practising

5. Copy each partial design and then complete it. Explain how you know each design is symmetrical.

a) 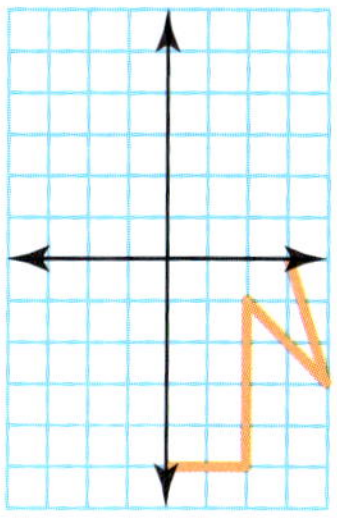

b) 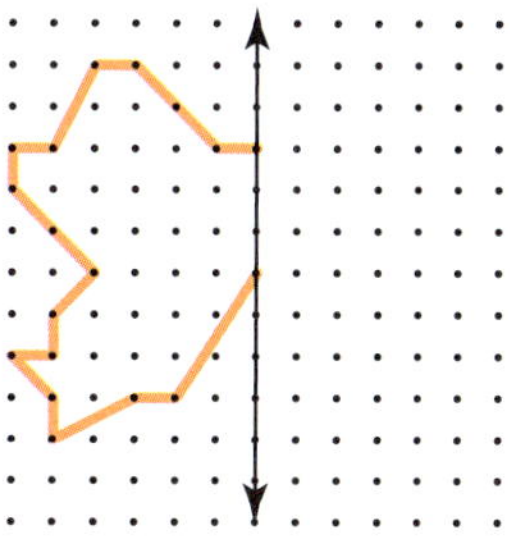

c) 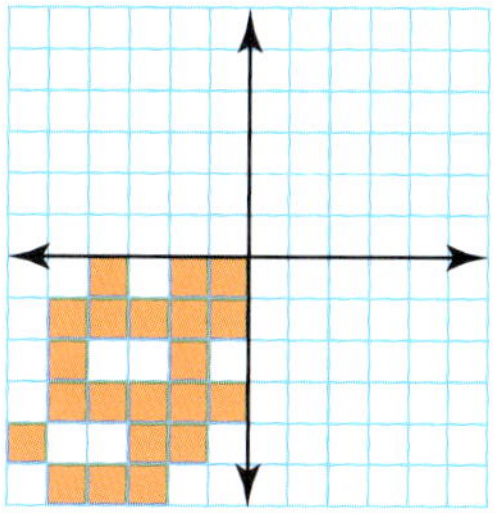

6. Is it possible to create different symmetrical designs for any partial design in Question 5? Explain.
7. Use a grid. Draw a design with 2 lines of symmetry.

8 Classifying 2-D Shapes

You will need

- sheet of shapes

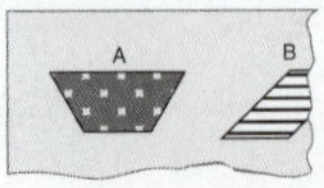

- a ruler

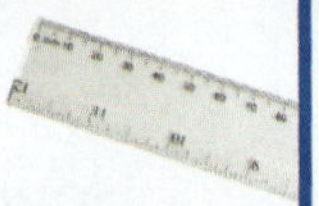

Goal **Identify and sort 2-D shapes.**

Think about attributes of some 2-D shapes.

- parallel sides
- number of sides
- equal sides
- symmetry
- perpendicular sides
- congruence
- angle size

? How many different ways can you sort these shapes?

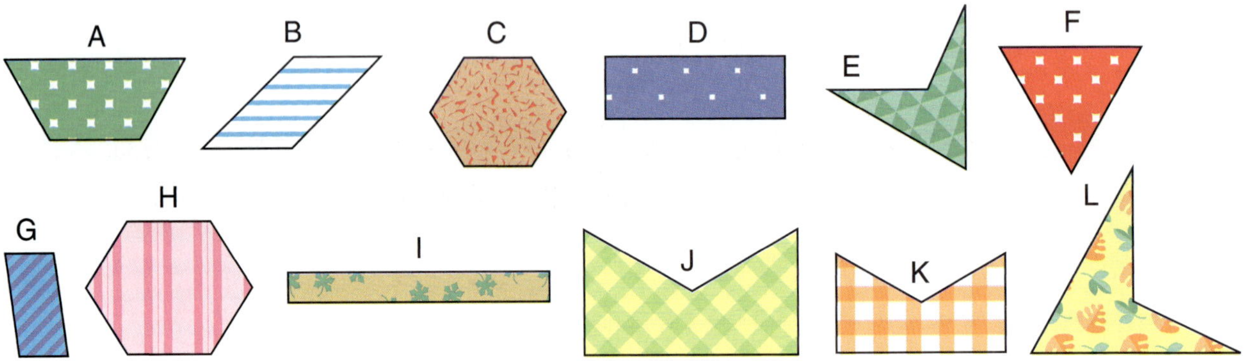

A. Sort the shapes using 2 or more attributes.

B. Show how you sorted the shapes.
You could make a chart or a list, or draw a Venn diagram or a Carroll diagram.

C. Find at least 4 other ways to sort the shapes.
Show how you sorted the shapes each time.

Reflecting

1. Which attributes do most of the shapes have?
2. Which attributes do only a few of the shapes have?

Tangram Quadrilaterals

This tangram puzzle is made up of 7 pieces, called tans.

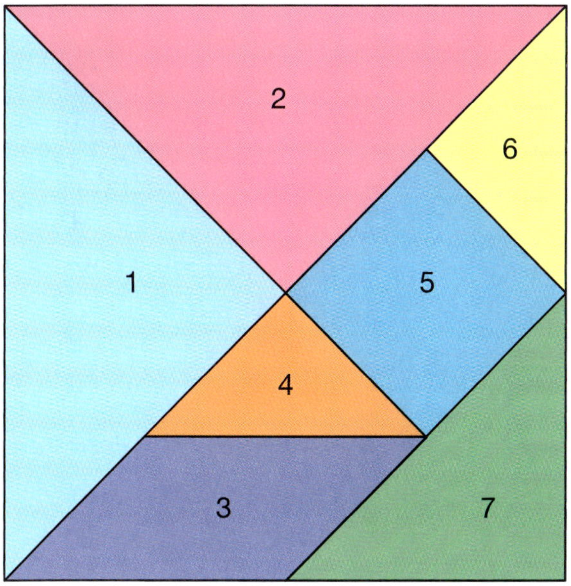

You will need

- a transparent mirror

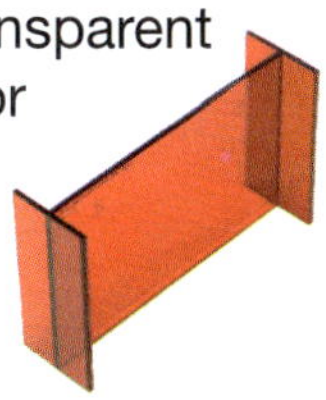

- a tangram

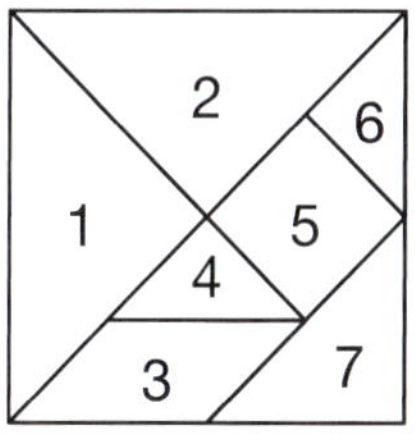

1 a) Identify which tans are quadrilaterals.
b) Name each quadrilateral.

2 You can join tans to make more quadrilaterals.
a) Draw quadrilaterals by tracing and combining tans.
b) Name each type of quadrilateral you make.

3 Use a transparent mirror to decide which quadrilaterals are symmetrical.
a) Name the quadrilaterals that are not symmetrical.
b) Name the quadrilaterals that have 1 line of symmetry.
c) Name the quadrilaterals that have more than 1 line of symmetry.

LESSON

Skills Bank

1 **1.** Is each angle a right angle, greater than a right angle, or less than a right angle?

a)

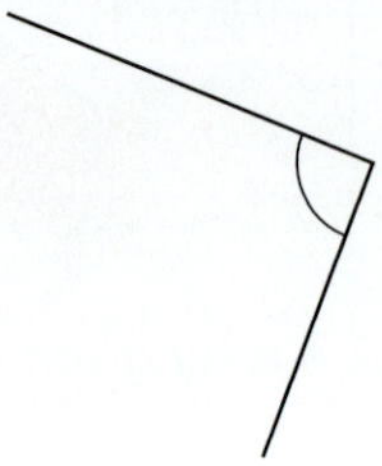

b)

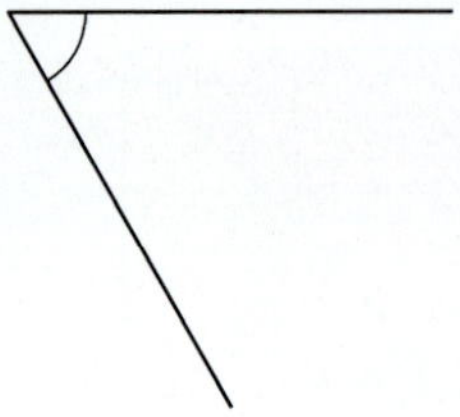

c)

2. Copy each shape.

- Mark each right angle with a **blue** square.
- Mark each angle that is less than a right angle with a **red** arc.
- Mark each angle that is greater than a right angle with a **green** arc.

a)

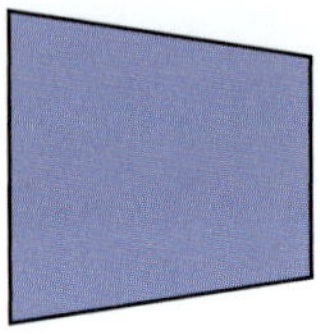

b)

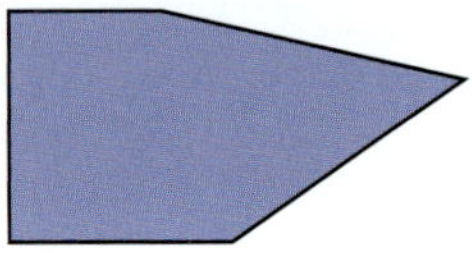

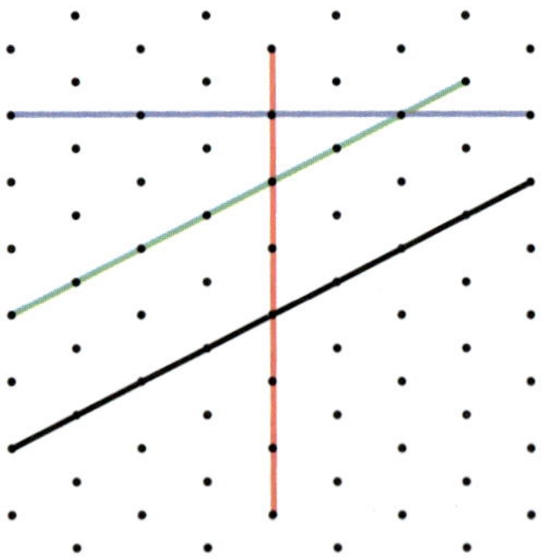

2 **3.** What colour lines match each description?

a) parallel

b) perpendicular

c) vertical

d) horizontal

3 **4.** Complete the chart. Find as many shapes as possible for each shape name.

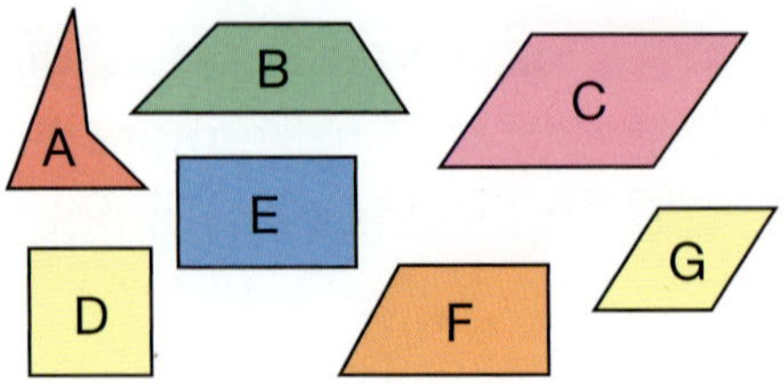

Shape name	Shape letter
parallelogram	
quadrilateral	
rectangle	
rhombus	
square	
trapezoid	

6 **5.** Draw each shape on dot paper.
Draw all the lines of symmetry.

a)

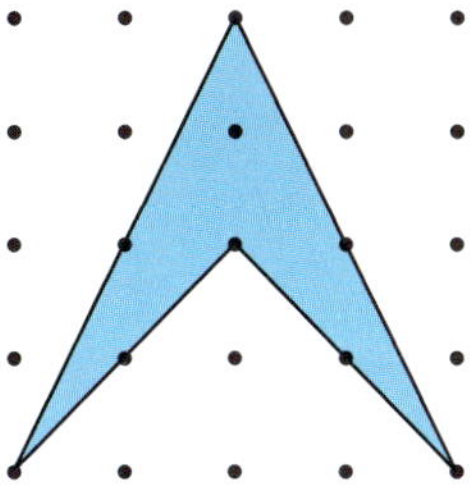

b)

c)

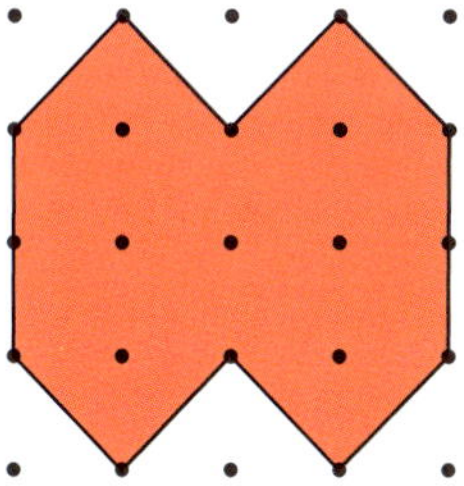

7 **6.** Complete the designs for the lines of symmetry.

a)

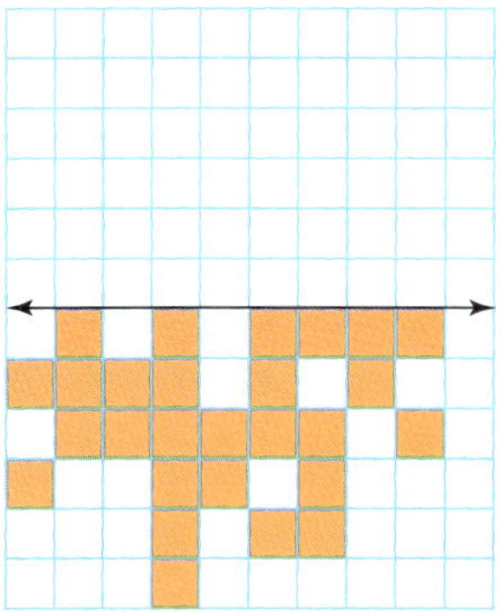

b)

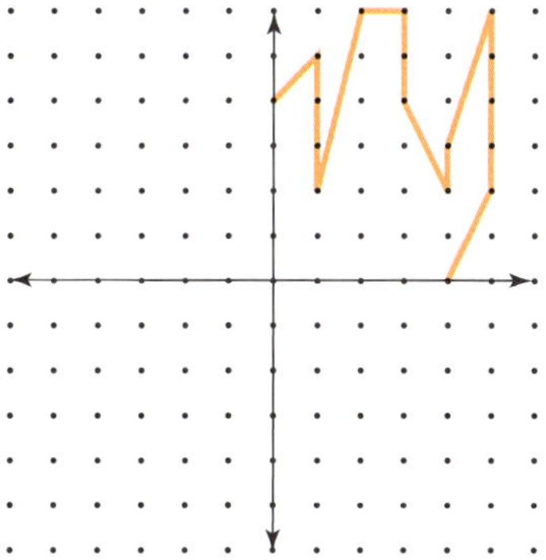

8 **7. a)** Complete the Venn diagram to sort the shapes.

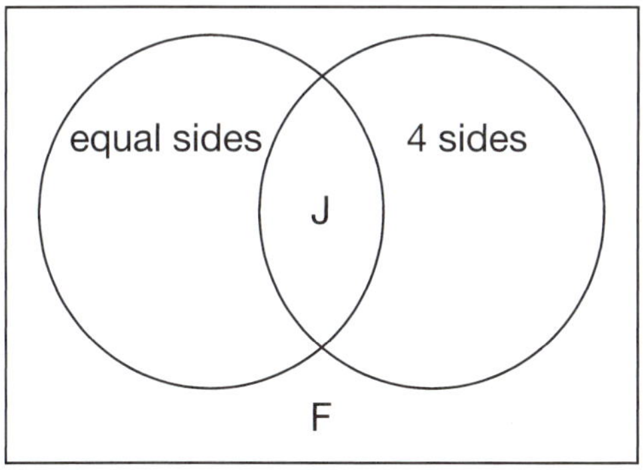

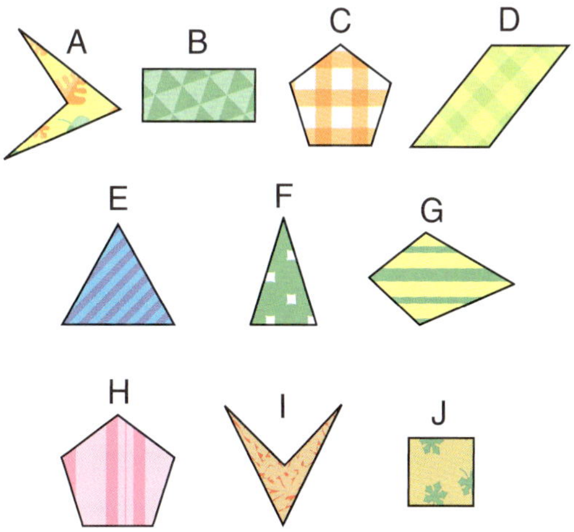

b) Sort the shapes another way.
Use a Venn diagram, a Carroll diagram, a chart, or lists. You can use 2 or more attributes such as

- number of sides
- congruence
- symmetry
- equal sides
- perpendicular sides
- angle size
- parallel sides

LESSON

Problem Bank

2 **1.** How many pairs of parallel lines are in the diagram? Show how you know.

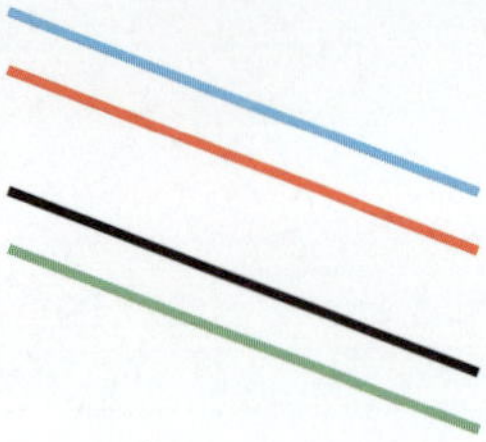

3 **2.** Shani found a board shaped like a rhombus. Show how she can cut the board and rearrange the pieces to make a rectangle.

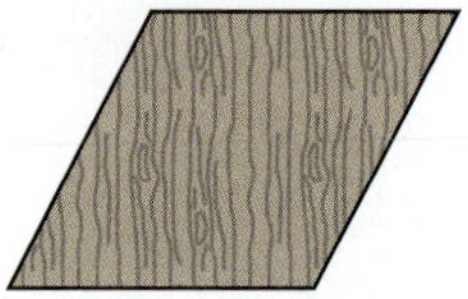

3. How many different sizes of parallelograms can you see in this figure? Draw pictures with labels to record your answer.

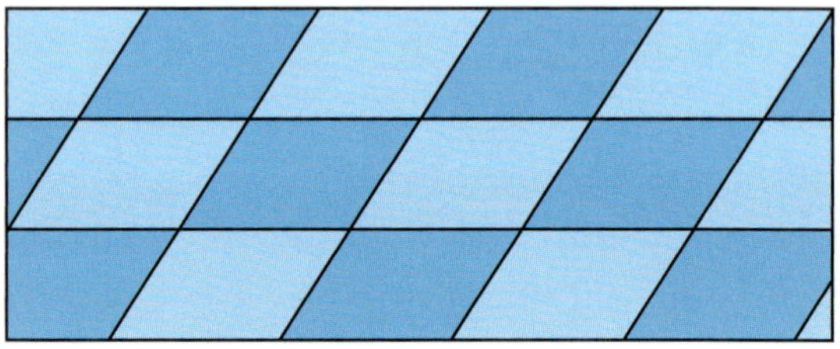

4. Imagine that you cut a quadrilateral into 2 pieces with a straight cut. How many sides can your resulting shapes have? Show all the possibilities you can think of.

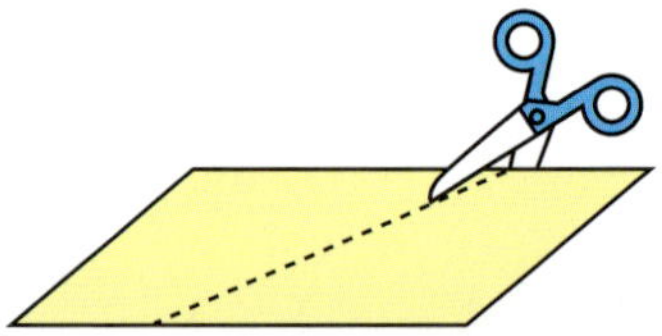

5 **5.** Trace these 2 triangles. Show how you can place 1 triangle on top of the other for each result.

a) Only 1 triangle and 1 trapezoid can be seen.

b) 4 congruent triangles can be seen.

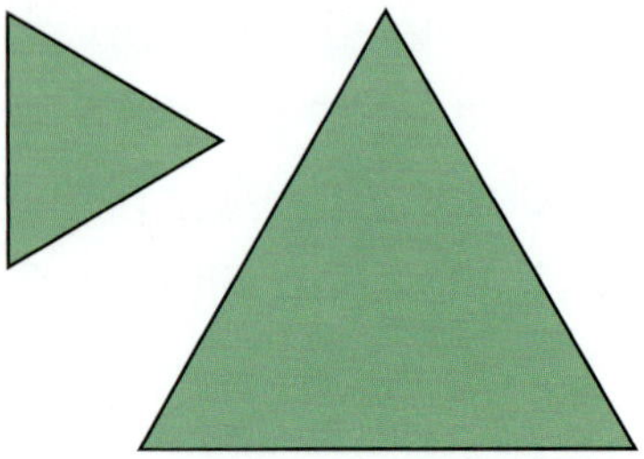

LESSON

Chapter Review

2 **1.** Name and sketch an example for each of these. Use objects you might see in a store.

a) parallel lines that are not horizontal

b) perpendicular lines with 1 vertical line

c) intersecting lines that are not perpendicular

3 **2.** Which of these shapes are parallelograms? Explain your thinking.

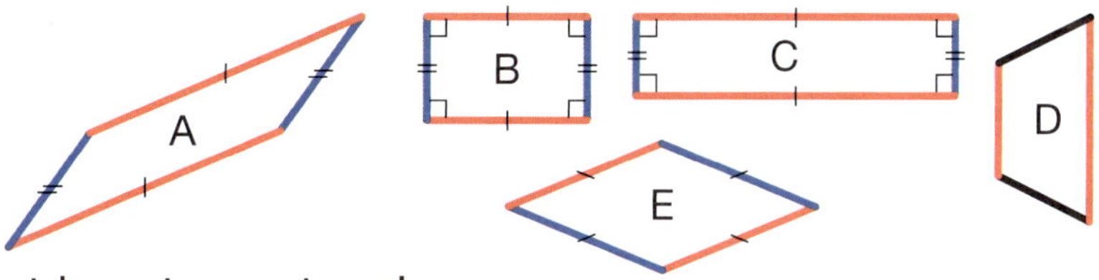

3. Draw a parallelogram that is not a rectangle.

a) How many angles are greater than a right angle?

b) How many angles are less than a right angle?

c) How many angles are right angles?

4. Name each quadrilateral.

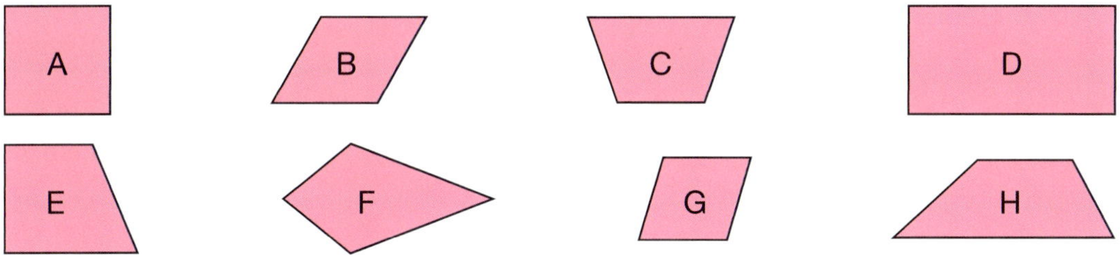

5. Draw a cake like this one on dot paper. Show how to cut the cake in 2 pieces with a straight cut to get each result.

a) Each piece is a parallelogram.

b) Each piece is a trapezoid.

7 **6.** Create a design with 2 lines of symmetry on grid paper. Explain how you know your design is symmetrical.

8 **7.** Sort the quadrilaterals in Question 3 using these rules. Use a Venn diagram, a Carroll diagram, a chart, or a list.

a) parallel sides and right angles

b) more than 1 pair of parallel sides and right angles

Chapter Task

Shape Names

Poly Gon is having a math party.
She wants a name tag for each of her guests,
but she has rules for the tags.

- More than half, but not all, of the name tags are quadrilaterals.
- 4 of the name tags have 1 angle greater than a right angle.
- More than half of the name tags have a line of symmetry.
- Only half of the name tags have sides that are parallel.
- 1 name tag has 2 angles greater than a right angle and 2 angles less than a right angle.

? How can you make and describe 10 different name tags for Poly's party?

A. Create 10 name tags to fit Poly's rules.

B. Completely describe each of the name tags you made.

Task Checklist

- ✓ Did you include diagrams?
- ✓ Did you use math language?
- ✓ Did you verify that your name tags agree with Poly's rules?
- ✓ Did you organize your work so it is easy to follow?

Cumulative Review

Cross-Strand Multiple Choice

1. Which calculations would help you find the missing digits?

```
  8 7
+ 9 ■
1 ■ 8
```

A. 8 + 9 and 7 + 8
B. 87 + 9 and 18 − 7
C. 80 + 90 and 70 + 80
D. 8 − 7 and 80 + 90

2. Josh paid for all of these items using two $20 bills and one $10 bill. How much change should he get?

blank CDs	$30.00
magazine	$6.95
CD labels	$8.25

E. $4.80
F. $4.90
G. $5.80
H. $5.87

3. Which measurements could describe the perimeter of a beach towel?

A. 40 mm
B. 400 cm
C. 4 m
D. 4 km

4. Which statement does not describe this array?

E. 42 − 7 − 7 − 7 − 7 − 7 − 7
F. 6 + 6 + 6 + 6 + 6 + 6
G. 6 × 7
H. 42 ÷ 6

5. How could you name the 2 groups on this sorting mat?

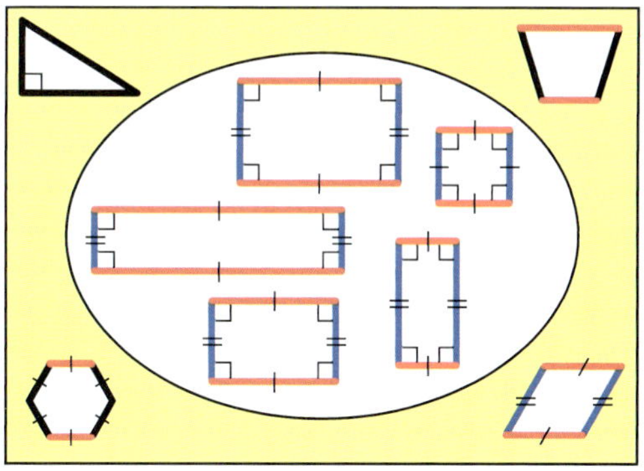

A. quadrilaterals/not quadrilaterals
B. polygons/quadrilaterals
C. rectangles/not rectangles
D. parallel sides/no parallel sides

6. Which angle is greater than a right angle?

E.
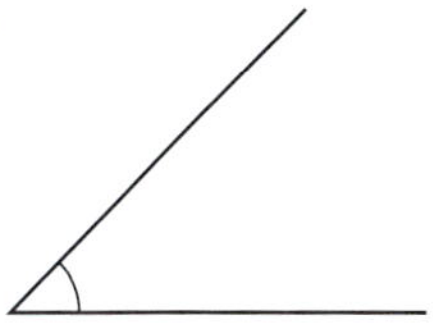

F.
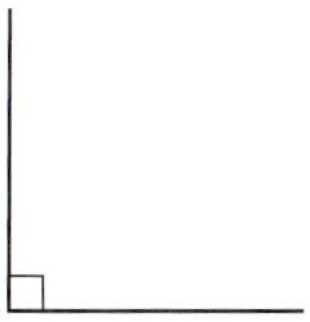

G.
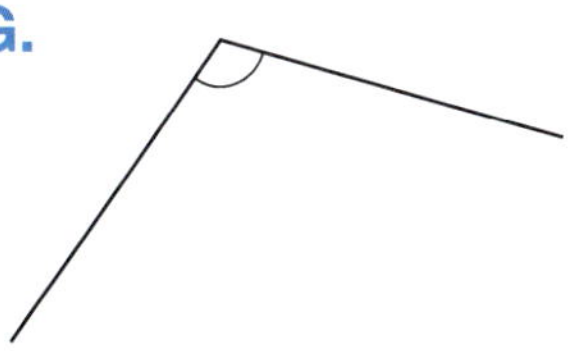

H.

Cross-Strand Investigation

Canada's Supreme Court

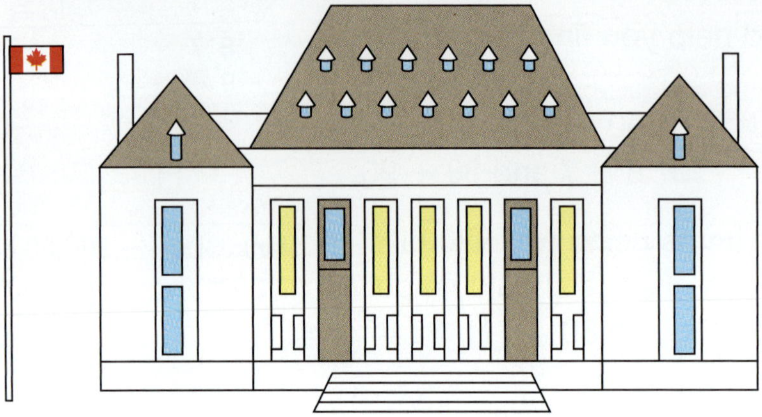

The Supreme Court of Canada in Ottawa

7. a) Identify 2 pairs of parallel lines in the picture. Explain how you know.
   b) Identify 2 pairs of perpendicular lines in the picture. Explain how you know.
   c) Why are parts of buildings often parallel? Explain using math words.
   d) Why are parts of buildings often perpendicular? Explain using math words.

8. Each lit window contains 6 rows of glass panes. There are 3 panes in each row. How many panes were needed to make all of the lit windows? Solve this problem using 2 different strategies. Explain your solution.

9. a) The Main Courtroom is a rectangle that is 16 m long and 12 m wide. How can the length and width help you find the perimeter? Explain using a labelled drawing.
   b) Draw a rectangle to fit exactly around the outside of the building on this page. Then draw a different rectangle with the same perimeter. Show how you know the perimeter is the same.

CHAPTER 8

Area and Grids

Goals

You will be able to

- use 2 standard area units
- measure area with the appropriate unit
- use maps and grids
- solve area problems using organized lists

Ploughed rectangular fields

Getting Started

You will need

- pattern block squares 
- grid paper 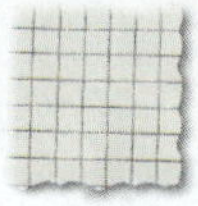
- a ruler 

Comparing Area

Your eyes can trick you into thinking something is a different size than it really is.

? How can you compare the areas of the red and blue shapes?

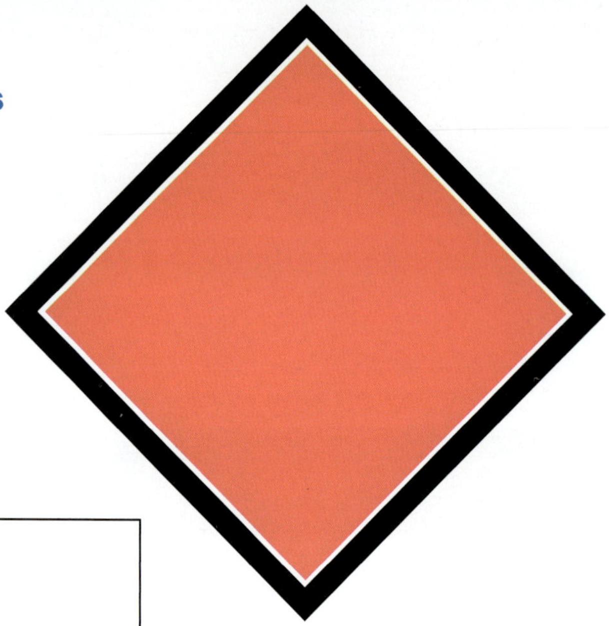

A. What do you think you will find when you compare the areas of the red and blue shapes? Give a reason for your answer.

B. Measure the areas of both shapes. What did you find?

C. The blue shape is inside a square.
   a) Estimate how much greater the area of the square is compared to the area of the blue shape. Measure to check.
   b) Estimate how much longer the perimeter of the square is compared to the perimeter of the blue shape. Measure to check.

Do You Remember?

1. Find the perimeter.

a)

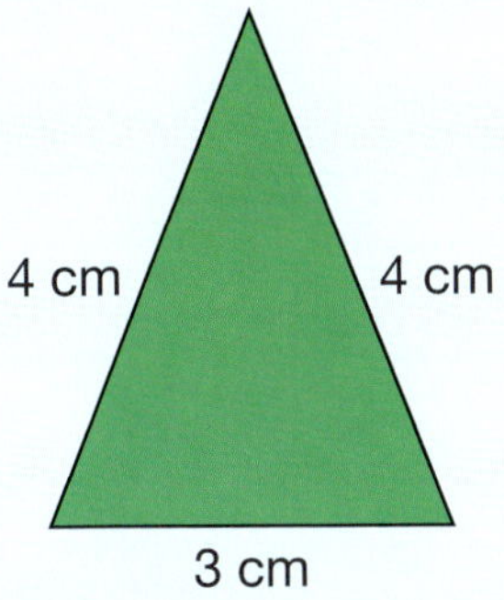

b)

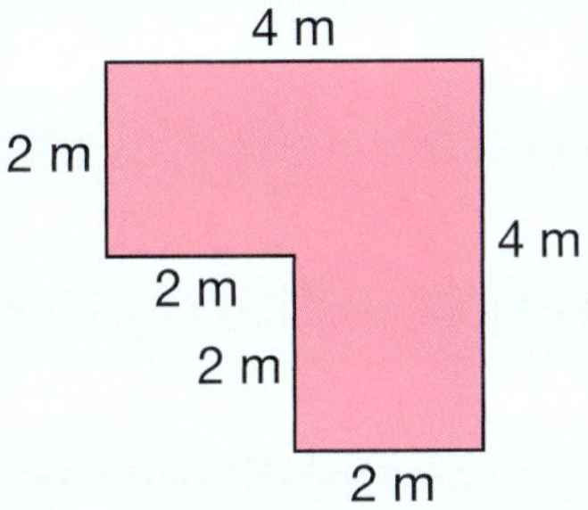

2. Which unit would you use to measure each perimeter, centimetres or metres?
   a) a library card
   b) a schoolyard

3. Which figures have the same area as  ?

A. 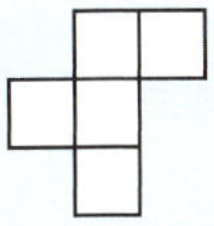

B. 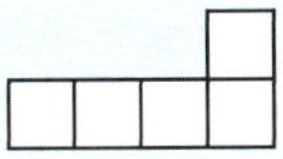

C. 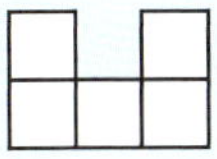

D. 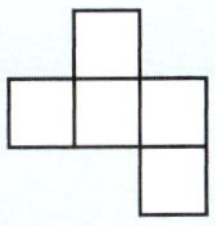

E. 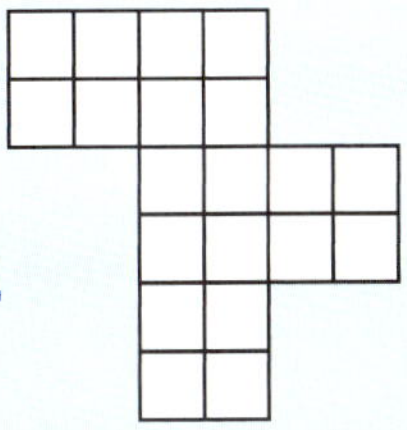

1 Square Centimetres

You will need

- a centimetre grid transparency

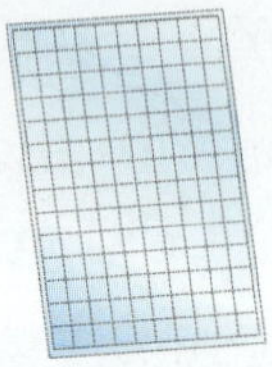

Goal Estimate, measure, and compare area using square centimetres.

Natalie and Calvin are making stars for art class.
They are using shiny gold paper.
The paper comes in 2 shapes, square and rectangular.

Natalie's Square

I think the square paper has a greater area.

Calvin's Rectangle

I think the rectangular paper has a greater area.

? How can Natalie and Calvin compare the areas of the 2 shapes of paper?

A. Who do you think is right?
Explain your reasoning.

B. Place a centimetre grid transparency over each shape.
Measure the area by counting the number of **square centimetres** it takes to cover each shape.

C. How do the areas compare?

square centimetre
A unit of measurement for area

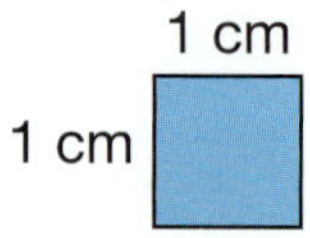

A square with sides of 1 cm has an area of 1 square centimetre.

Reflecting

1. Why was it difficult to compare the areas of Natalie's square and Calvin's rectangle without measuring?
2. Sometimes area is measured to the nearest whole square centimetre. The area of this triangle is about 7 whole square centimetres. Why does that make sense?

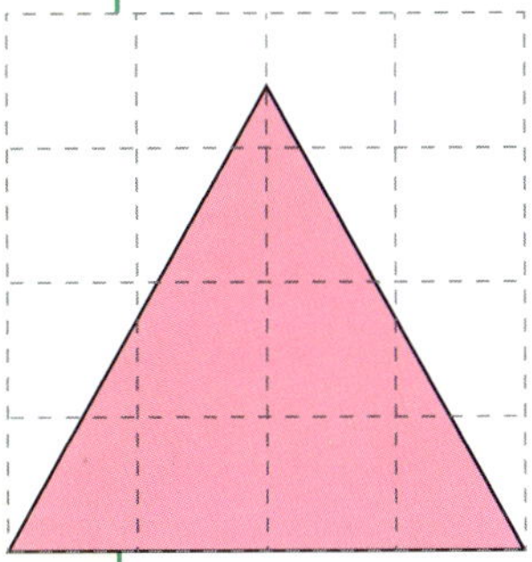

Checking

3. a) Which shape do you think has the greater area?
   b) Estimate the area of each shape.

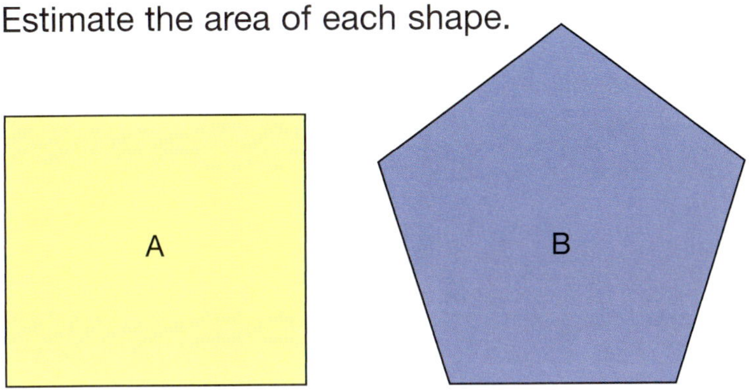

   c) Measure the area of each shape to the nearest whole square centimetre. Use a centimetre grid transparency.

Practising

4. Estimate the area of each shape. Measure the area of each shape to the nearest whole square centimetre.

a)

b)

5. Estimate the area of each shape.
Measure each area to the nearest whole square centimetre.

a)

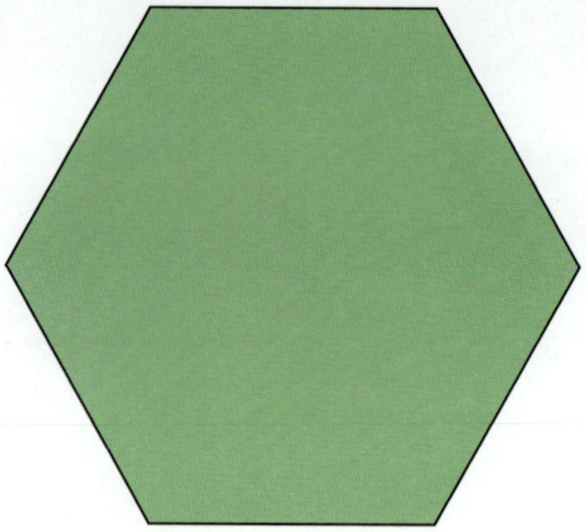

b)

c)

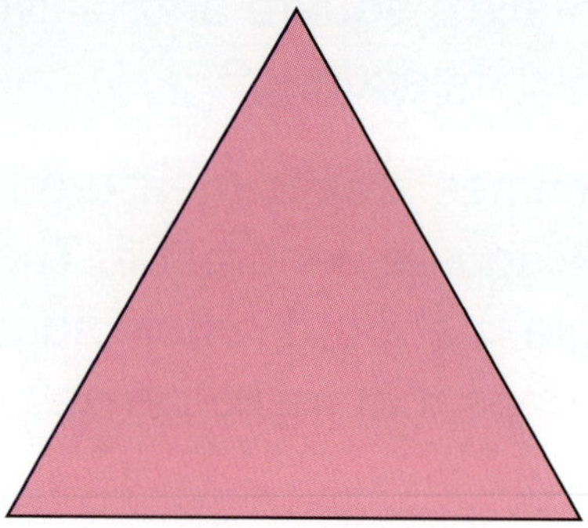

d)

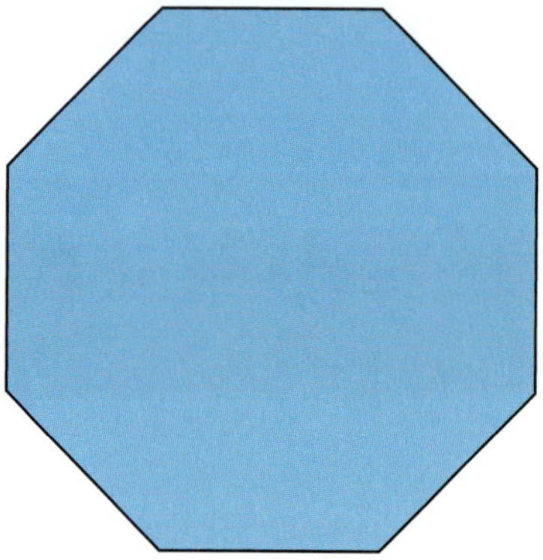

6. List the shapes in Question 5 in order from the least to the greatest area.

7. Choose 3 surfaces you can measure with a square centimetre grid.
   a) Estimate each area in square centimetres.
   b) Measure each area to the nearest whole square centimetre.

8. Is a shape with an area of 1 square centimetre always a square? Use an example to help you explain.

Area Logic

Number of players: 2 or 3
How to play: Colour areas on a grid.
Each player uses a different coloured pencil crayon.

Step 1 Roll 2 dice. Use the product of the 2 numbers as the area of a rectangle.

Step 2 Players try to colour their rectangles on the centimetre grid paper.
If there is no space left to colour a rectangle with this area, lose your turn.

Step 3 Continue taking turns until no player can go.

Count the coloured squares to see which player has coloured the greatest area.

You will need

- 2 dice

- grid paper

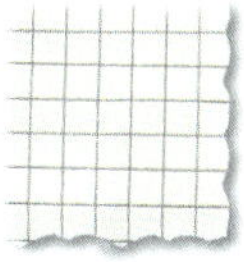

- pencil crayons or crayons

Paulette's Turn

I rolled 2 and 4.

I will need to colour a rectangle with an area of 8 square centimetres.

The rectangle could be 2 cm by 4 cm or 1 cm by 8 cm.

I will colour a 2 cm by 4 cm rectangle.

2 Square Metres

Goal **Estimate, measure, and compare area using an appropriate area unit.**

When you buy carpet, you need to know how many square metres it will take to cover the floor.

You will need

- a metre stick

- scissors

- chart paper
- tape

? About how many square metres of carpet would you need to cover the floor in your classroom?

A. Use chart paper to make a model of a **square metre**.
Each side of the square should be 1 m long.

B. Look at your model. Estimate the area of the floor of your classroom.

C. Use your model to measure the area of the floor in your classroom to the nearest square metre.

D. About how many square metres of carpet would you need to cover the floor?

square metre
A unit of measurement for area

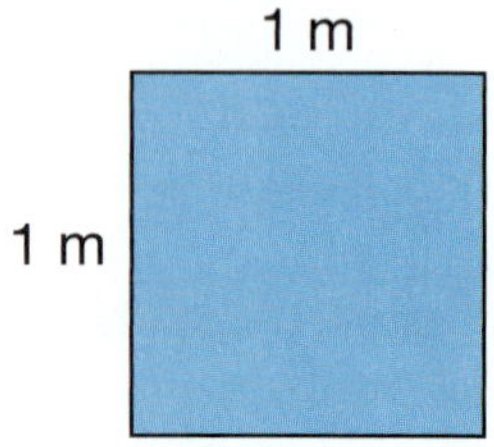

A square with sides of 1 m has an area of 1 square metre.

Reflecting

1. a) Explain how you measured the area of the floor in part C.

b) Explain why your measurement could be called an estimate.

2. Suppose you used square centimetres. Would your measurement be a larger or a smaller number of units? Explain your thinking.

Checking

3. Find a surface that you estimate has an area greater than 5 square metres.
   Use your model to measure the area to the nearest square metre.

Practising

4. Find 3 different surfaces that you estimate have an area between 6 square metres and 12 square metres.
   a) Use your model to measure each surface to the nearest square metre.
   b) Order the surfaces from the greatest to least area.

5. Which unit would you use to measure each area, square metres or square centimetres? Explain each answer.
   a) a baseball field
   b) a book cover
   c) a backyard
   d) an envelope

6. What could be made from each material?
   a) about 1 square metre of cloth
   b) about 1 square centimetre of metal
   c) about 1 square metre of plastic

7. Josh cut his model of a square metre. He used the pieces to make a new shape.

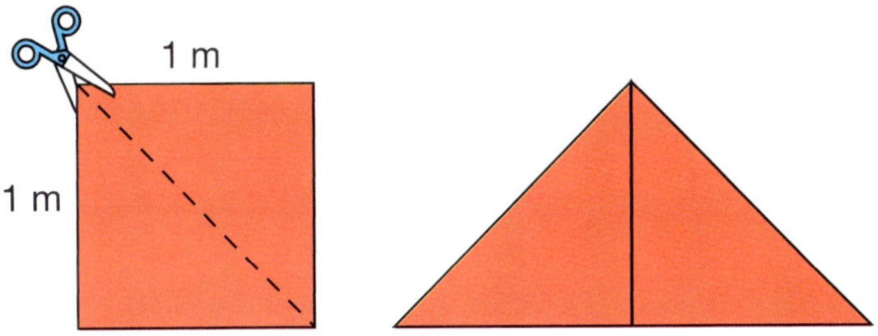

   a) What is the area of Josh's new shape? Explain how you know.
   b) What other shapes can you make with your model of a square metre? What is the area of each shape?

CHAPTER 8

3 Solve Problems Using Organized Lists

You will need

- square tiles

- grid paper

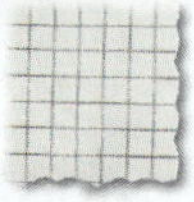

Goal **Use an organized list to solve area problems.**

? How many different rectangles with an area of 12 square tiles can you make?

Rey's Solution

Understand

I know I can make a rectangle that is 1 row of 12 tiles.

I can also make a rectangle that is 2 rows of 6 tiles.

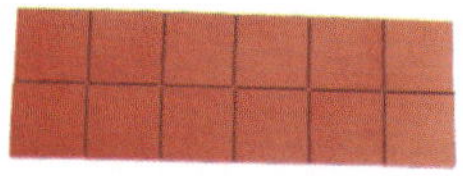

Make a Plan

If I keep trying different numbers of rows in order, I should be able to find all the rectangles that are possible.

I can keep track of what I do in an **organized list**.

organized list The strategy of following a certain order to find all possibilities

Carry Out the Plan

I started with 1 row and then tried all the numbers of rows to 12.

I crossed out the rectangles that were the same as another rectangle.

I can make 3 different rectangles with 12 square tiles.

Organized List of Rectangles

1 row of 12 ✓

2 rows of 6 ✓

3 rows of 4 ✓

~~4 rows of 3~~ (same as 3 rows of 4)

5 rows (rectangle impossible)

~~6 rows of 2~~ (same as 2 rows of 6)

7, 8, 9, 10, 11 rows (rectangle impossible)

~~12 rows of 1~~ (same as 1 row of 12)

Reflecting

1. How did Rey's organized list help him find all the different rectangles?
2. Suppose each tile has a design. Would there still be only 3 different rectangles? Explain your thinking.
3. Explain why a rectangle with 3 rows of 4 tiles and a rectangle with 4 rows of 3 tiles both have the **dimensions** 3 units by 4 units.

dimensions
Measurements such as length and width

Checking

4. Use grid paper. What are the dimensions of all the rectangles with whole number sides that have an area of 36 square centimetres? Show how you know.

Practising

5. Use grid paper. What are the dimensions of all the rectangles with whole number sides that have an area of 40 square centimetres? Show how you know.
6. This square has an area of 16 square centimetres. Complete the organized list. Find the total number of squares.

Size of square	Area of square	Number of squares with the area
4 by 4	16 square centimetres	1 square
3 by 3	■ square centimetres	4 squares
2 by 2	■ square centimetres	■ squares
1 by 1	■ square centimetres	■ squares

7. Write a problem that can be solved with an organized list. Give your problem to a partner to solve.

LESSON

Mid-Chapter Review

1

1. Which shape has the same area as the red square? How do you know?

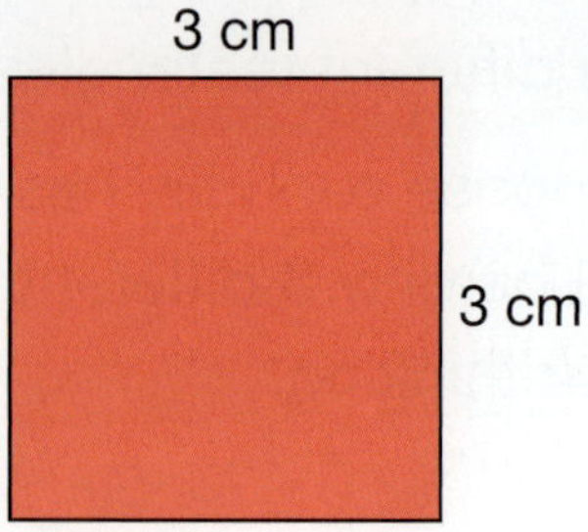

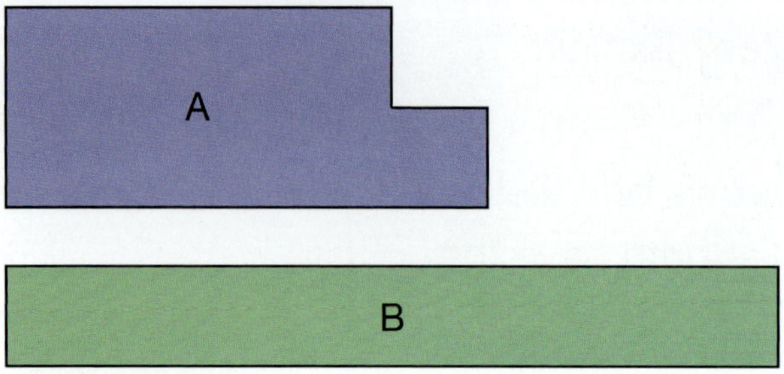

2. Measure each area to the nearest whole square centimetre. Use a centimetre grid transparency.

a)

b)

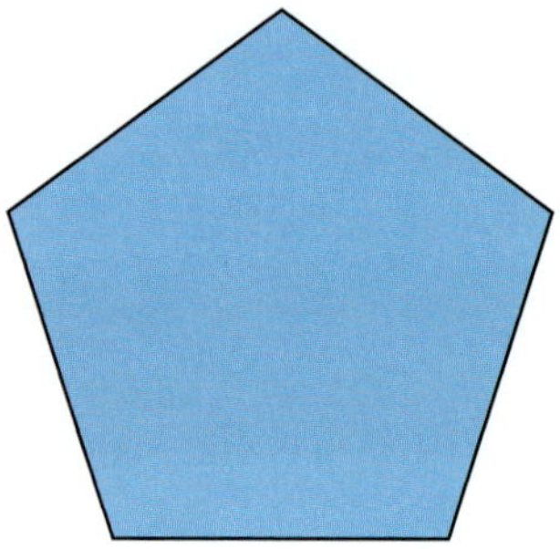

3. Trace your hand or shoe on grid paper.

a) Estimate the area in square centimetres.

b) Measure the area to the nearest whole square centimetre.

c) Compare the measurement with your estimate.

2

4. Find 3 surfaces each with an estimated area between 10 square centimetres and 1 square metre.

a) Use a centimetre grid transparency to measure each area.

b) Order the surfaces from the least to greatest area.

5. Which unit would you use to measure each area, square centimetres or square metres? Explain each answer.

a) a photograph

b) the library ceiling

c) the surface of an ice rink

d) the palm of your hand

e) the school parking lot

f) the chalkboard

Area on Board

You will need

- a geoboard

- elastics

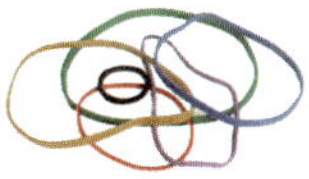

1 This geoboard is divided in half.

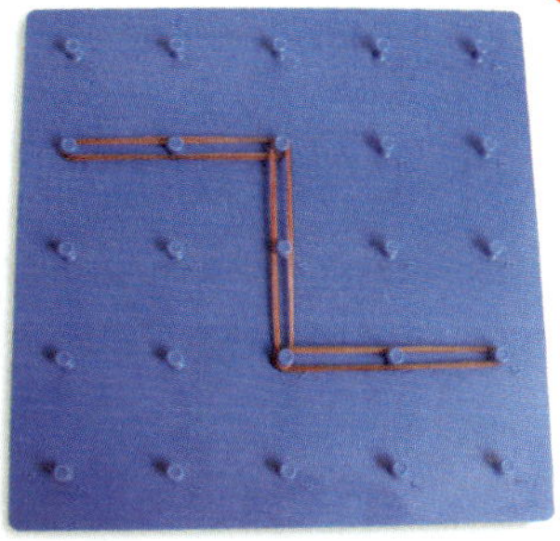

a) Find 5 more ways to divide a 5 by 5 geoboard in half.

b) How do you know the halves have the same area?

c) Are the halves always congruent?

2 Do the parallelogram and the rectangle have the same area? Explain how you know.

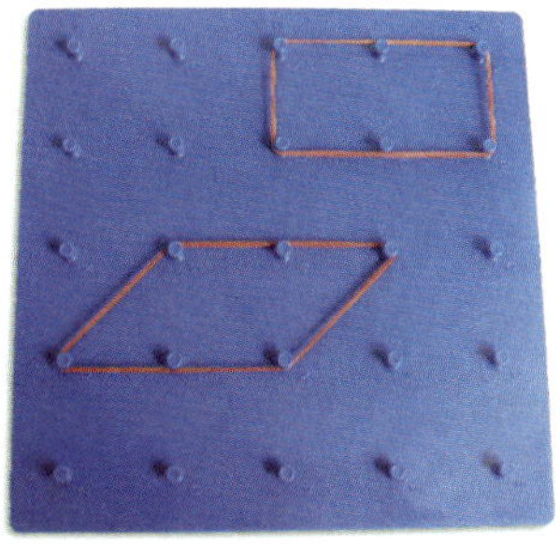

3 Make as many squares as you can on a 5 by 5 geoboard. How many squares with a different area can you make?

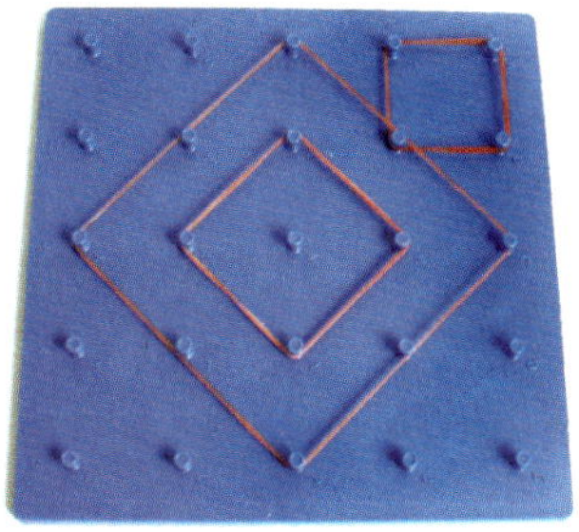

4 Maps and Paths

Goal **Use directions with north, south, east, and west for a map.**

Mandy phoned Shani to give her instructions to get to her house.

You will need

- pencil crayons

- Mandy's map

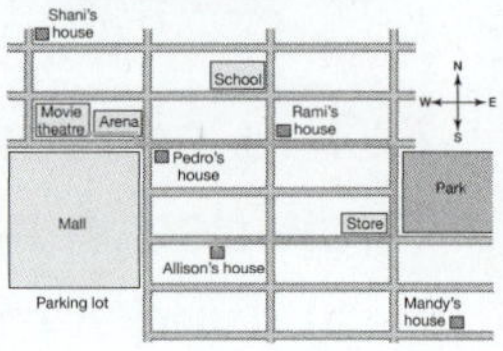

Mandy's Map

This is the way from your house to mine.

Go east from your house for 2 blocks.

Then walk south.

You'll go past our school, then past Rami's house.

After 6 blocks, turn to go east.

Walk 1 block, and then another half block to my house.

I live about the middle of the block on the north side of the street.

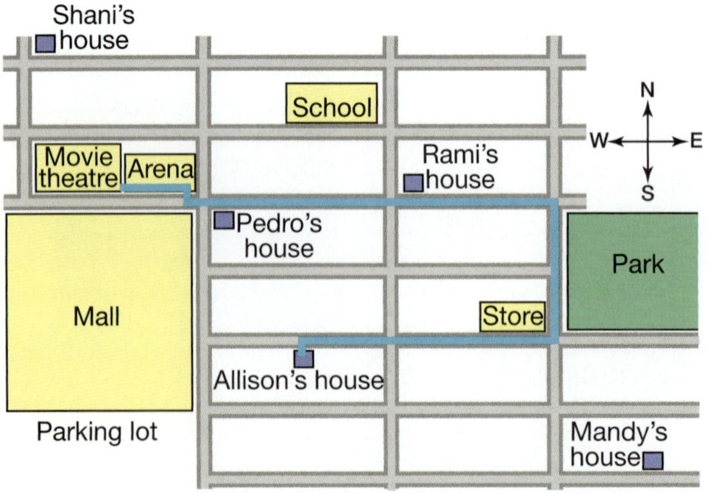

? How can you describe paths using north, south, east, and west?

Use Mandy's map.

A. Draw the path that Mandy described. Use a **red** pencil crayon.

B. Trace a different path from Shani's house to Mandy's house. Use **green**.

C. Write instructions for your **green** path from Shani's house to Mandy's house. Remember to include directions such as north, south, east, and west.

D. Describe the **blue** path on the map from Allison's house to the movie theatre.

E. Rami went from school to the southwest corner of the park, to the arena, and then home. Draw a path he could take. Use **orange**. Then write the instructions for the path.

F. Choose 2 places on the map. Write instructions to get from one place to the other. Use each word, north, south, east, and west.

Then pretend you are talking on the phone. Read your instructions for your partner to trace in **purple** on a map.

Reflecting

1. Suppose you could only use the direction words left and right instead of north, south, east and west.
   a) Describe Mandy's instructions.
   b) Which direction words do you prefer? Explain your reasons.
2. How would you change Mandy's instructions for Shani to go home? How are the instructions the same?
3. Compare Mandy's map with other maps you have seen.

CHAPTER 8

5 Coordinate Grids

Goal **Identify and describe locations on a grid using columns and rows.**

You will need
- a coordinate grid

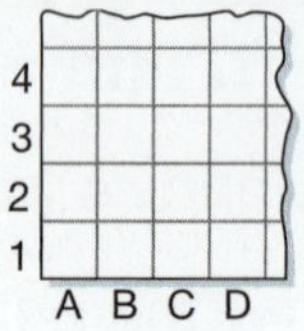

- counters

? How can you locate spaces on a grid?

Manitok's Grid

I got this grid for playing BINGO.
To get BINGO, I have to cover 5 squares to complete a row, a column, or a diagonal.

The green counter is under the G and it covers the number 2. Its **coordinates** are G2.

coordinates
Numbers or letters that describe a location on a grid

A. Name the coordinates of the other counters on Manitok's grid.

B. What are the coordinates of the FREE square?

C. What coordinates could be covered to get BINGO on Manitok's grid?

This grid shows another way to label coordinates. The red counter is at C4.

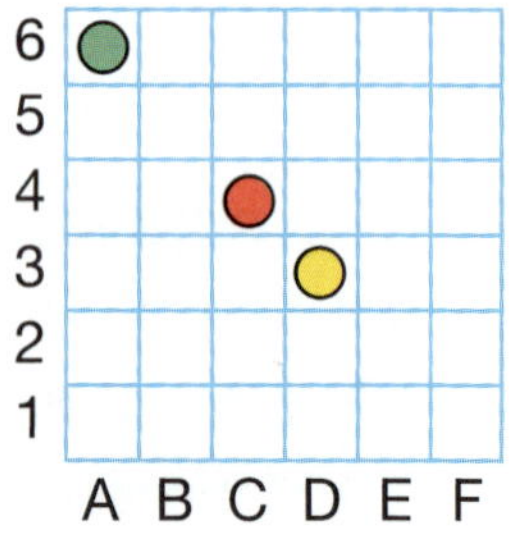

D. Name the coordinates of the other counters.

E. What coordinates would complete the diagonal?

Reflecting

1. Can the same coordinates describe 2 different spaces on a grid? Explain your answer.

Checking

2. The coordinates G4, O5, B2, I3, N4, and G1 have been called.

Would these coordinates make BINGO on Manitok's grid?
Explain your answer.

3. Suppose B2, E2, D2, and A2 are covered.

a) What other spaces could be covered to complete a row on the coordinate grid? Show your work.

b) What other spaces would be covered to complete the column for D on the coordinate grid? Show your work.

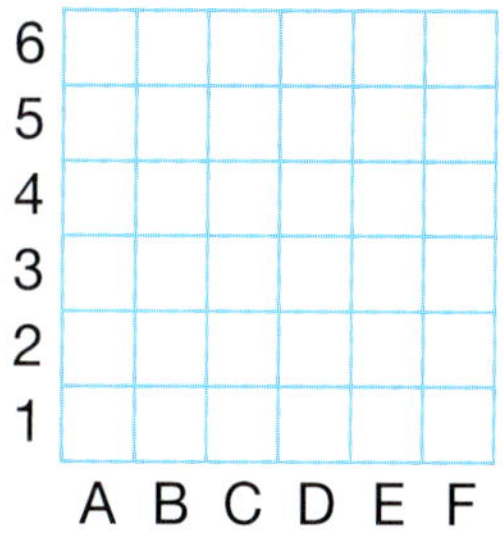

Practising

4. a) What are the coordinates of the red counters on this BINGO card?

b) What coordinates could be called to get BINGO for a row? Give 2 possible answers.

c) What coordinates could be called to get BINGO for a column? Give 2 possible answers.

5. In this game, each rectangle covers 3 spaces on a coordinate grid.
Players take turns guessing the coordinates of the other player's rectangles.

The shaded spaces show correct guesses that have already been made. What coordinates would you guess to complete the rectangles?

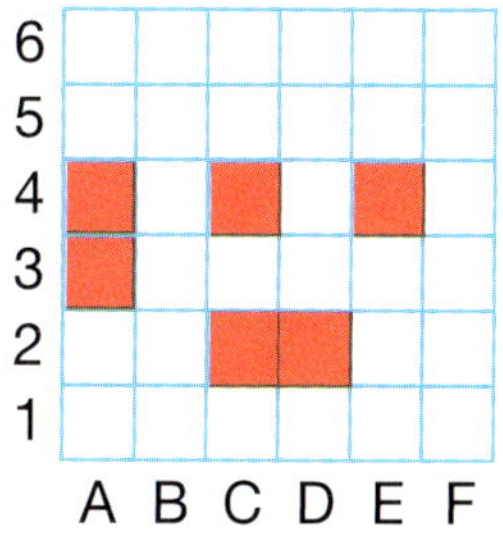

6. Use a 6 by 6 coordinate grid like the one in Question 3. Write the coordinates of 4 spaces that would make a square on this grid.

Optical Illusions

You will need

- a transparent mirror

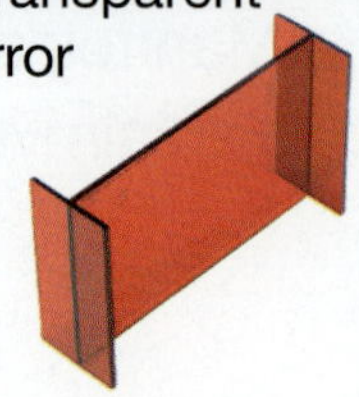

- tracing paper

Optical illusions can make shapes that are congruent appear not congruent.
Optical illusions can also make shapes that are not congruent appear congruent.

Try These

1. Do you think the **blue** circles are congruent?
   Use a transparent mirror or tracing paper to check.

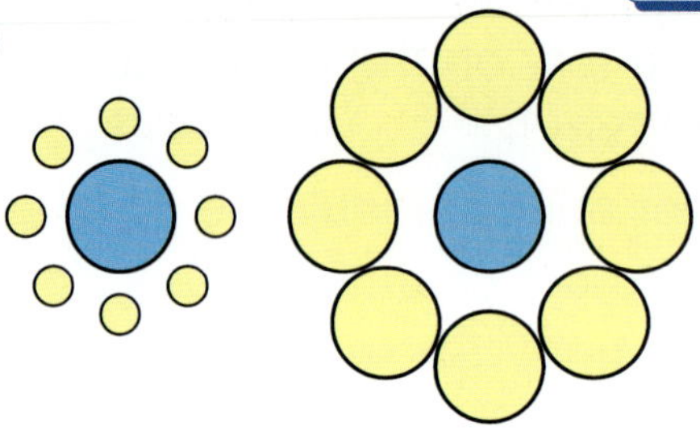

2. Do you think the **red** hexagons are congruent?
   Check your prediction.

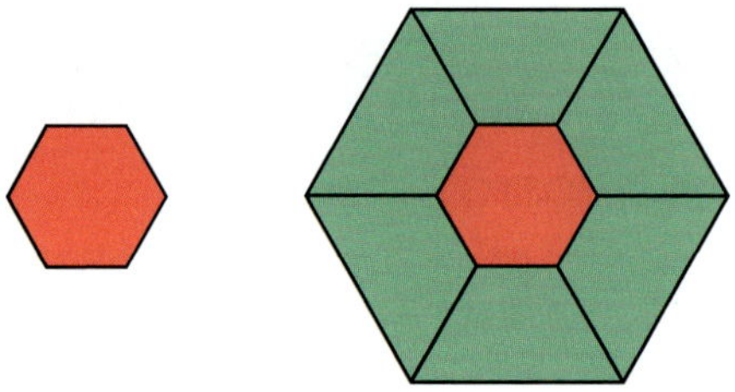

3. Do you think the **yellow** triangles are congruent?
   Check your prediction.

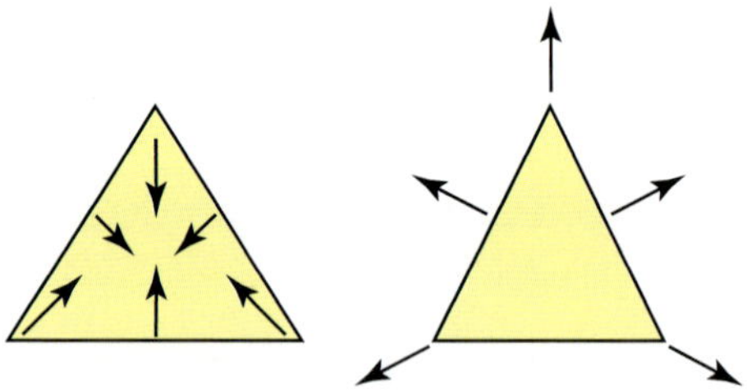

4. Do you think the **green** squares are congruent?
   Check your prediction.

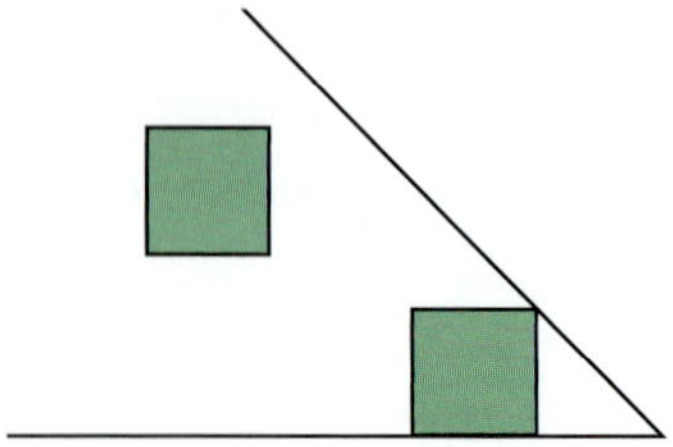

5. Create your own optical illusion.
   Explain why it is an optical illusion.

Grid Hide and Seek

Number of players: 2

How to play: Players guess the coordinates of another player's counters on a grid.

Step 1 Each player gets 1 coordinate grid and 3 counters. Set up a divider so the other player cannot see your grid.

Step 2 Place 3 counters on the spaces of your grid.

Step 3 Take turns guessing the coordinates of the other player's counters.
If a guess is correct, say, "Found."
If a guess is incorrect, say, "Keep looking."
If a guess is 1 space away from a counter, say, "Very close."

Continue playing until a player guesses all of the other player's coordinates.

You will need

- counters

- a coordinate grid

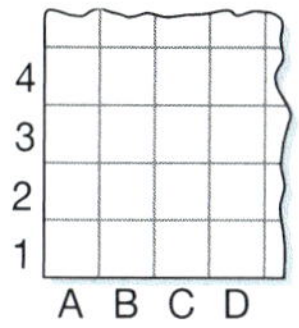

- a divider

Skills Bank

LESSON

1

1. Estimate the area of each shape. Then measure the area to the nearest whole centimetre. Use a centimetre grid transparency.

a)

b)

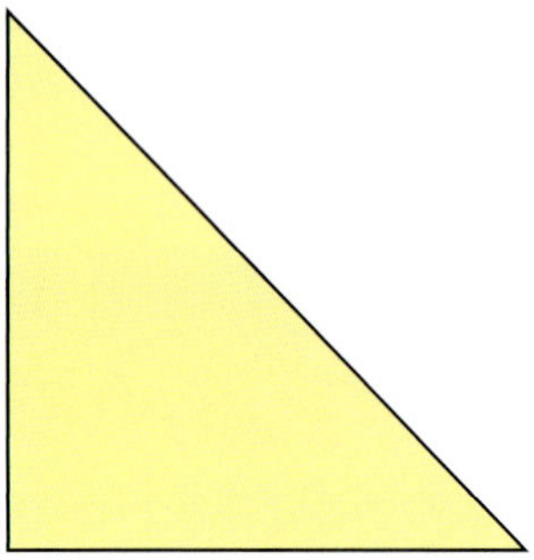

c)

d)

e)

f)

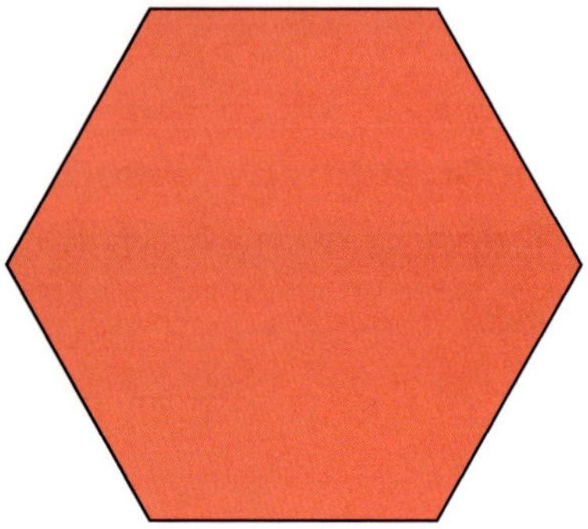

g)

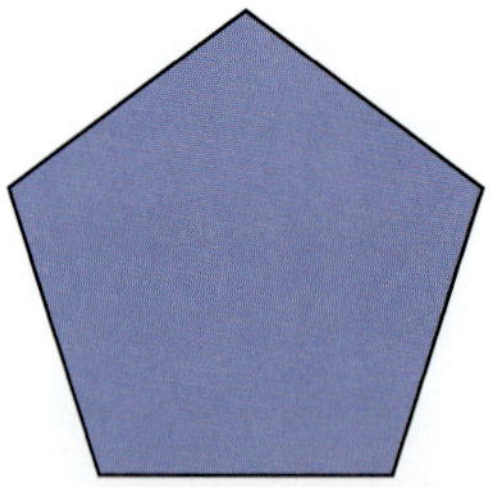

h)

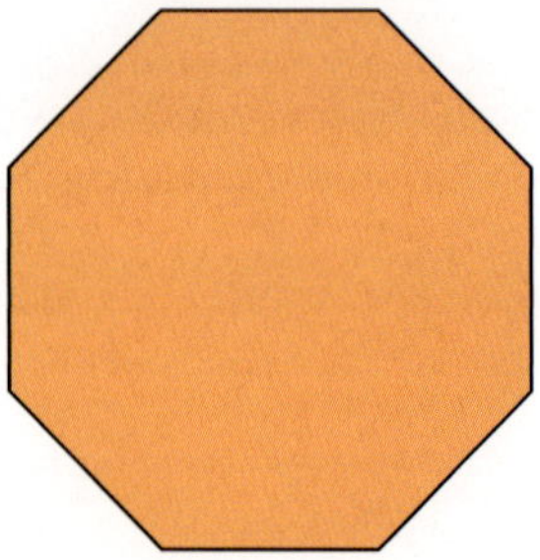

2 **2.** Find an object that you estimate has each area. Then use a model of a square metre. Measure each area.

a) between 1 square metre and 4 square metres
b) between 4 square metres and 9 square metres
c) between 9 square metres and 16 square metres
d) greater than 16 square metres

3. Name at least 5 things that you would measure using each area unit.

a) square metres
b) square centimetres

4 **4.** Each square represents a block. Describe the path from A to B to C and back to A.

Use directions such as north, south, east, and west.

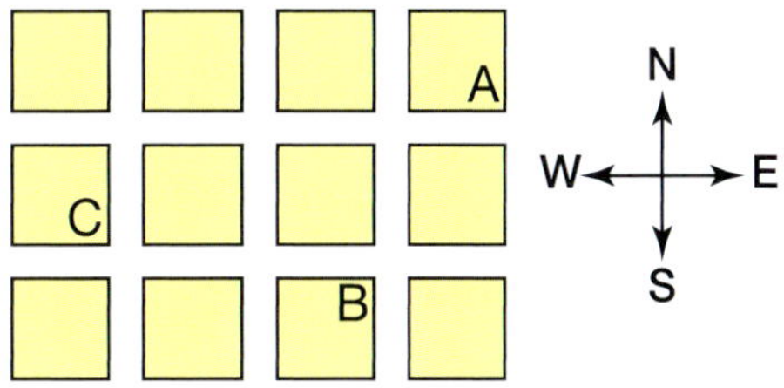

5 **5.** What are the coordinates for each colour?

a) red
b) green
c) blue

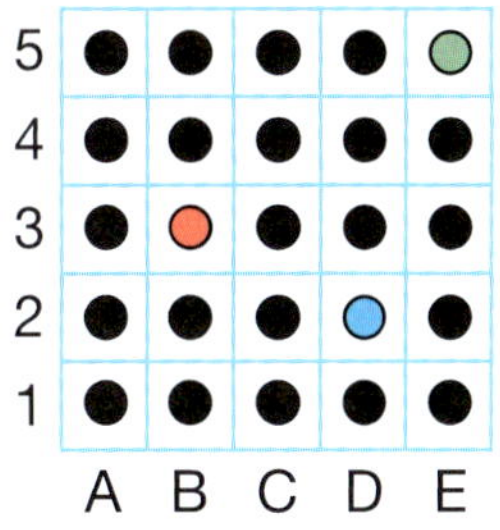

6. What are the coordinates of each trapezoid?

a) green
b) yellow
c) red
d) blue

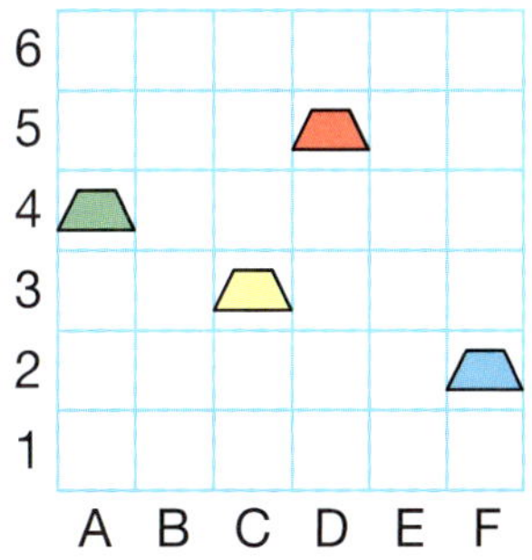

LESSON

Problem Bank

1 **1.** Justin designed this tile on grid paper.
His tile is 4 squares by 4 squares.
Each small square is 1 square centimetre.
He laid 3 tiles side by side.

a) What is the perimeter of the 3 tiles?

b) How much area is **green** in the 3 tiles?

c) How much area is **orange** in the 3 tiles?

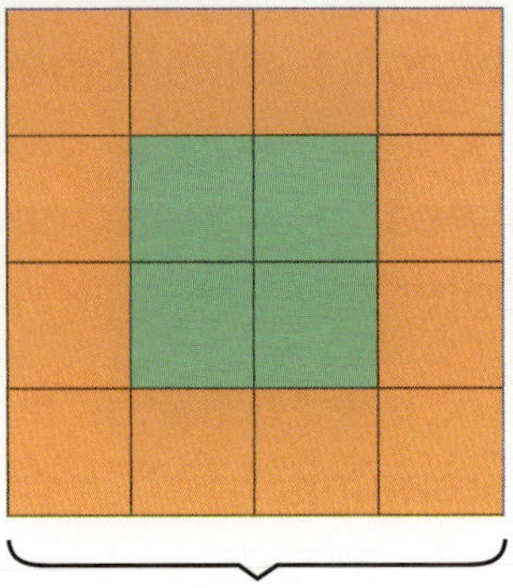

1 tile

3 **2.** There are 5 possible tetrominoes.
Here are 2 of them.

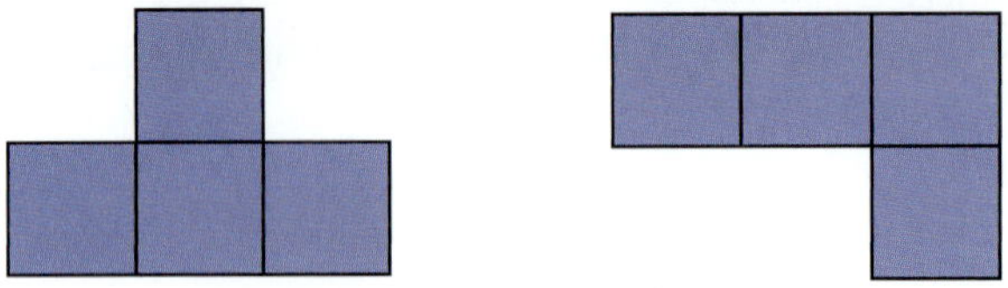

tetrominoes (4 squares)

a) Draw the other 3 tetrominoes.

b) What is the total area of all 5 tetrominoes?

c) Can the 5 tetrominoes fit together to make a rectangle? Show your work.

4 **3.** Brad can walk from his home to school in 2 different ways.

a) Describe each path. Use directions such as north, south, east, and west.

b) Which path is shorter? Explain your answer.

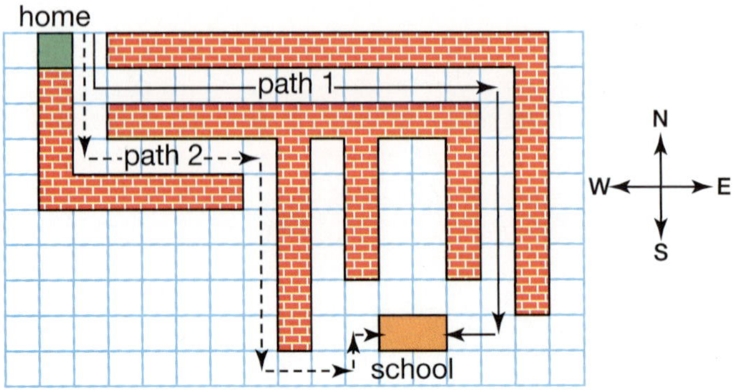

LESSON

Chapter Review

1 **1.** Measure the area of each shape to the nearest whole square centimetre.

a)

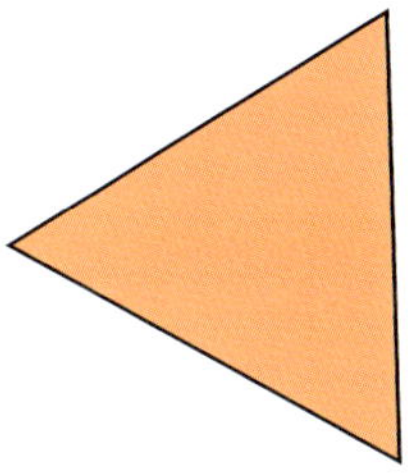

b)

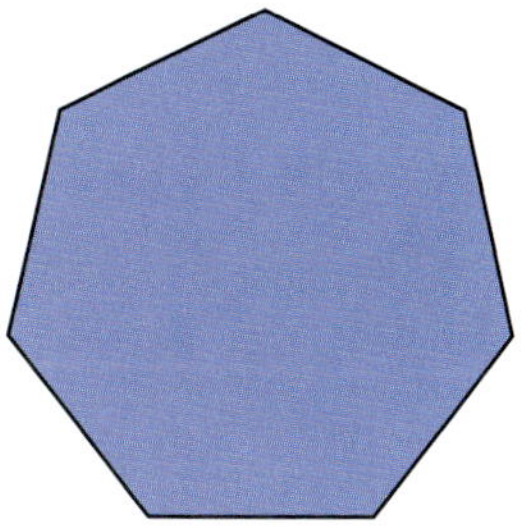

2 **2.** What unit of measurement would you use for each? Explain each answer.

a) the amount of space taken by a stamp

b) the amount of wallpaper needed to cover a wall

3. Shani said that the number of square centimetres to measure an area is more than the number of square metres. Use drawings or examples to show that this is true.

3 **4.** **a)** Use grid paper. What are the dimensions of all the rectangles with whole number sides that you can make with an area of 32 square centimetres? Show your work.

b) Did you find all the rectangles? Explain how you know.

4 **5.** Name a place such as your school, a friend's home, or a store.

a) Write instructions to go from your home to this place. Use directions such as north, south, east, and west.

b) Write instructions to go back to your home.

5 **6.** **a)** What are the coordinates of the counters on this grid?

b) Where would you put new counters to complete a row? Name the coordinates. Give 2 possible answers.

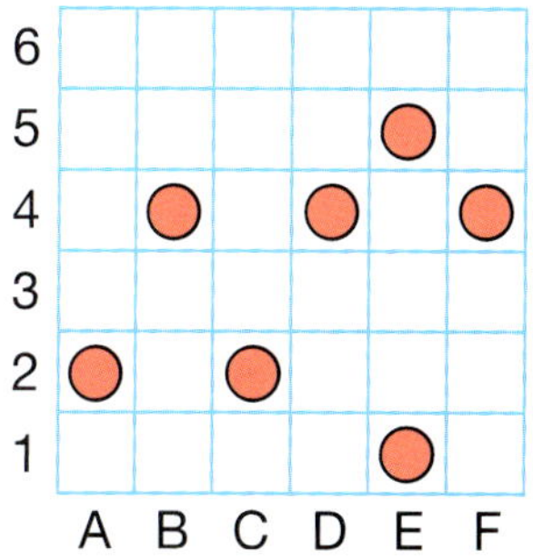

Chapter Task

Design a Petting Zoo

You are designing a petting zoo for the Spring Fair. Each type of animal should be in its own pen. You have only 80 m of fencing.

- The pony and calf each need a large pen.
- The chickens and rabbits need small pens.
- The lambs and piglets need medium pens.
- The pens must have some space between them.

? How can you design the petting zoo so that it is well planned, and doesn't take up too much space?

Understand

A. How many pens are needed?

B. What do you know about the areas and the perimeters of the pens?

Make a Plan

C. Use square tiles to model the pens. Each square represents 1 square metre.

Carry Out the Plan

D. Draw your model on grid paper.

Look Back

E. Does your design meet the requirements?

Communicate

F. Explain why you think your design is a good one.

Task Checklist

- ☑ Did you use a model?
- ☑ Did you include diagrams?
- ☑ Did you show the right amount of detail?
- ☑ Did you explain your thinking?
- ☑ Did you use math language?

CHAPTER 9

Multiplying Greater Numbers

Goals

You will be able to

- **recognize a variety of multiplication situations**
- **multiply 2-digit and 3-digit numbers by 1-digit numbers**
- **create and solve multiplication problems using different strategies**

Ukrainian eggs at the Canadian Museum of Civilization

Getting Started

Making Multiplication Facts

A single **array** represents 2 multiplication facts. Some arrays can be rearranged to show other multiplication facts.

You will need

- dice

- tiles

- a multiplication table

×	1	2	3
1	1	2	3
2	2	4	6
3	3	6	9

? How many multiplication facts can you find by making and rearranging arrays?

A. Roll 2 dice. Record the results.

B. Use tiles to make an array. Use the numbers you rolled as the numbers of rows and columns.

C. Write 2 multiplication sentences for the array. Circle the products on a multiplication table.

Josef and Pedro's Array

Pedro and I rolled a 2 and a 6.
We made this array.

We wrote our multiplication sentences.

$2 \times 6 = 12$ $6 \times 2 = 12$

Then we circled 12 for 2×6 and 12 for 6×2 on a multiplication table.

We think we can make a different array with the 12 tiles.

D. If possible, rearrange the array to make other arrays.

E. Write 2 multiplication sentences for each new array. Circle the products on a multiplication table.

F. Repeat parts A to E 6 times to see how many different facts you can make.

Do You Remember?

1. Multiply. Explain the strategy that you used.

a) Use $3 \times 4 = 12$ to find 3×8.

b) Use $2 \times 10 = 20$ to find 4×5.

c) Use $5 \times 3 = 15$ to find 6×3.

d) Use $3 \times 5 = 15$ to find 6×5.

e) Use $8 \times 9 = 72$ to find 9×8.

2. Multiply.

a) 7×8 d) 2×8 g) 0×0

b) 9×3 e) 9×7 h) 8×8

c) 6×0 f) 6×5 i) 1×9

3. Multiply.

a) 7×10 c) 9×100 e) 2×1000

b) 3×50 d) 7×20 f) 5×200

4. Write the **fact family** for this array.

1 Exploring Multiplication

You will need
- base ten blocks

Goal Solve multiplication problems using models.

A grade 4 class is planning a pizza party.
There are 28 students in the class.
They want 3 slices of pizza each.

? How many pizzas should the class order, and how much will the pizzas cost?

Use base ten blocks to model the problem.

A. How many slices of pizza will the class need?

B. How many slices will they have when the pizzas arrive?

C. How much will the pizzas cost?

Reflecting

1. How do you know parts A, B, and C are multiplication questions?
2. Is using base ten blocks a reasonable method to solve this problem? Explain.
3. How can changing 28 groups of 3 to 3 groups of 28 help you to model the solution?

Curious Math

Persistent Numbers

You will need

- a calculator

To find the persistence of a number, you multiply the digits of the number until you get a 1-digit number.

	Step 1		Step 2
723 →	$7 \times 2 \times 3 = 42$	→	$4 \times 2 = 8$

The persistence of 723 is 2 because it takes 2 steps to get a 1-digit number.

1 Use a calculator to find the persistence of each number.

a) 54 c) 328 e) 5063

b) 12 d) 4528

2 What is the least number with a persistence of 2?

Curious Math

Sum and Product

1 What 3 numbers have a sum equal to their product?

■ + ▲ + ● = ■ × ▲ × ●

2 5 different numbers have a sum of 15. Those same 5 numbers have a product of 120. What are the 5 numbers?

CHAPTER 9

2 Multiplying with Arrays

Goal **Use easier numbers to simplify multiplication.**

You will need

- grid paper

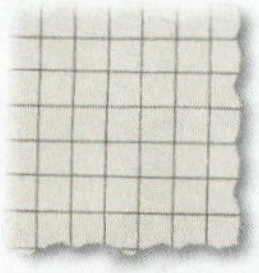

- pencil crayons

- a calculator

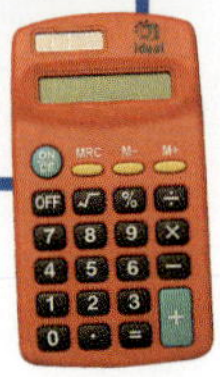

? How many squares are on Natalie's game?

Natalie's Solution

There are 8 rows of squares in my game. There are 12 columns of squares. I can multiply 8 × 12 to find how many squares there are altogether.

First, I estimate that 12 is about 10.

8 × 10 = 80. There are more than 80 squares.

I can model the game on grid paper.

I noticed there are 2 arrays *inside* the 8 × 12 array. I colour the 8 × 10 array red. I colour the 8 × 2 array blue. I find the products for the 2 smaller arrays and add them together.

8 x 12 = ■

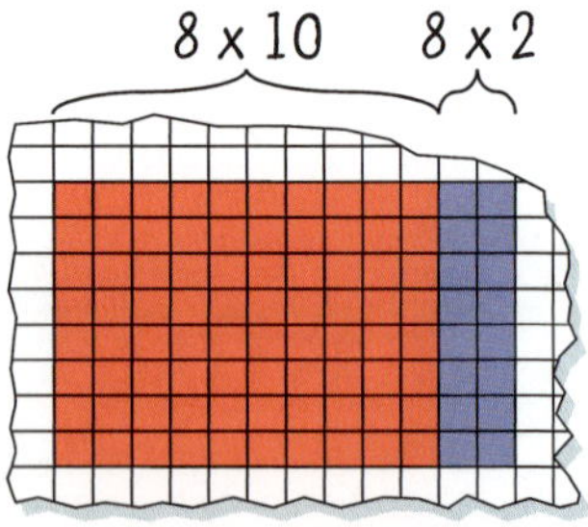

8 x 12 = 8 x 10 + 8 x 2

A. Use 8 × 10 + 8 × 2 to find 8 × 12.

The array can also be broken into other smaller arrays.

B. Draw an 8 × 12 array on grid paper.
Starting in the 1st column, colour in an 8 × 8 array.
Then colour in an 8 × 4 array.

C. Write 8 × 12 as the sum of 2 products to represent what you coloured.
8 × 12 = 8 × ■ + 8 × ■
8 × 12 = ■ + ■
8 × 12 = ■

D. Draw another 8×12 array on grid paper. Find and colour 2 arrays inside it. Use different arrays than in parts A and B. Write 8×12 as the sum of 2 products to represent what you coloured.

Reflecting

1. a) How does using smaller arrays help you to multiply?
   b) Which arrays were easiest to multiply? Why?
2. Describe the pattern in the multiplication sentences from parts A, C, and D.

Checking

3. A game has 7 rows and 14 columns.
   a) Estimate the number of squares on the game.
   b) Draw the array on grid paper. Find and colour smaller arrays that have easier numbers to multiply.
   c) Write 7×14 as the sum of 2 products to represent what you coloured.

Practising

4. A tree planter has planted 7 rows of 18 trees. How many trees did he plant? Solve using arrays.
5. Complete.
   a) $6 \times 21 = 6 \times 20 + 6 \times 1$
      $6 \times 21 = \blacksquare$
   b) $4 \times 16 = 4 \times 8 + 4 \times \blacksquare$
      $4 \times 16 = \blacksquare$
   c) $5 \times 32 = 5 \times \blacksquare + 5 \times \blacksquare$
      $5 \times 32 = \blacksquare$
   d) $5 \times 28 = \blacksquare \times \blacksquare + \blacksquare \times \blacksquare$
      $5 \times 28 = \blacksquare$
6. Multiply. Use a calculator to check.
   a) 9×17
   b) 2×15
   c) 8×11

CHAPTER 9

3 Multiplying in Expanded Form

Goal **Multiply 1-digit numbers by 2-digit numbers using expanded form.**

You will need

- base ten blocks

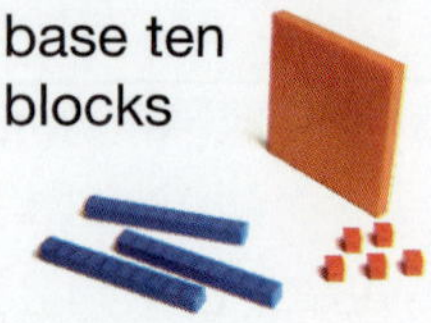

- a place value chart

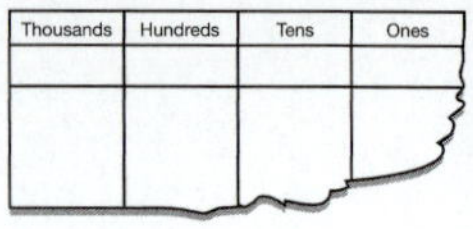

Paulette is making 54 leather bags for gifts at a celebration.
She sews 3 medallions on each bag.

? How many medallions will Paulette need?

Paulette's Solution

54 bags with 3 medallions on each bag is 54 groups of 3.
When I have equal groups, I can multiply.
54×3 and 3×54 have the same product, so I'll multiply 3×54 instead.
I'll multiply using expanded form and check my work using blocks.

Step 1 I write 54 in expanded form. I make 3 groups of 54.

$$\begin{array}{r} 54 \\ \times\ 3 \\ \hline \end{array} \qquad \begin{array}{r} 50 + 4 \\ \times\ 3 \\ \hline \end{array} \qquad \begin{array}{r} 5 \text{ tens} + 4 \text{ ones} \\ \times\ 3 \\ \hline \end{array}$$

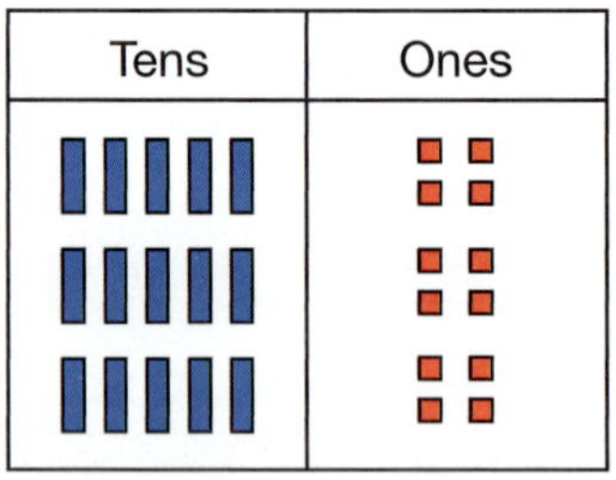

Step 2 I multiply 5 tens by 3. I see 3 groups of 5 tens, or 15 tens.

$$\begin{array}{r} 54 \\ \times\ 3 \\ \hline \end{array} \qquad \begin{array}{r} 50 + 4 \\ \times\ 3 \\ \hline 150 \end{array} \qquad \begin{array}{r} 5 \text{ tens} + 4 \text{ ones} \\ \times\ 3 \\ \hline 15 \text{ tens} \quad \end{array}$$

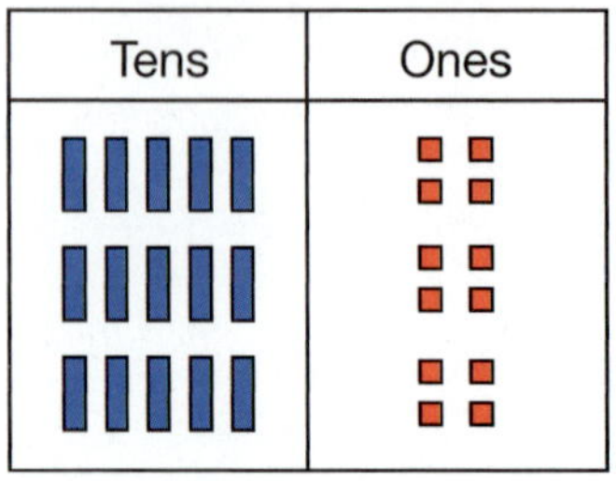

Step 3 I multiply 4 ones by 3. I see 3 groups of 4 ones, or 12 ones.

$$\begin{array}{r} 54 \\ \times\ 3 \\ \hline \end{array} \qquad \begin{array}{r} 50 + 4 \\ \times\ 3 \\ \hline 150 \\ +\ 12 \\ \hline \end{array} \qquad \begin{array}{r} 5 \text{ tens} + 4 \text{ ones} \\ \times\ 3 \\ \hline 15 \text{ tens} + 12 \text{ ones} \end{array}$$

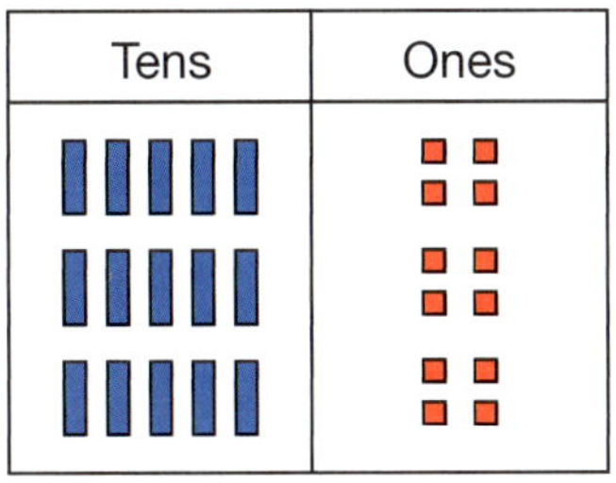

Step 4 I add the 2 products. I regroup the blocks.

$$\begin{array}{r} 54 \\ \times\ 3 \\ \hline \end{array} \qquad \begin{array}{r} 50 + 4 \\ \times\ 3 \\ \hline 150 \\ +\ 12 \\ \hline 162 \end{array} \qquad \begin{array}{r} 5 \text{ tens} + 4 \text{ ones} \\ \times\ 3 \\ \hline 15 \text{ tens} + 12 \text{ ones} \\ 162 \end{array}$$

1 hundred + 5 tens + 1 ten + 2 ones

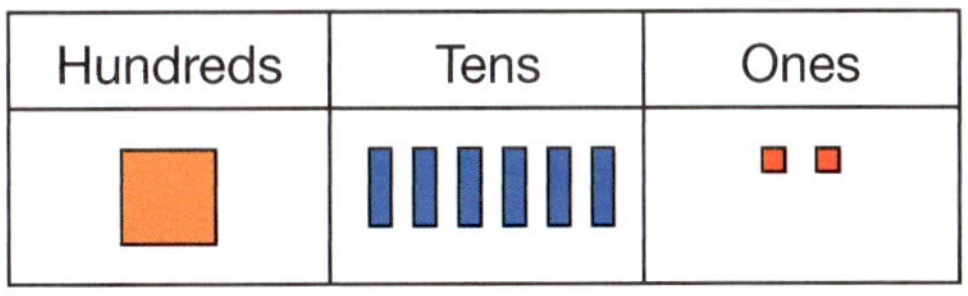

1 hundred + 6 tens + 2 ones

Reflecting

1. How did writing 54 in expanded form help Paulette to multiply?

Checking

2. Stanley serves 4 trays of salmon at a feast. Each tray holds 32 pieces of salmon. How many pieces of salmon does Stanley serve? Use expanded form to multiply. Use blocks to check.

Practising

3. Alasie made a necklace with 6 rows of 64 beads. How many beads did she use altogether?

4. Estimate. Then calculate. Use estimates to check.
   a) 17×5 b) 14×7 c) 56×6

LESSON

Mid-Chapter Review

1 **1.** Which problem can be solved using multiplication? Explain how you know.

a) A basketball player played in 3 games. In each game, he scored 22 points. How many points did he score altogether?

b) A basketball player scored 27 points. Each basket was worth 3 points. How many baskets did she make?

c) A basketball player scored 26 points. Another player scored 32 points. How many points did they score together?

2 **2.** Calculate each product. Show your work.

a) 3×18 **b)** 6×11 **c)** 7×14

3. Olivia made a 6×16 tray of Nanaimo bars. Write a set of smaller arrays to find the number of bars she made.

$6 \times 16 = ■ \times ■ + ■ \times ■$

4. Kaytan is helping to make a brick pathway. It is going to have 19 rows with 7 bricks in each row. How many bricks will he need? Show your work.

3 **5.** Ryan is making 27 bean-bag creatures. Each creature will have 3 buttons on its face. How many buttons will Ryan need? Show your work.

6. Complete.

a)
7 tens + 9 ones
× 6 ones
42 tens + ■

b)
60 + 7
× 8
480 + ■

c)
3 tens + 2 ones
× 3
■ + 6 ones

7. Justine estimates that 8×72 is more than 560. Explain her thinking.

Estimating with Multiples of 10

You will need

- a calculator

Rey's Estimating

I estimated 35×36 by rounding this way.

I rounded each factor.	$35 \times 36 \longrightarrow 30 \times 40$
I renamed 30 as 3×10.	$3 \times 10 \times 40$
I changed the order.	$3 \times 40 \times 10$
I multiplied 3×40.	120×10
I multiplied 120×10.	1200

35×36 is about 1200.

A. Why could Rey round 35×36 to 40×40 instead of 30×40?

B. Which is closest to the product for 35×36?
30×40 or 40×40
Use a calculator to help you decide.

C. Estimate 53×45. Explain your steps.

Try These

1. Choose the factors that give the closest estimate.
Use a calculator to help you decide.

a)	63×35	60×30	60×40	70×30	70×40
b)	25×58	20×60	30×60	20×50	30×50

2. Estimate each product.

a) 52×67	**c)** 19×39	**e)** 45×44	**g)** 65×56
b) 25×25	**d)** 61×86	**f)** 73×92	**h)** 22×45

4 Communicate About Solving Problems

Goal **Explain your thinking when solving a problem.**

? How can Chantal explain how she solved a problem?

Chantal's Problem

I heard that horses age differently.
A 1-year-old horse is like a 3-year-old human.
I wondered how old our 16-year-old horse would be in human years, so I figured it out.

Chantal's Explanation

I asked Vinh to help me improve my explanation.

I made sure I understood the problem.

A horse this old ...	... is like a human this old
1	3
2	6
3	9

I made a plan. I will multiply.
I carried out the plan. The answer was 48,
so I knew my horse would be like a 48-year-old human.
Then I looked back to check.
I think 48 years old is a reasonable answer.

You showed the problem-solving steps.

How did you know you could multiply?

What numbers did you multiply?

You looked back to check your answer.

How do you know 48 is reasonable?

A. What strengths did Vinh see in Chantal's explanation?

B. What did Vinh think was missing from Chantal's explanation?

C. How would you answer Vinh's purple questions?

Reflecting

1. How did Chantal's table help to show what she was thinking?
2. Which question in the Communication Checklist was each of Vinh's purple questions about?

Communication Checklist

- ✓ Did you show the right amount of detail?
- ✓ Did you explain your thinking?

Checking

3. A 1-year-old bear is like a 4-year-old human. Vinh calculated the age of a 19-year-old bear in human years.
   a) Identify at least 1 strength in Vinh's explanation.
   b) What questions would you ask Vinh to help him improve his explanation?

Vinh's Explanation

I made sure I understood the problem.

A bear this old ...	... is like a human this old
1	4
2	8
3	12

I made a plan. I multiplied 19 x 4.

I carried out the plan. 19 x 4 = 76

I looked back to check. 76 is right because 20 x 4 = 80, so 19 x 4 must be a bit less.

Practising

4. A 1-year-old deer is like an 8-year-old human.
   a) Calculate the age of a 13-year-old deer in human years.
   b) Explain how you solved and checked the problem. Use the Communication Checklist.

5 Multiplying 3 Digits by 1 Digit

Goal **Multiply 3-digit numbers by 1-digit numbers using expanded form.**

You will need

- base ten blocks

- a place value chart

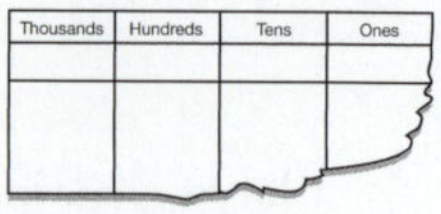

Miki lives near a beach.
She collects 114 shells each week.

? How many shells will Miki have in 4 weeks?

Miki's Solution

Collecting 114 shells each week for 4 weeks is like 4 groups of 114.

When I have equal groups, I can multiply.

Step 1 First, I estimate.
4×114 is about $4 \times 100 = 400$.
I will have more than 400 shells.

Step 2 I write 114 in expanded form.

$$\begin{array}{r} 114 \\ \times\ 4 \\ \hline \end{array} \qquad \begin{array}{r} 100 + 10 + 4 \\ \times\ 4 \\ \hline \end{array}$$

Step 3 I multiply 4 by the hundreds digit.

$$\begin{array}{r} 114 \\ \times\ 4 \\ \hline \end{array} \qquad \begin{array}{r} 100 + 10 + 4 \\ \times\ 4 \\ \hline 400 \end{array}$$

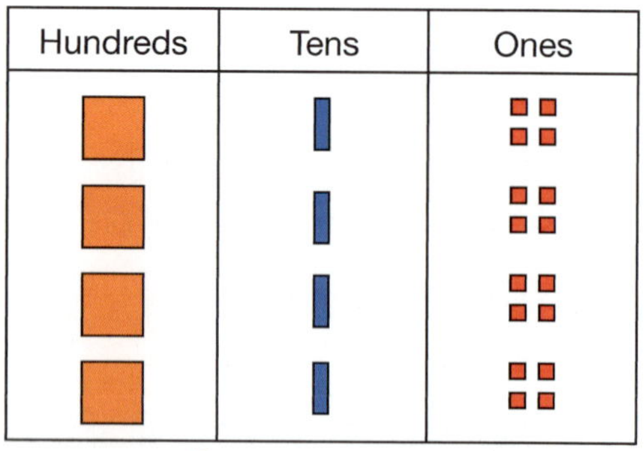

A. Complete Miki's multiplication.

B. Check your answer with base ten blocks on a place value chart.

Reflecting

1. Suppose you solved Miki's problem by multiplying this way. Would the product be the same? Explain.

$$\begin{array}{r} 114 \\ \times\ 4 \\ \hline \end{array} \qquad \begin{array}{r} 100 + 10 + 4 \\ \times\ 4 \\ \hline 16 \end{array}$$

2. Was your final product in Question 1 greater or less than Miki's estimate? Explain why.

Checking

3. Craig polishes small rocks in a rock tumbler. He polishes 115 rocks each month. How many rocks will Craig polish in 6 months?
   a) Estimate how many rocks Craig will polish in 6 months.
   b) Calculate how many rocks Craig will polish in 6 months. Multiply using expanded form.
   c) Check your answer using base ten blocks.

Practising

4. Kristi and her family grow carrots to sell at the farmer's market. They plant 350 carrots each month from May to July. How many carrots do they plant altogether?

5. In a school parade, 485 students will each carry 3 balloons. 125 students will each carry 2 flags. How many balloons and flags will the students carry?

6. Create a 3-digit by 1-digit multiplication problem about collecting cards. Estimate the answer. Then calculate. Use your estimate to check.

7. Estimate. Then calculate each product.
   a) $\begin{array}{r} 986 \\ \times\ 3 \\ \hline \end{array}$ b) $\begin{array}{r} 181 \\ \times\ 5 \\ \hline \end{array}$ c) $\begin{array}{r} 332 \\ \times\ 7 \\ \hline \end{array}$

6 Multiplying with an Algorithm

Goal **Multiply using a procedure.**

You will need

- base ten blocks

- a place value chart

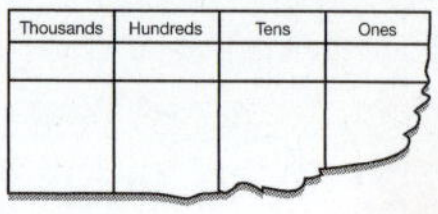

- a calculator

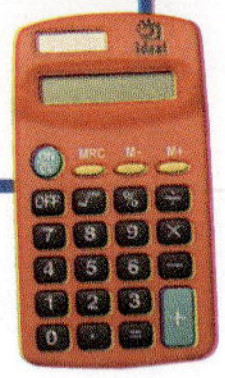

Terry has 256 hockey cards. Pedro has twice as many.

? How many hockey cards does Pedro have?

Terry's Calculations

Twice as many means 2 times as many.
I'll multiply to find how many cards Pedro has.

Step 1 First, I estimate.
256 is about halfway between 200 and 300.
$2 \times 200 = 400$ $\quad$ $2 \times 300 = 600$
500 is halfway between 400 and 600.
Pedro has about 500 cards.

Step 2 I calculate by making 2 groups of 256.
I see 2×6 ones blocks.

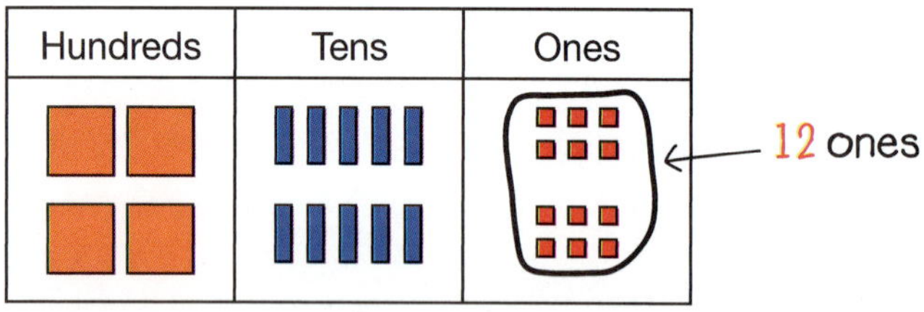

$$\begin{array}{r} 256 \\ \times\ 2 \\ \hline \end{array}$$

Step 3 I regroup 10 ones as 1 tens block.

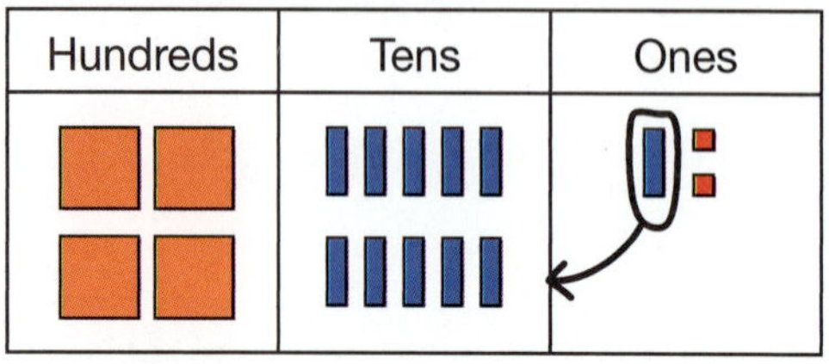

1 ten
2 ones

$$\begin{array}{r} 1 \\ 256 \\ \times\ 2 \\ \hline 2 \end{array}$$

Step 4 I see 2 × 5 tens blocks + 1 tens block.
10 tens + 1 ten = 11 tens

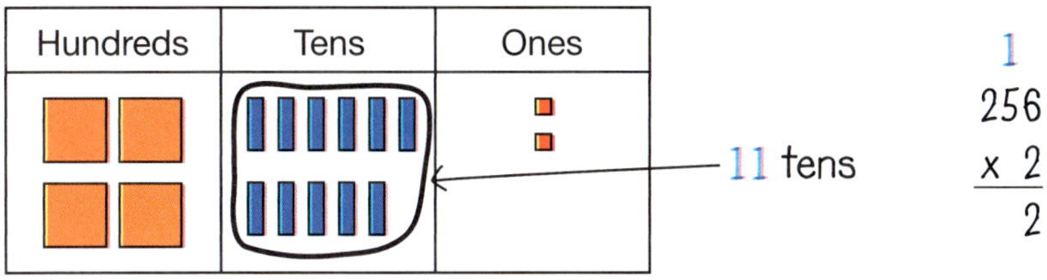

Step 5 I regroup 10 tens as 1 hundreds block.

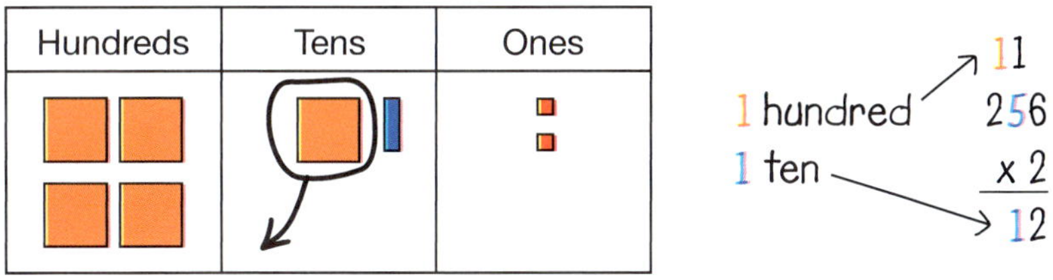

Step 6 I see 2 × 2 hundreds blocks + 1 hundreds block.
4 hundreds + 1 hundred = 5 hundreds

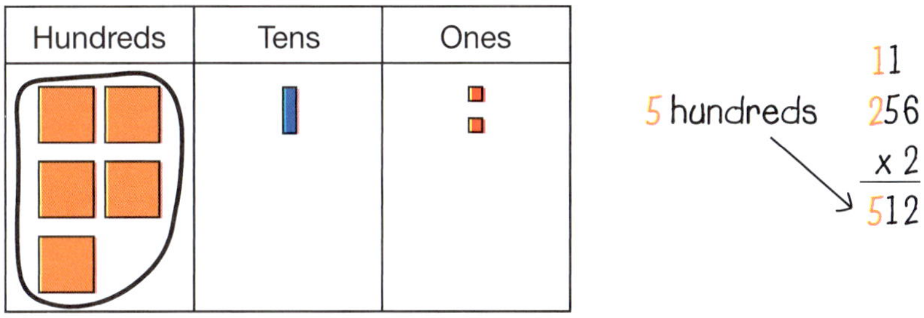

512 is just a bit more than my estimate of 500, so my answer is reasonable. Pedro has 512 hockey cards.

Reflecting

1. In Step 3, why did Terry record a 1 above the 5?
2. Terry calculates that Shani has 2 × 756 cards.
   a) What is the next step of his new multiplication?
   b) How is this next step different from when he multiplied 2 × 256?

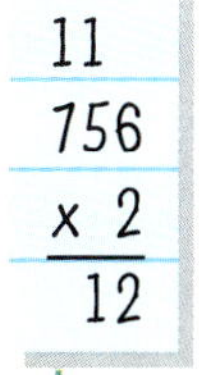

Checking

3. Chantal likes horses. She has 145 plastic horses. She says she has 3 times as many other animals as she has horses.
   How many other animals does she have?
   Give an estimate. Then calculate the answer.

Practising

4. 2 art classes are making mugs.
   The 1st class made 154 mugs.
   The 2nd class made 4 times as many mugs.
   How many mugs did the 2nd class make?

5. Jamal multipied 384×4.
   He used 4×400 for his estimate.
   a) Was his estimate high or low?
   Explain your answer.
   b) Write 3 other 3-digit by 1-digit multiplications that would have the same 4×400 estimate.

6. Write the multiplication sentence for this model.

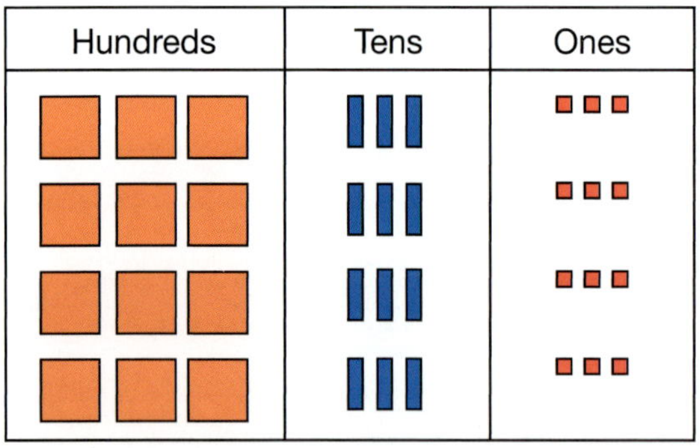

7. Estimate. Then calculate. Use estimates to check.
   a) $\begin{array}{r} 225 \\ \times\ 9 \\ \hline \end{array}$ b) $\begin{array}{r} 863 \\ \times\ 3 \\ \hline \end{array}$ c) $\begin{array}{r} 594 \\ \times\ 7 \\ \hline \end{array}$ d) $\begin{array}{r} 943 \\ \times\ 8 \\ \hline \end{array}$

8. Make up a 3-digit by 1-digit multiplication problem about a hobby or game that interests you.
   Solve your problem. Check using a calculator.

Math Game

Greatest Product

Number of players: 2 to 4
How to play: Create the greatest product.

Step 1 Each player rolls 1 die and records the digit in 1 of the boxes of their own game card.

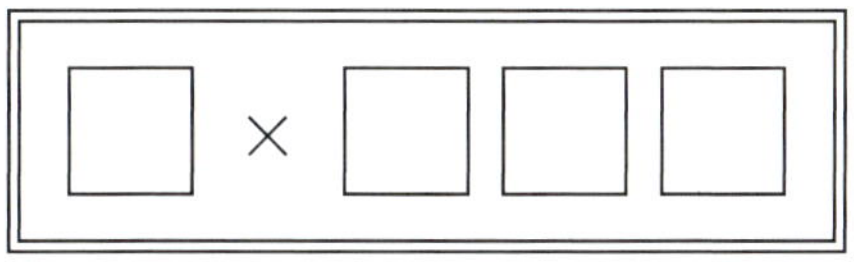

Step 2 Players keep rolling and recording until all 4 boxes of the game card are filled. Once a digit is recorded in a box, it cannot be changed.

Step 3 Calculate the product. The player with the greatest product wins.

You will need

- dice

- game cards

Mandy's Turn

I rolled 2, 5, 1, 4.
My product is
5 × 412, or 2060.

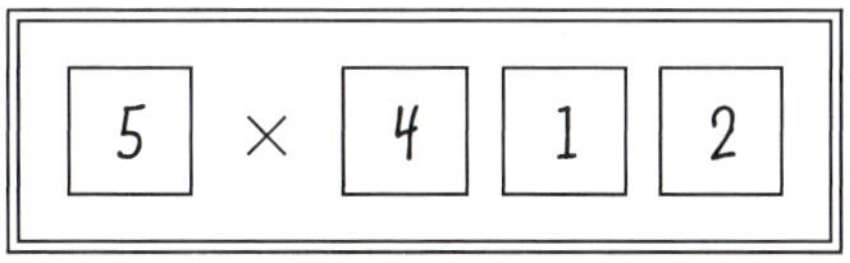

Other ways to play:

- Find the least product.
- Use a 1-digit by 2-digit game card.
- Use a spinner or digit cards instead of dice.

7 Choosing a Method to Multiply

You will need

- base ten blocks

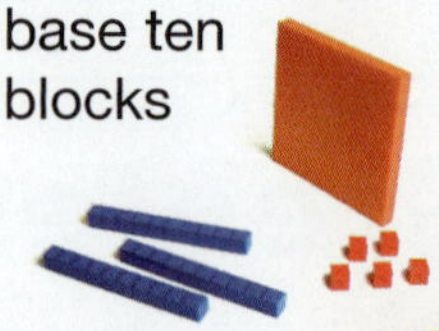

- a calculator

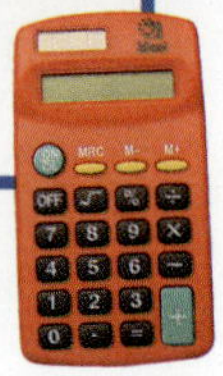

Goal **Choose and justify a multiplication method.**

Did you know that in 1 minute ...

- 250 babies are born in the world
- a cartoon uses 1430 animated pictures
- Mary can walk 67 m

? How can you choose the best multiplication method for calculating each?

For each situation, would you

- estimate or find an exact answer?
- use mental math, base ten blocks, pencil and paper, or a calculator?

Explain your choices.

A. You want to find out how many babies will be born in the next 4 minutes.

B. You want to find out how many babies will be born in the next 15 minutes.

C. You want to find out how many pictures are needed to show a cartoon that is 15 minutes long.

D. You want to find out how many pictures are needed to show a cartoon that is 2 minutes long.

E. You want to find out how far Mary walks in each of the times below.

a) 2 minutes
b) 10 minutes
c) 17 minutes
d) 20 minutes

Reflecting

1. Suppose the number of babies born had been 247 in 1 minute instead of 250. Would you have answered parts A and B the same way? Explain.
2. a) Why might the number of minutes change the calculation method you choose?
   b) Why might the number of times in 1 minute change the calculation method you choose?

Curious Math

Egyptian Multiplication

Here is a curious way to multiply.
It is called Egyptian multiplication.

Suppose you want to multiply 314 by many different numbers.
Begin by making a list starting with $314 \times 1 = 314$.
Double the red number and the product to make each new line in the list.
Decide what number you want to multiply 314 by.
Find red numbers that add up to your value.
Add the numbers to the right of the equal sign from those rows.

$314 \times 1 = 314$
$314 \times 2 = 628$
$314 \times 4 = 1256$
$314 \times 8 = 2512$

For example, $9 = 8 + 1$, so
$9 \times 314 = 8 \times 314 + 1 \times 314$
$9 \times 314 = 2512 + 314$
$9 \times 314 = 2826$

1. Calculate. Show your work.
   a) 7×314 b) 6×314

LESSON

Skills Bank

2

1. Complete.

a)

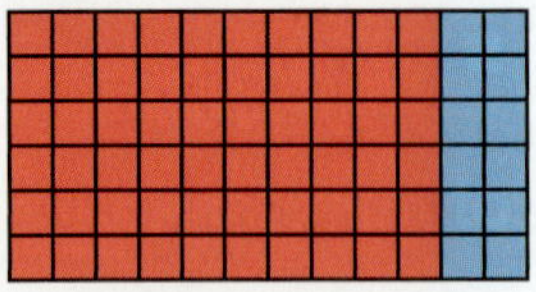

$6 \times 12 = 6 \times 10 + 6 \times$ ■

$6 \times 12 =$ ■

b)

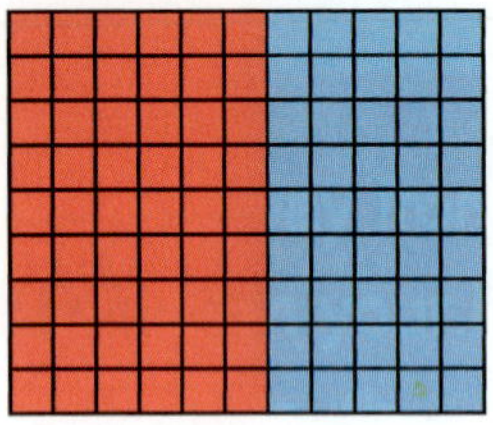

$9 \times 11 =$ ■ $\times 6 +$ ■ $\times$ ■

$9 \times 11 =$ ■

c)

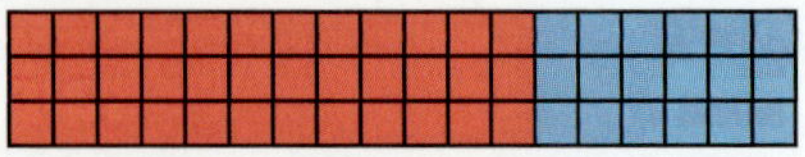

$3 \times 18 = 3 \times$ ■ $+ 3 \times$ ■

$3 \times 18 =$ ■

d)

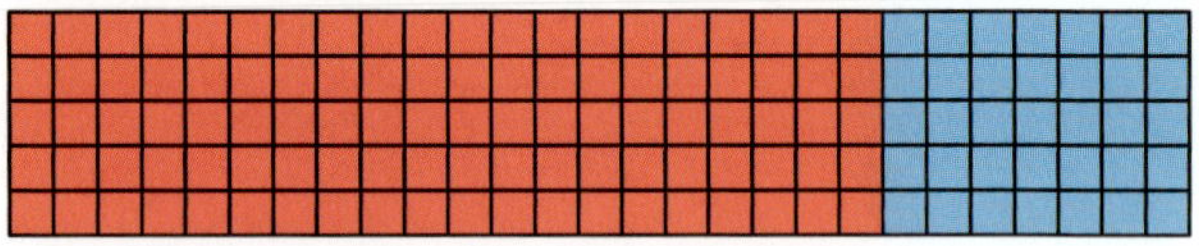

$5 \times 27 =$ ■ $\times$ ■ $+$ ■ $\times$ ■

$5 \times 27 =$ ■

2. Multiply to find how many in each array.

a) 3 rows of 19 stamps
b) 5 rows of 21 carrots
c) 6 rows of 18 trading cards
d) 4 rows of 12 books
e) 9 rows of 17 new cars
f) 8 rows of 16 bottles

3. Multiply. Check using a calculator.

a) $3 \times 17 =$ ■
b) $2 \times 16 =$ ■
c) $4 \times 15 =$ ■
d) $7 \times 11 =$ ■
e) $7 \times 29 =$ ■
f) $8 \times 14 =$ ■
g) $9 \times 14 =$ ■
h) $9 \times 23 =$ ■
i) $9 \times 19 =$ ■

3

4. Estimate. Then complete using expanded form.
Use estimates to check.

a)
6 tens + 2 ones
× 8
48 tens + ■ ones
■ hundreds + ■ tens + ■ ones

b)
70 + 4
× 6
420 + ■
■

c)
30 + 8
× 4
■
■

3 **5.** Estimate. Then multiply using expanded form.

a) 73×4 c) 72×2 e) 61×3 g) 92×2

b) 29×5 d) 16×3 f) 14×7 h) 87×3

6. Solve each problem. Show your work.

a) There are 5 pencils in each box.
How many pencils are in 85 boxes?

b) There are 4 bookshelves and 21 books on each shelf.
How many books are there altogether?

5 **7.** Complete each multiplication.

a)
$$\begin{array}{r} 300 + 20 + 7 \\ \times\ 5 \\ \hline 1500 \\ 100 \\ +\ ■ \\ \hline ■ \end{array}$$

b)
$$\begin{array}{r} 200 + 80 + 3 \\ \times\ 7 \\ \hline 1400 \\ ■ \\ +\ ■ \\ \hline ■ \end{array}$$

c)
$$\begin{array}{r} 400 + 60 + 9 \\ \times\ 9 \\ \hline ■ \\ ■ \\ +\ ■ \\ \hline ■ \end{array}$$

8. Estimate. Then multiply using expanded form.
Use estimates to check.

a) 361×7 c) 421×4 e) 618×3 g) 333×6

b) 125×8 d) 753×5 f) 618×7 h) 275×3

9. Solve each problem. Show your work.

a) A jumbo jet travels 885 km in 1 hour.
How many kilometres does the jet travel in 5 hours?

b) Kia earns $678 a week working at a greenhouse.
How much will she earn in 5 weeks?

c) There are 341 mL of orange juice in each litre of punch. How many millilitres of orange juice will be needed for 7 L of punch?

6 **10.** Estimate. Then calculate. Use estimates to check.

a) 762×7

c) 491×3

e) 611×9

g) 383×5

b) 145×7

d) 952×6

f) 937×9

h) 268×7

11. Solve each problem. Show your work.

a) An art class made 267 plates.
Another art class made 4 times as many plates.
How many plates did the 2nd class make?

b) Some jets can carry 131 passengers.
A jumbo jet can carry about 4 times as many passengers.
About how many passengers can a jumbo jet carry?

7 **12.** Estimate or find an exact answer.
Explain why you chose your multiplication method.

a) 1 bucket contains 5000 nails.
How many nails are in 3 buckets?

b) 1 houseboat costs $18 325.
How much do 6 houseboats cost?

c) 72 minutes of music can be burned on 1 CD.
How many minutes of music can be burned on 7 CDs?

d) The drugstore ordered 76 cases of shampoo.
Each case has 36 bottles of shampoo.
How many bottles did they order altogether?

e) Jonah has $387 in his bank account.
His brother saved about 5 times as much money.
About how much money has Jonah's brother saved?

f) Ms. Stein drives the school bus about 217 km each day. How many kilometres does she drive in 5 days?

Problem Bank

LESSON

2 **1.** A greenhouse has
3 rows of 11 vegetables,
3 rows of 13 vegetables,
3 rows of 18 flowers, and
3 rows of 16 flowers.
Are there more vegetables or flowers? Show your work.

3 **2.** Ava is playing Scrabble.
She can get a triple word score with F I X
or a double word score with F R I E N D.
Which word would score more points?

6 **3.** Talima and Mark are making dreamcatchers.
Mark uses 880 cm of thread.
Talima uses 3 times as much thread.
How much thread does Talima use?

4. Create the greatest product you can by putting the digits 2, 4, 6, and 8 once into this multiplication. Show your work.

■■■
× ■

5. What is the greatest possible product that can be made by putting the digits 6, 7, 8, and 9 once into this multiplication? Use a calculator. Show your work.

■ × ■■■

6. Find all the pairs of numbers with the product 252. Use a calculator.

LESSON

Chapter Review

2 **1.** Matthew multiplied using this array.

a) What multiplication does the array model?
b) Show the array as 2 smaller arrays.
c) Write the arrays as the sum of 2 products.

2. Laura has 12 books on each of her 7 shelves.
How many books does Laura have?
Show your answer as arrays.
Write the arrays as the sum of 2 products.

3 **3.** Which estimate is closer to the actual product?
Explain each choice.

a) 79×9	**A.** 80×10	**B.** 80×9
b) 47×7	**A.** 50×7	**B.** 47×10
c) 18×4	**A.** 20×4	**B.** 18×5

4. There are 28 students in a class.
Each student brought in 5 items for a food drive.
a) Estimate the number of items the class collected.
b) Solve the problem. Show your work.
c) Is your answer reasonable?
Check your answer using your estimate.

5 **5.** An illustrator of science books is drawing 7 centipedes with 354 legs each.
How many legs will she need to draw altogether?
Multiply using the expanded form of 354.

6 **6.** Sound travels 344 m in 1 second.
How far does sound travel in 7 seconds?

6 **7.** Pierre is fencing off a square field for his horse. Each side of the field is 152 m long. What is the length of the fence?

8. 750 can be represented as 3×250. Represent 750 in 3 more ways.

9. Estimate. Then calculate each product. Use estimates to check.

a) $\begin{array}{r} 35 \\ \times\ 9 \\ \hline \end{array}$ b) $\begin{array}{r} 654 \\ \times\ 3 \\ \hline \end{array}$ c) $\begin{array}{r} 185 \\ \times\ 4 \\ \hline \end{array}$

10. Determine the missing digits. Show your calculations.

a) $\begin{array}{r} \blacksquare\,8 \\ \times\ 8 \\ \hline 3\,0\,\blacksquare \end{array}$ b) $\begin{array}{r} \blacksquare\,2\,4 \\ \times\ 5 \\ \hline 2\,6\,\blacksquare\,\blacksquare \end{array}$ c) $\begin{array}{r} \blacksquare\,8\,\blacksquare \\ \times\ 4 \\ \hline 1\,1\,2\,8 \end{array}$

11. Chris is multiplying 348×7. He estimates that the product is more than 2100 because 7×300 is 2100.

a) Is Chris's answer reasonable? Why or why not?
b) Do the exact multiplication.
c) Explain how you did the calculation.

Chris's Multiplication

$\begin{array}{r} 348 \\ \times\ 7 \\ \hline \end{array}$ $\begin{array}{r} 300 + 40 + 8 \\ \times\ 7 \\ \hline 56 \\ 280 \\ +\ 210 \\ \hline 546 \end{array}$

7 **12.** Heather says that she can find the answer for 2×104 faster using mental math than using paper and pencil or a calculator. Explain why this could be true.

13. Multiply. Explain your choice of method.

a) $\begin{array}{r} 50 \\ \times\ 4 \\ \hline \end{array}$ b) $\begin{array}{r} 312 \\ \times\ 6 \\ \hline \end{array}$ c) $\begin{array}{r} 479 \\ \times\ 8 \\ \hline \end{array}$ d) $\begin{array}{r} 2896 \\ \times\ 3 \\ \hline \end{array}$

Chapter Task

Describing a School Year

There are 3 recesses each day, including lunch.
There are about 190 days in 1 school year.
So, there are about $3 \times 190 = 570$ recesses in 1 year.

? How can multiplication be used to describe other things that happen in a school year?

A. List 4 other things about school that can be described using multiplication.

B. How many times does each thing happen in a day, a week, or a month?

C. What 2 pieces of information do you need to determine how often each thing happens in a year?

D. Describe each thing using multiplication.

E. Explain why you chose your multiplication methods.

Task Checklist

- ✓ Did you explain how you estimated and calculated?
- ✓ Did you use math language?
- ✓ Did you organize your work so it is easy to follow?

CHAPTER 10

Dividing Greater Numbers

Goals

You will be able to

- use the relationship between multiplication and division to solve problems
- use different methods to divide 2-digit numbers by 1-digit numbers
- explain the strategies you use to solve division problems

Dividing a pizza

Getting Started

You will need
- counters

Planning a Play Day

The students are planning a Play Day.
There will be 54 students participating.

Think of an event that would be fun for 54 students to do at the Play Day.

? How many groups would be in your event? How many people would be in each group for your event? How many items need to be shared?

A. Miki is planning a relay race for the 54 students. She wants 2 equal teams. How many students should she put on each team? Explain your thinking.

B. Rey is planning a beanbag toss. If he puts the 54 students in teams of 6, how many teams can he make? Explain your thinking.

C. Sometimes you divide to find the number of groups. Sometimes you divide to find the number in each group. Which type of division did you use in part A? Which type of division did you use in part B?

D. Write a division sentence to represent what you did in part A. Write another division sentence for part B.

E. Make up a Play Day event for 54 students. Describe your event and tell how it will involve equal groups. Explain how you used division or multiplication.

Do You Remember?

1. Complete.

a) $9 \times 6 =$ ■
b) $18 \div 2 =$ ■
c) $64 \div 8 =$ ■
d) $35 \div 5 =$ ■
e) $8 \times 8 =$ ■
f) $18 \div 3 =$ ■
g) $54 \div 6 =$ ■
h) $12 \div 4 =$ ■
i) $3 \times 7 =$ ■

2. Write the multiplication and division **fact family** to describe this **array**. Remember to include 2 multiplication sentences and 2 division sentences.

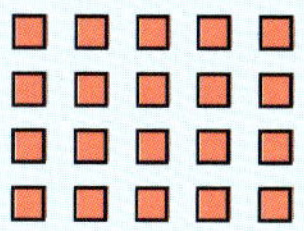

3. Use the multiplication sentence to divide.

a) $8 \times 7 = 56$, so $56 \div 7 =$ ■.
b) $6 \times 8 = 48$, so $48 \div 8 =$ ■.

1 Exploring Division

Goal Solve division problems using models.

The students in the Science Club made kaleidoscopes. Each kaleidoscope had a reflector shaped like a triangle-based prism.
Each reflector is made of 3 plastic rectangles.
The students used between 70 and 90 plastic rectangles.
There were no plastic rectangles left over.

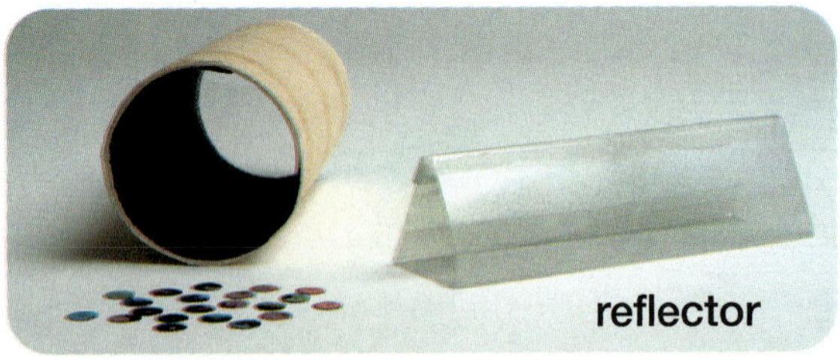

? How many kaleidoscopes could the students have made?

A. Why can you solve this problem using division?

B. What will the **divisor** be? Explain how you know.

C. Use any materials available to solve the problem. Find more than 1 answer. Show your work.

Reflecting

1. Why is there more than 1 answer to this problem?
2. Did you divide every number between 70 and 90? Why or why not?
3. Write a division sentence and a multiplication sentence to describe 1 of your answers.

You will need

- How to Make a Kaleidoscope
- counters

- base ten blocks

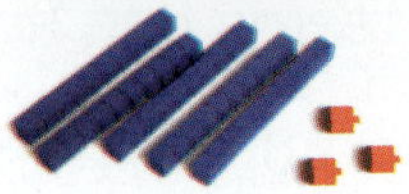

- grid paper

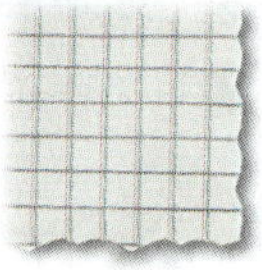

- a 100 chart

1	2	3	4
11	12	13	14
21	22	23	24
31	32	33	34

divisor
The number you divide by in a division sentence

$6 \div 2 = 3$
↑ divisor

Adding in Steps

When you want to add numbers in your head, it sometimes helps to add the numbers in steps.

Manitok's Mental Addition

I want to add 23 + 18.

First I'll add 10. That's easy.

$23 + 10 = 33$

Then I'll add 7 to get to 40.

$33 + 7 = 40$

I have 1 left because 18 = 10 + 7 + 1.

So 23 + 18 = 41.

A. Why do you think Manitok added 10 first?

B. What other way can you add 23 + 18 in steps? Explain how your way makes the addition easier.

C. How can you add 23 + 18 + 12 in your head?

Try These

1. Use mental math to add. Describe your steps.
   - a) 38 + 11
   - b) 25 + 15
   - c) 42 + 21
   - d) 16 + 18
   - e) 45 + 25
   - f) 51 + 19
   - g) 12 + 18 + 11
   - h) 26 + 12 + 18
2. Show the steps you could use to subtract 63 − 15 using mental math.

2 Using Repeated Subtraction to Divide

You will need

- counters

Goal Use repeated subtraction to divide.

Zola is making decorations for a banquet.
She is tying balloons together in bunches of 4.

? How many bunches of 4 balloons can Zola make with 77 balloons?

Zola's Division

I can model the balloons with 77 counters.

Then I can use repeated subtraction to subtract bunches of 4 until there aren't enough balloons to make any more bunches.

```
4)77          ← I start with 77 balloons.
 -40     10 ← I can make at least 10 bunches
                because 10 x 4 = 40.
  37     ←      Now I have 37 balloons left.
 -■■      5 ← I can make 5 more bunches because ...
```

remainder
The number left over after an amount is divided into equal parts

When you divide 7 into 3 groups, you get 2 in each group and a remainder of 1.

$7 \div 3 = 2$ **R1**
↑ remainder

A. Why does Zola think she can make 5 bunches next? Show how you would continue Zola's recording to complete the division.

B. How many bunches of 4 balloons can Zola make in all? Explain how you know.

C. What will the **remainder** be? Explain how you know.

Reflecting

1. Why did Zola start by subtracting 10 groups of 4?
2. Zola checked her division.
   She multiplied the number of bunches of balloons by the number of balloons in each bunch. Then, she added the number of balloons left over.
   Explain why she can check division this way.
3. Why does it make sense to subtract multiples of 4 to find out how many 4s there are in 77?

Checking

4. Zola has 99 balloons to put in bunches of 6.
   a) How many bunches could she subtract first? Why? How many balloons would that be?
   b) How many bunches can she make in all? Show how to check by multiplying, then adding.
   c) What will the remainder be? Explain how you know.

Practising

5. Zola has 62 balloons to put in bunches of 5.
   How many bunches can she make?
   What will the remainder be? Show your steps.
6. Use repeated subtraction to divide.
   Multiply, then add to check.
   a) $3\overline{)64}$ b) $4\overline{)50}$ c) $6\overline{)62}$ d) $2\overline{)95}$ e) $4\overline{)87}$
7. a) Write a problem about balloons for 75 ÷ 8.
   b) Solve your problem.
   c) Check your solution by multiplying and then adding.
8. Zola had about 70 balloons. When she made groups of 4 balloons, she had no balloons left over. How many balloons could Zola have had? Explain how you know.

3 Interpreting Remainders

You will need

- a calculator

Goal **Decide how to treat the remainder in a division problem.**

Each student wrote a problem that can be solved by calculating $6\overline{)51}$.

? How does the problem help you decide what to do with the remainder?

$$\begin{array}{r l} 6\overline{)51} & \\ -42 & 7 \\ \hline 9 & \\ -6 & 1 \\ \hline 3 & \end{array} \quad 7 + 1 = 8$$

$51 \div 6 = 8 \text{ R}3$

Chris's Problem

51 students went for a ride on the Super Coaster. Each car holds 6 people. How many cars did they need?

Solution: $51 \div 6 = 8 \text{ R}3$

Answer: 9 cars

Miki's Problem

A restaurant served 51 slices of pepperoni pizza. There are 6 slices in each pizza. How many pizzas did they serve?

Solution: $51 \div 6 = 8 \text{ R}3$

Answer: 8 and $\frac{1}{2}$ pizzas

Jon's Problem

Sarah had 51 stickers to give to 6 friends. She wanted to give the same number to each friend. How many stickers did each friend get?

Solution: $51 \div 6 = 8 \text{ R}3$

Answer: 8 stickers each

Reflecting

1. Describe what happened to the remainder in each problem. Explain why it happened.

Checking

2. a) Chantal has 58 photographs to arrange in an album. She wants to put 4 pictures in each spread. How many spreads will she need?
   b) Explain what you did with the remainder. Tell why.

3. Write a new problem about 58 ÷ 4 where you would do something different with the remainder. Explain what you would do differently.

Practising

4. a) 4 friends will share 25 licorice strings. How many strings will each friend get if they share equally and all the strings are eaten?
   b) Explain what you did with the remainder. Tell why.

5. a) Raffle tickets cost 5¢. How many tickets can Michael buy with 72¢?
   b) Explain what you did with the remainder. Tell why.
   c) Multiply, then add to check.

6. a) The rafts for the Scary Tunnel ride hold 4 people. If 62 children go on the ride at the same time, how many children will be on the last raft?
   b) Explain what you did with the remainder. Tell why.

7. Use a calculator to divide. What do you notice about the result when there is a remainder?
   a) 24 ÷ 2 b) 27 ÷ 2 c) 40 ÷ 4 d) 27 ÷ 8

8. Use a calculator to check your answer for how many tickets Michael can buy for Question 5.

9. Write a problem about 53 ÷ 3. Show how to solve it. Explain what you did with the remainder. Tell why.

Mid-Chapter Review

LESSON

1

1. Melissa has 40 counters.
   How many groups of 5 can she make?
   Show your work.

2

2. a) Tony has 43 jars of jam to pack in boxes of 6. How many boxes can he fill? Show your steps.
   b) Check your answer by multiplying and then adding. Show your work.

3. Use repeated subtraction to divide. Show your steps. Multiply and then add to check.
   a) $5\overline{)71}$ b) $7\overline{)82}$ c) $4\overline{)93}$

4. a) Zola has 74 balloons to put in bunches of 8.
   How many bunches can she make?
   How many balloons will be left over?
   b) Multiply and then add to check.

3

5. a) Miki has 52 jingle cones to sew on her dress.
   She wants rows of 8 cones.
   How many rows can she have?
   b) Explain what you did with the remainder.
   Tell why.

6. a) Today 41 students are going to the H.R. MacMillan Space Centre.
   Each van can take 6 students.
   How many vans will they need?
   b) Explain what you did with the remainder.
   Tell why.

7. a) Write a problem about 92 ÷ 8.
   b) Show how to solve your problem.
   c) Explain what you did with the remainder.
   Tell why.

Math Game

Remainder Hunt

You will need

- number cards

Number of players: 2 to 5
How to play: Arrange 3 cards in a division frame to get the least possible remainder.

Step 1 Each player makes a division frame that looks like $\square\overline{)\square\square}$. The boxes should be large enough to fit the number cards.

Step 2 Deal 3 cards to each player.

Step 3 Arrange your cards on your division frame. Try to make the division calculation that will have the least possible remainder.

Step 4 Complete the division and find the remainder.

Step 5 Have the other players check your remainder.

The player with the least remainder wins 1 point.
The game ends when a player has 5 points.

Pedro's Turn

I can't make a division sentence with a remainder of 0.
I'm going to make $3\overline{)25}$ because the remainder is 1.

Another way to play: Try to make the greatest possible remainder.

4 Estimating with 2-Digit Dividends

Goal **Use multiplication and division facts to estimate quotients.**

Allison made a list of the books she would like to read.

Allison's Reading Plan

I want to read *Cam Jansen and the Birthday Mystery* in 3 days.

I plan to read the same number of pages every night.

Title	Author	Pages
The Girl Who Loved Wild Horses	Paul Goble	31
The Missing Sun	Peter Eyvindson	47
Dolphins at Daybreak	Mary Pope Osborne	67
Jacob Two-Two and the Dinosaur	Mordecai Richler	85
Cam Jansen and the Birthday Mystery	David A. Adler	58
Esio Trot	Roald Dahl	62
The Good, the Bad, and the Goofy	Jon Scieszka	70

? About how many pages does Allison need to read each night?

A. How do you know that Allison has to read more than 10 pages each night?

B. How do you know that Allison has to read fewer than 20 pages each night?

C. How does knowing $6 \div 3 = 2$ help Allison estimate?

D. About how many pages should Allison read each night?

Reflecting

1. Why might it be useful for Allison to think of 58 as about 6 tens?
2. Suppose Allison had only 2 nights to read the book.
   a) What division sentence would Allison have to estimate?
   b) What multiplication or division fact would help Allison find the number of pages she would need to read each of the 2 nights?

Checking

3. Allison wants to read *Esio Trot* in 5 days. She plans to read the same number of pages each day.
   a) Why will she read more than 10 pages each day?
   b) What fact would help Allison estimate the number of pages she needs to read each day?
   c) About how many pages should she read each day?

Practising

4. Allison wants to read the first 2 books on her list in 3 days. She plans to read the same number of pages each day.
   a) What fact would help Allison estimate the number of pages she should read each day?
   b) About how many pages should she read each day?
   c) About how many pages would she have to read each day to finish in 6 days instead of 3?
5. Estimate each quotient. Explain your thinking.
   a) $6\overline{)92}$ b) $8\overline{)71}$ c) $7\overline{)89}$
6. Gerry can finish his book if he reads about 30 pages every day for 3 days. About how long is the book?

5 Dividing in Parts

Goal **Divide in steps using simpler numbers.**

Josef and Shani both collect postcards. Josef has 3 times as many postcards as Shani.

? If Josef has 84 postcards, how many postcards does Shani have?

Calvin's Solution

I can rename 84 as 60 + 24.

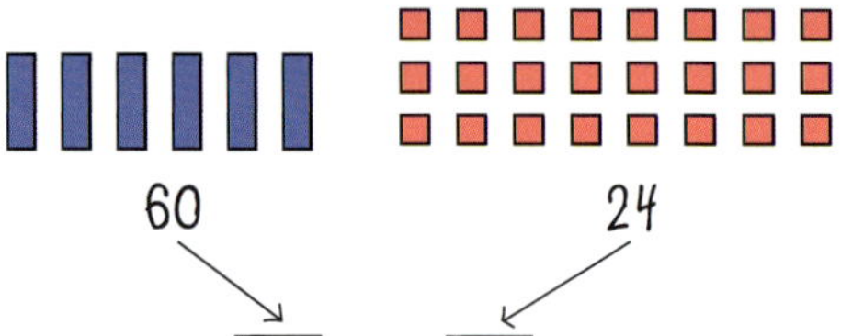

$3\overline{)84}$ is the same as $3\overline{)60}$ + $3\overline{)24}$.

Shani has 20 + ■ postcards.

A. Why did Calvin divide 84 by 3?

B. Why is dividing 84 into 2 parts the same as dividing 60 by 3 and then dividing 24 by 3?

C. Finish Calvin's division. Why did Calvin rename 84 as 60 + 24 and not as 80 + 4?

Reflecting

1. Would this be a good way to rename 84 to solve the problem? Explain your answer.
   $3\overline{)84} = 3\overline{)30} + 3\overline{)30} + 3\overline{)24}$
2. How else could you rename 84 to solve the problem?
3. How did knowing multiplication facts for 3 help Calvin divide?

Checking

4. Shani and Josef both collect hockey cards. Shani has 4 times as many cards as Josef. If Shani has 92 cards, how many does Josef have?
   a) How could you rename 92 to solve the problem?
   b) How many cards does Josef have? Show your work.

Practising

5. Shani has 96 stickers in her sticker book. How many stickers does Josef have if Shani has this many stickers? Show your work.
   a) 4 times as many b) 8 times as many
6. Calvin used this method to share 87 postcards equally in 3 albums.
   a) Is his method correct? Explain.
   b) Use his method to divide 72 by 4.

Calvin Divides Again

I can rename 87 as 90 − 3.
$3\overline{)87}$ is the same as $3\overline{)90} - 3\overline{)3}$.
I can put 30 − 1 = 29 postcards in each album.

7. Would you divide in parts? Why or why not?
   a) $5\overline{)85}$ b) $7\overline{)91}$ c) $9\overline{)90}$
8. Show 2 ways to rename 78 to make it easier to divide by 2.

6 Dividing 2 Digits by 1 Digit

You will need

- base ten blocks

Goal **Use base ten blocks and pencil and paper to divide a 2-digit number by a 1-digit number.**

There are 94 stickers for 4 people to share equally.

? How many stickers will each person get?

Rami's Division

I think each person will get about 25 stickers because $4 \times 25 = 100$.

To find out, I can model the **dividend** with base ten blocks and divide the blocks into 4 equal groups.

dividend

The number that is divided into equal parts

$6 \div 2 = 3$ (6 ← dividend)

Step 1 I model 94 stickers.
I need to put them in 4 equal groups.

4)94

Step 2 First I share the tens in 4 groups.

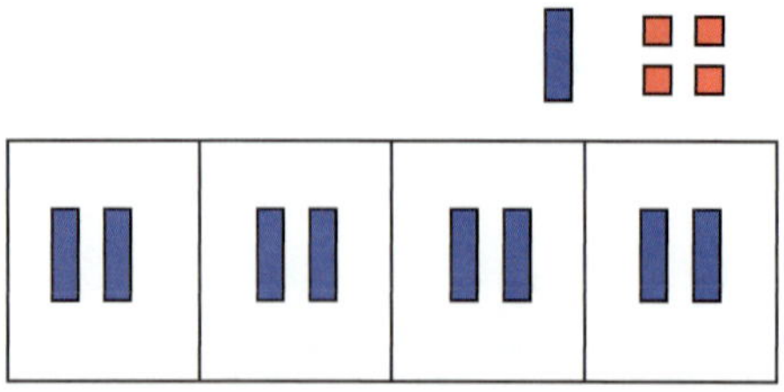

I put 2 tens in each group so I write 20 in the quotient.

```
  20
4)94
 -80   ← I've shared 80 stickers.
  14   ← I have 14 stickers left.
```


Step 3 Now I regroup the ten and share the 14 ones.
I can put 3 ones in each group.

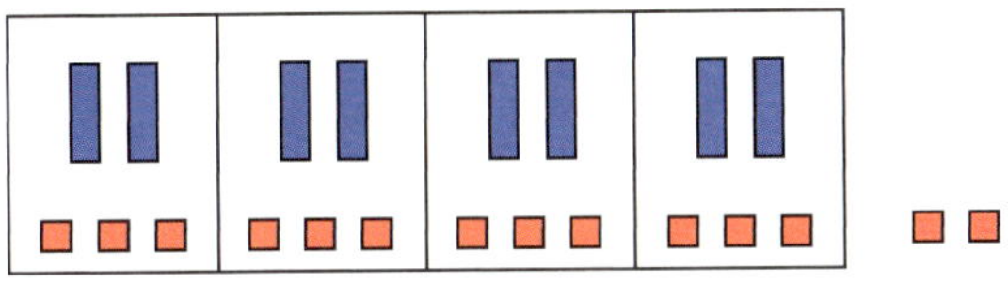

Each person will get 23 stickers with 2 stickers left over.

This is reasonable because 23 is close to my estimate of 25.

```
   3    ← I record 3 in the quotient.
  20    } The quotient is 20 + 3 = 23.
4)94
 -80
  14
 -12    ← I've shared 12 more stickers.
   2    ← The remainder is 2.
```

Reflecting

1. Why did Rami regroup 1 ten as 10 ones in Step 3?
2. Why did Rami subtract 12 in Step 3?

Checking

3. 6 people will share a bowl of 75 grapes.
   a) Model the sharing of the grapes. Record the division.
   b) How many grapes will each person get?
   c) Will there be any grapes left over? Explain.

Practising

4. There are 83 jellybeans in a jar.
   5 people will share the jellybeans equally.
   a) Will any jellybeans be left over? Explain.
   b) Estimate the number of jellybeans each person will get. Explain your method.
   c) Divide to find out how many jellybeans each person will get.
   d) Is your answer reasonable? Explain.

5. Divide. Multiply, then add to check.
   a) $6\overline{)78}$ b) $3\overline{)95}$ c) $7\overline{)93}$ d) $4\overline{)65}$

7 Solve Problems by Guessing and Testing

Goal **Use a guess-and-test strategy to solve problems.**

Terry put the candles on his mother's birthday cake in an array.
He used 1 candle for each year of her age.
He made 5 equal rows of candles.
He put 6 candles around the edges because there weren't enough candles to make another row.

? How old could Terry's mother be if she is younger than 50?

Mandy's Solution

Understand

Since there are the same number of candles in 5 rows, I can think about multiplying or dividing.

The number of candles in each row must be more than 6 or Terry could make another row with the 6 extra candles.
The least number of candles that could be in each row is 7.

Make a Plan

I'll try multiplying 5 rows by 7 candles. Then I'll add the 6 extra candles to see if the total number is less than 50.

Carry Out the Plan

5 rows of 7 = 35
35 + 6 extra candles = 41
41 is less than 50, so Terry's mother could be 41.

Could there be another answer?

5 rows of 8 = 40
40 + 6 = 46
Terry's mother could be 46.

5 rows of 9 = 45
45 + 6 = 51
She is younger than 50, so this array isn't a solution.

There are 2 possible answers.
Terry's mother could be 41 years old or 46 years old.

Reflecting

1. How did Mandy know there were at least 7 candles in each row? How did she know there were fewer than 9 candles?
2. How did Mandy use information from the problem to make sure her guesses were reasonable?

Checking

3. Terry packed cookies in equal-sized bags. He packed more than 40 cookies but less than 50. He filled 4 bags and had 5 cookies left over. How many cookies could he have packed?

Practising

4. Terry had between 50 and 60 square tiles. He made a rectangle that was 6 units long. He had 4 tiles left over. Draw Terry's rectangle. Label the side lengths.

5. Create your own guess-and-test problem. Give your problem to a partner to solve.

8 Choosing a Method to Divide

You will need

- base ten blocks

- a calculator

Goal Choose and justify a division method.

Paulette is making crafts for a sale next year. She will use the same number of beads, feathers, or quills for each item.

She is planning how many items she can make.

Paulette has these materials.
- beads: 88 red, 72 yellow, 69 blue, 762 white
- feathers: 25 white, 90 blue, 53 yellow
- quills: 92 green, 1002 purple, 1241 blue

? How can she choose the best division method to plan her crafts?

For each situation, would you
- estimate or find an exact answer?
- use mental math, base ten blocks, paper and pencil, or a calculator?

Explain your choices.

A. Paulette is trimming 47 boxes with purple quills. Can she put 20 purple quills on each box?

B. How many bracelets can Paulette make with 9 yellow beads on each?

C. Can Paulette make 30 pairs of moccasins with 40 blue quills and 25 white beads on each moccasin?

D. How many dreamcatchers can Paulette make with 3 white feathers and 9 blue beads on each?

E. How many key chains can Paulette make with 6 red beads and 4 yellow feathers on each?

F. Paulette is trimming 4 pairs of mitts with green quills.
She is using as many of the 92 green quills as she can.
Will she have any green quills left over to sew on a jacket?

G. How many clocks can Paulette decorate with each number of blue feathers on each clock?
a) 10 b) 8 c) 19 d) 90

Reflecting

1. For part E, suppose the number of yellow feathers for each key chain was 14 instead of 4.
Would you have chosen the same division method?
Explain the reason for your answer.

2. Paulette wants to decorate some dresses with beads.
She needs to know how many dresses she can decorate.
   a) How might the number of beads on each dress affect the division method you choose?
   b) What else might change the division method you choose? Explain your thinking.

3. Make up a division problem about crafts that you would solve using mental math.
Explain why you would use mental math.

LESSON

Skills Bank

1 **1.** Students are arranged in groups of 4 for a class trip. The number of students on the trip is between 25 and 35.

a) How many students could be on the trip?

b) Write a multiplication sentence and a division sentence to describe 1 of the possible answers.

2 **2.** Use repeated subtraction to divide. Show your steps.

a) 2)80	**c)** 6)78	**e)** 3)46	**g)** 8)87
b) 4)56	**d)** 5)58	**f)** 4)53	**h)** 7)89

3. Solve each problem. Show your work.

a) Carlos has 65 balloons to put in bunches of 3. How many bunches can he make? What will the remainder be?

b) Carlos had about 80 balloons. He put them in bunches of 6 and had none left over. How many balloons could he have had? Find more than 1 answer.

4. a) Explain how you know this division is incorrect.

b) Show how to correct the division.

c) Why do you think the mistakes happened?

5)73	
-25	5
52	
-25	5
33	
-25	5
8	
-5	1
3	16R3

3 **5.** Divide to solve each problem. Show your work. For each problem, explain what you did with the remainder.

a) Nadine uses 4 pieces of wood for each picture frame.
How many frames can she make with 58 pieces of wood?

b) Cupcakes come in packages of 6.
Jessie needs 57 cupcakes.
How many packages should he buy?

c) Lisa and 3 friends have earned $25.
They have to share the money equally.
How much money will each person get?

4 **6.** Estimate each quotient.

a) $91 \div 5$ d) $78 \div 8$ g) $87 \div 4$
b) $7\overline{)90}$ e) $8\overline{)63}$ h) $9\overline{)71}$
c) $74 \div 4$ f) $6\overline{)93}$ i) $61 \div 5$

7. Which quotient is closest to 10?

A. $92 \div 6$ B. $71 \div 3$ C. $68 \div 5$ D. $83 \div 9$

8. Vilay has a book with 94 pages.
He plans to read the same number of pages each day.
Estimate how many pages he should read each day if he wants to finish the book in each number of days.

a) 3 days b) 4 days c) 8 days

5 **9.** Estimate. Then divide in parts.
Show your work.

a) $5\overline{)75}$ d) $9\overline{)99}$ g) $2\overline{)58}$
b) $4\overline{)56}$ e) $7\overline{)91}$ h) $6\overline{)84}$
c) $3\overline{)81}$ f) $5\overline{)95}$ j) $7\overline{)98}$

5 **10.** Solve each problem. Show your work.

a) Greg has 87 marbles.
He has 3 times as many marbles as Kelly.
How many marbles does Kelly have?

b) Oliver has read 92 pages today for the read-a-thon.
Oliver has read about 4 times as many pages as Kyle.
About how many pages has Kyle read?

6 **11.** Estimate. Then divide.
Use your estimates to check.

a) $3\overline{)43}$	**d)** $5\overline{)91}$	**g)** $6\overline{)89}$
b) $6\overline{)73}$	**e)** $7\overline{)96}$	**h)** $9\overline{)97}$
c) $7\overline{)82}$	**f)** $4\overline{)73}$	**i)** $8\overline{)95}$

12. There are 95 collector cards to be shared among 6 people.

a) Estimate the number of cards each person will get.

b) How many cards will each person get?

c) Is your answer reasonable? Explain.

13. Solve each problem. Show your work.

a) Terry baked 78 cookies for 6 families to share.
How many cookies will each family get?

b) Suppose Terry baked 96 cookies instead of 78.
How many more cookies would each of the 6 families get?

8 **14.** Would you use pencil and paper, base ten blocks, mental math, or a calculator for each division? Why?
Use your chosen method to answer each question.
Show your work.

a) $150 \div 5$	**c)** $539 \div 8$	**e)** $89 \div 7$
b) $6\overline{)48}$	**d)** $37\overline{)1813}$	**f)** $60 \div 3$

LESSON

Problem Bank

1 **1.** When plush toy elephants are shipped to stores, the manufacturer can pack up to 6 elephants in 1 box. How many boxes will be needed to ship 95 elephants?

2. Choose any 4 numbers in this pattern on a 100 chart. Divide each number by 4. How many of the numbers have no remainder? Try some others. What pattern can you find?

14	15	16	17	18	19
24	25	26	27	28	29
34	35	36	37	38	39

6 **3.** Find at least 1 way to complete this division so the digits are all different. ■■ ÷ 3 = ■■ R5
Explain the strategy you used.

4. Each box represents the same digit.

$$\begin{array}{r} ■6 \\ ■\overline{)5■} \end{array}$$

a) What is the digit?
b) Explain how you solved the problem.

5. The triangle and the square each have a perimeter of 84 cm.
a) How much longer is 1 side of the triangle than 1 side of the square?
b) Explain how you solved the problem.

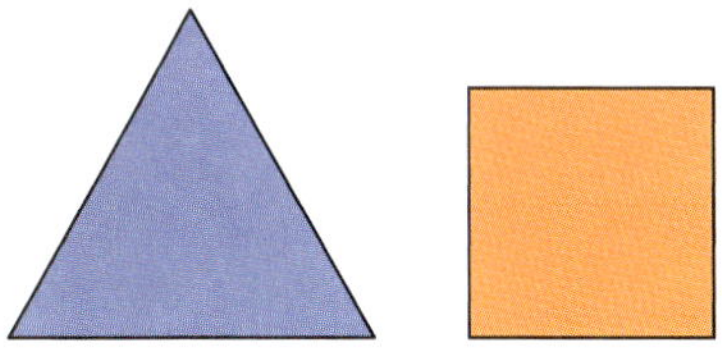

7 **6.** Find the missing digits.
a) Find a divisor for ■$\overline{)81}$ that has no remainder.
b) Find a divisor for ■$\overline{)81}$ that has a remainder of 3.

7. Find all the whole numbers that divide into 176 without a remainder. Use a calculator.

LESSON

Chapter Review

1 **1.** How does knowing $4 \times 2 = 8$ help you to find $80 \div 4$?

3 **2.** Solve. Explain what you did with the remainder. Tell why.

a) 63 students went on a trip to the Art Gallery. The students toured the gallery in groups of 5. How many groups went to the Art Gallery?

b) Hong is mixing batter for a Pancake Lunch. He needs 5 scoops of flour to make the batter for 1 batch of pancakes. If he has enough flour for 82 scoops, how many batches of pancakes can he make?

c) Alex put 98 quarters in piles of 4. How much money does she have?

4 **3.** Which quotient is greatest? Explain how you know.

A. $92 \div 4$ **B.** $83 \div 9$ **C.** $49 \div 2$ **D.** $93 \div 5$

5 **4.** How does knowing $78 = 60 + 18$ help you to find $78 \div 6$?

5. Adrian has 4 times as many coins in his collection as Sunita. If Adrian has 76 coins, how many coins does Sunita have?

6 **6.** Estimate and then divide. Show your work.

a) $74 \div 8$ **c)** $99 \div 7$ **e)** $68 \div 3$

b) $6\overline{)86}$ **d)** $4\overline{)65}$ **f)** $5\overline{)76}$

6

7. Mary plans to pack 85 cookies in bags of 4.
   a) Mary says she will be able to make more than 20 bags. Is she correct? Explain your answer.
   b) Mary says there will be a remainder. Is she correct? Explain your answer.
   c) How many full bags can Mary pack?

8. Devon prepared 93 m of yarn for the Craft Club. He cut the yarn into pieces that are 5 m long.
   a) How many 5 m pieces did he cut if he used all the yarn? Show your work.
   b) How much yarn was left over? Explain how you know.

9. Julia got a quotient of 21 when she divided. What numbers might she have been dividing? Find 3 possible answers.

7

10. There were more than 40 and less than 60 people at a team banquet. There were 4 people at the head table and 6 people at each of the other tables.
    a) How many tables of 6 were there? Find more than 1 answer. Show your steps.
    b) Explain how you know you found all the possible answers.

8

11. Which of these divisions would you solve using mental math? Explain your reasons.

    A. $42 \div 6$ B. $39 \div 3$ C. $57 \div 3$

12. Would you use pencil and paper, base ten blocks, mental math, or a calculator for each division? Explain your thinking.

    a) $35 \div 7$ c) $51 \div 3$ e) $140 \div 7$
    b) $68 \div 8$ d) $80 \div 5$ f) $3669 \div 3$

13. Vinh says that he can calculate $35 \div 8$ faster using mental math than using a calculator or pencil and paper. Explain why this could be true.

Chapter Task

Printing Pages

Book pages are printed on large sheets of paper called *forms*. All pages for a book are on the forms. This includes blank pages at the beginning and end of the book. Each form contains the same number of pages, so the number of pages in a book will always divide by that number with no remainder. After the forms are printed, they are cut apart so the pages can be bound.

Title	Pages
The Girl Who Loved Wild Horses	32
The Missing Sun	48
Dolphins at Daybreak	72
Jacob Two-Two and the Dinosaur	88
Cam Jansen and the Birthday Mystery	64
Esio Trot	64
The Good, the Bad, and the Goofy	72

? How many pages can be on a form? Explain your answer.

Task Checklist

- ✓ Did you show all your steps?
- ✓ Did you use math language?
- ✓ Did you explain your thinking?

CHAPTER 11

3-D Geometry and 3-D Measurement

Goals

You will be able to

- **draw, build, and describe 3-dimensional (3-D) shapes**
- **describe attributes of 3-D shapes**
- **estimate, measure, and compare the capacity and the mass of 3-D shapes and relate units**
- **model volume**
- **make connections between 2-D and 3-D geometry and between geometry and measurement**

Cree Village at Moose Factory Island

Getting Started

Describing Packages

These students are describing and comparing packages of different sizes and shapes.

? How can you describe the sizes and shapes of the packages?

A. Examine 2 packages in the picture. Make a list of math words that describe each package.

B. Compare the 2 packages. Write at least 1 thing that is the same and 1 thing that is different about the packages.

C. List 2 things that might come in each package.

D. Circle the words in your list that have to do with shape. Underline the words that have to do with measuring.

Do You Remember?

1. Match each name with a shape.
 a) rectangle-based prism
 b) triangle-based prism
 c) triangle-based pyramid
 d) cone

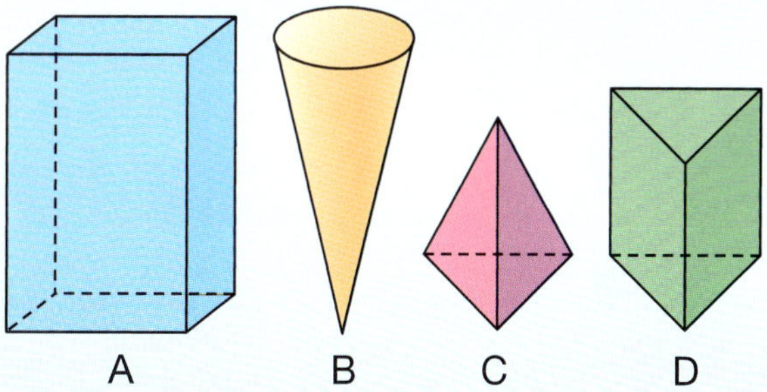

2. How are the rectangle-based prism and triangle-based prism the same? How are they different?

3. Which shapes in Question 1 have congruent bases?

4. Complete the chart for 1 pyramid and 1 prism.

3-D shape	Number of faces	Number of edges	Number of vertices
____ prism			
____ pyramid			

5. Would you use a spoon, a cup, or a large pail to measure the amount of water each object holds?
 a) a kitchen sink
 b) a bath tub
 c) a soup can
 d) a lid of a jar

1 Designing Nets

Goal **Make nets for pyramids and prisms.**

You will need

- 3-D models

- paper
- stickers
- scissors

- tape

Allison is building 3-D models of homes through the years that are in the shapes of prisms and pyramids.

Allison's Models

I want to make a **net** for each model.

net
A 2-D pattern that can be folded into a 3-D shape

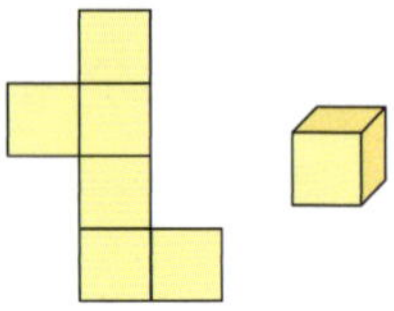

? How can Allison make nets for prisms and pyramids?

A. Choose a prism model.

B. Name the shapes of the faces.
How many faces are there of each shape?
How many vertices does the model have?
How many edges does the model have?

C. Trace 1 face.
Put a sticker on the face of the model to show that it has been traced.
Roll the prism over an edge so that you can trace a face connected to the face you just traced. Trace this face.
Put a sticker on this face of the model.

D. Repeat part C until you have traced each face once. You may need to roll the prism back on a traced face.

E. Cut out the net.
Fold and tape it to make the 3-D shape.

F. Repeat parts A to E for another prism.

G. Choose 2 pyramid models.
Repeat parts B to E for each pyramid.

Reflecting

1. Look at each net and its 3-D model.
How are they the same? How are they different?
2. Why is it important to pay attention to which faces are connected when you design a net?
3. The tracings of the 2 bases of a prism never touch on its net. Explain why this makes sense.

2 Building 3-D Shapes with Congruent Faces

Goal Build 3-D shapes and describe relationships between faces and vertices.

You will need

- triangles and squares

- tape

Some geodesic domes are made of congruent triangles.

Geodesic dome at Ontario Place

? What 3-D shapes can you build using only congruent triangles with equal sides?

A. Join different numbers of triangles to create nets. Try 2 faces, 3 faces, and so on up to 10 faces. Fold your nets to see if they make complete 3-D shapes.

B. For each successful 3-D shape, count and record these attributes:
- the number of faces
- the number of vertices
- the number of faces that meet at each vertex

C. Repeat parts A and B using congruent squares.

Reflecting

1. Did you build more 3-D shapes with the triangles or the squares? Why do you think that is?
2. Compare the number of faces to the number of vertices in each shape. What do you notice?
3. What did you notice about the number of faces that meet at the vertices of each shape?

Curious Math

Faces, Edges, and Vertices

1 a) If you know the shape of the base of a pyramid, how can you predict the number of edges?

b) If you know the shape of the base of a prism, how can you predict the number of edges?

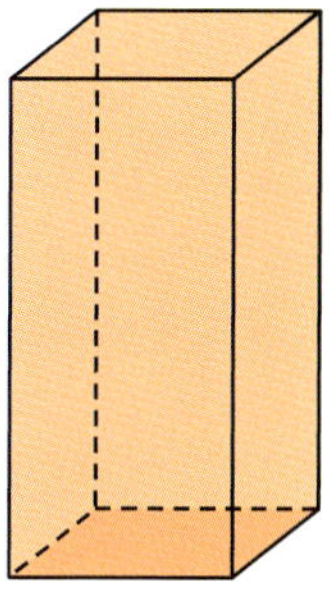

2 A historical statement says the sum of the faces and vertices of a prism or pyramid is 2 more than the number of edges.

number of faces + number of vertices − number of edges = 2

Check the statement for yourself by completing a chart for 2 prisms and 2 pyramids.

3-D shape	Number of faces	Number of vertices	Number of edges

3 Comparing Nets and Shapes

You will need

- 3-D models

Goal Relate nets to prisms and pyramids, and compare prisms and pyramids.

Josef's Nets

Josef's class made nets for 3-D shapes, but the nets got mixed up.

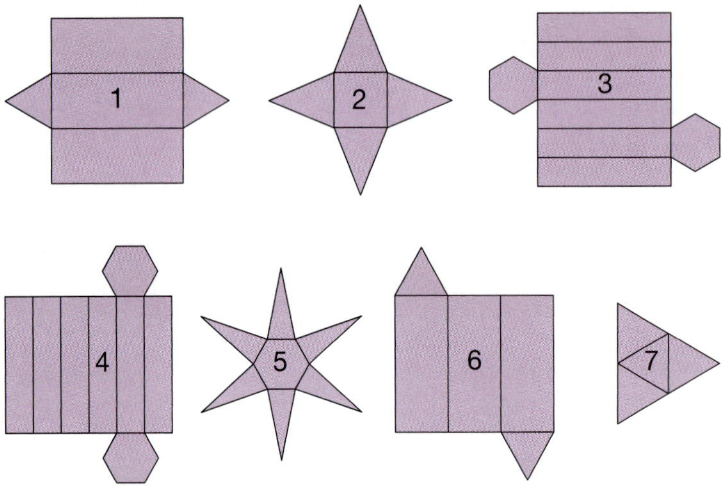

? How can Josef match each net with a 3-D shape?

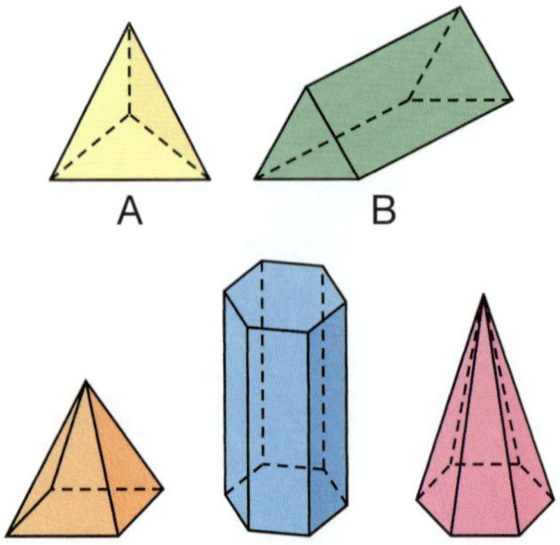

A. 2 of the nets match shape B. Which nets are they? How do you know? Name the shape.

B. 2 of the nets match another shape. Which nets are they? Which shape is it? Name the shape.

C. Match each of the other nets with 1 of the shapes. Name each of these shapes.

Reflecting

1. Explain the clues you used to match the nets with the 3-D shapes.
2. a) What is the difference between a net for a pyramid and a net for a prism with the same base?
   b) How does this show the difference between a pyramid and a prism?

Checking

3. a) Which is a net for this shape?
   b) Name the shape.

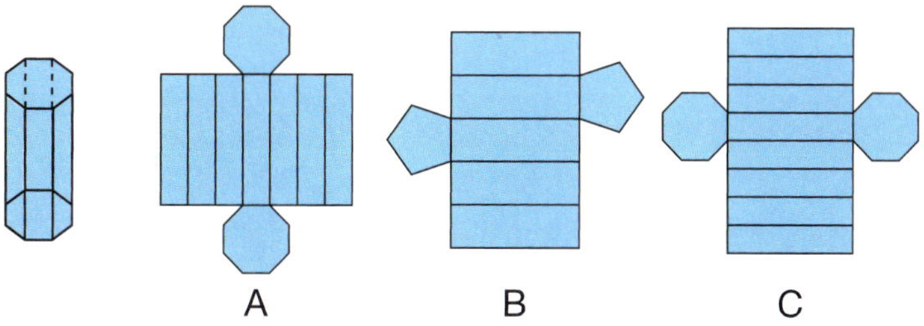

Practising

4. a) Which shape can be made with this net? Explain how you know.
   b) Name the shape. Explain how the net shows the name of the shape.

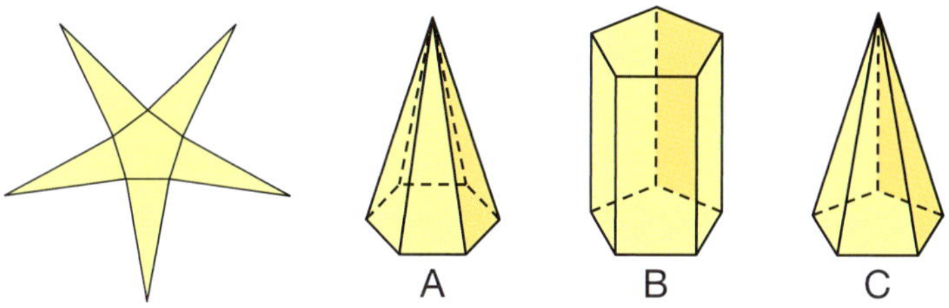

5. Explain how you know that each drawing cannot be a net for a cube.

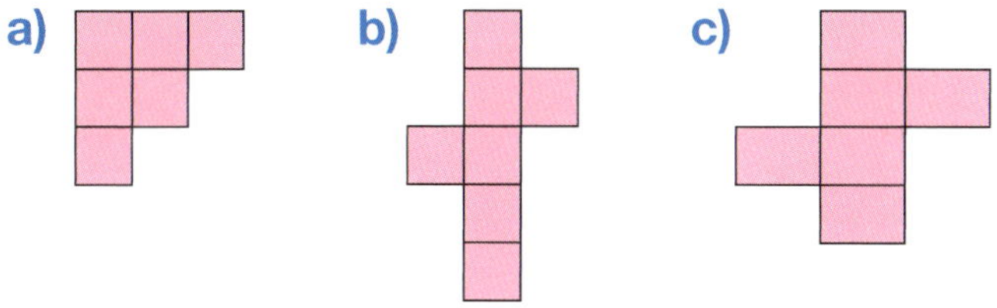

6. Use 2 congruent triangles and 3 congruent rectangles to draw a net. What shape can you make with the net?

CHAPTER 11

4 Communicate an Understanding of Geometric Concepts

You will need
- 3-D models

- paper
- stickers
- scissors

- tape

Goal Use math language to show what you know about a 3-D shape.

The famous inventor Alexander Graham Bell had a cabin shaped like a triangle-based pyramid, or **tetrahedron**. Calvin wrote about what it would be like to live in a tetrahedron.

? **How can Calvin show what he knows about tetrahedrons in his writing?**

Calvin's Writing

First I used a net to make a tetrahedron model. Then I talked with a classmate about my ideas. After that, I wrote my rough copy. Then I looked for ways to improve it for my good copy.

Living in a Tetrahedron (Rough Copy)

If I lived in a tetrahedron house, the pictures would hang down away from the walls because the walls are slanted. There are only 2 walls to hang things on. I would only be able to stand up in the middle because of the slanted walls.

To make my house, I would only need 6 pieces of wood. I wouldn't want to live in a tetrahedron house, but a tent would be fun.

I like the way I explained about the pictures hanging.

I should use math words to explain why there are only 2 walls.

I should explain why I only need 6 pieces of wood using math words and a picture.

A. Use the Communication Checklist to decide how to improve Calvin's rough copy.
How would you answer each question in the checklist? Explain your thinking.

B. What changes could Calvin make when he writes his good copy?

Communication Checklist

- ☑ Did you explain your thinking?
- ☑ Did you use a model?
- ☑ Did you use math language?

Reflecting

1. Why is it helpful to talk with someone about your ideas?
2. What do Calvin's rough copy and comments show you about his knowledge of tetrahedrons?

Checking

3. Write about what it would be like to live in 1 of these buildings. Create a net to make a model of the shape. Talk with a classmate about your ideas before you write your rough copy.

Cree Eco-Lodge on Moose Factory Island

Cube house in Toronto

Practising

4. Look at the rough copy you wrote for Question 3.
   a) What do you like about your writing?
   b) What could you do to improve your writing?
   c) Write your good copy.

Cross-Sections

Rey's Cube

I tried to imagine what new faces I would get if I cut a cube diagonally. I also wondered what 3-D shapes I would get.

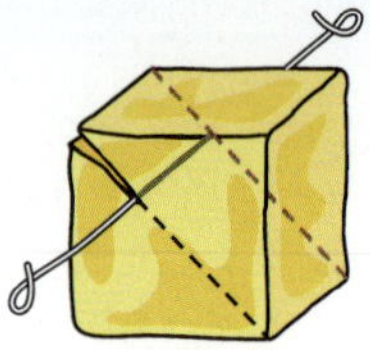

I guessed that I would get rectangle faces because the shape would still have square corners.

I checked my prediction by cutting a clay cube with dental floss.

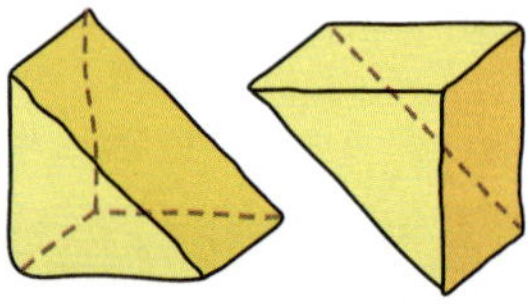

The new faces are rectangles and the cut made 2 triangle-based prisms.

You will need

- modelling clay

- dental floss

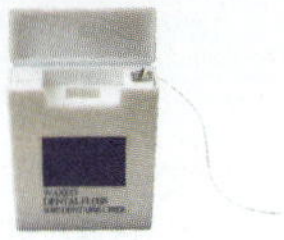

Predict the shapes of the faces for each cut. Check your predictions. Describe the 2 new 3-D shapes.

A. Cut a cube vertically.

B. Cut a cube horizontally.

C. Cut a corner off a cube.

Try These

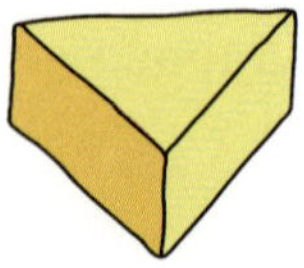

1. What new faces will you get if you cut a triangle-based prism horizontally?

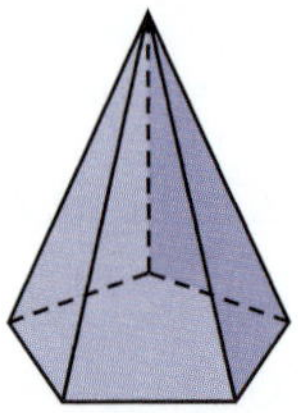

2. Describe the new face and the shapes you get when you slice the top off a pentagon-based pyramid.

Mid-Chapter Review

LESSON 3

1. Complete the chart.

3-D shape and name	Net	Number of vertices	Number of edges
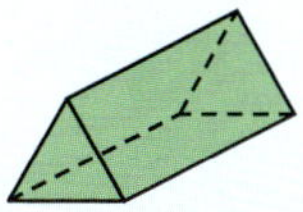 triangle-based prism			
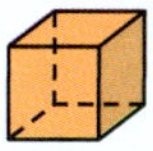	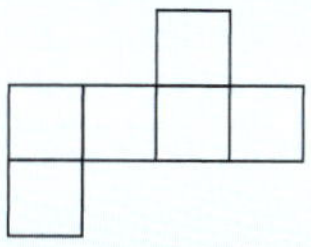		
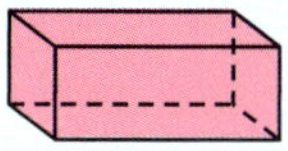			12
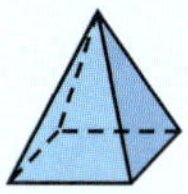		5	

2. Use your completed chart from Question 1.
   a) What do you notice about the number of edges and vertices in any shape?
   b) How are prisms and pyramids the same? How are they different?
   c) Which 3-D shapes have all congruent faces?

3. Look at the 2 nets for pyramids in your completed chart from Question 1.
   a) How are they the same?
   b) How are they different?

4. Show that there is more than 1 possible 3-D shape with 6 faces.

5 Modelling Volume

Goal **Model 3-D shapes to measure volume.**

An object is a 3-D shape because it has 3 dimensions: length, width, and height.

? How do we measure the size of 3-D shapes?

A. Find a small box. Use linking cubes. Make a model that is about the same size as the box. Count the linking cubes you used. Record the **volume** of your model.

B. Use ones blocks and the same box. Make a model that is about the same size as the box. Count the ones blocks you used. Record the volume.

C. Repeat parts A and B with a different small box.

Reflecting

1. If you turn a 3-D shape upside down or sideways, does its volume change? Give a reason for your answer.
2. Zola made 3 different shapes. Each shape was made with 12 linking cubes. Do the shapes have the same volume? Explain how you know.
3. For each box, are your models the same size? Does each model have the same number of cubes? Are the volumes the same? Explain your thinking.

You will need

- linking cubes

- ones blocks

- small boxes

volume

The amount of space taken up by an object

This shape has a volume of 5 cubes.

6 Measuring Mass

Goal Estimate, measure, and compare the mass of objects, and relate units.

Shani's Backpack

My brother and I are going on a bicycle hike.
We need food for 3 meals and extra clothing.
I can't carry more than 2 kg in my backpack.
My brother can't carry more than 6 kg.

You will need

- balance scales and masses

- list of food and clothing items
- a calculator

? What can Shani and her brother pack so the total mass in their backpacks is 8 kg or less?

A. Find an object with each **mass**.
Measure using a balance scale.
a) about 5 g b) about 500 g c) about 1 kg

B. Make a list of items Shani and her brother might take.

C. Find the mass of each item on your list.

D. Choose what to pack in each backpack. Explain.

mass
The amount of matter in an object. Mass is measured in grams (g) and kilograms (kg).
1000 g = 1 kg

Reflecting

1. Why does it make sense that 1000 g is called 1 kilogram?

2. Are more kilograms or more grams needed to balance Shani's backpack? Explain your thinking.

3. Name an object with each mass.
a) 5 g b) 600 g c) 1 kg d) 2 kg

CHAPTER 11

7 Measuring Capacity

Goal **Estimate, measure, and compare the capacity of containers, and relate units.**

You will need

- empty containers, including a 250 mL measuring cup and a 1 L bottle
- water
- a funnel

Pedro's Water Intake

I measured how much water I drank each day.
On Saturday, I drank 4 mugs of water.
On Sunday, I drank 4 large glasses of water.
On Monday, I drank a 1 L bottle of water.
On Tuesday, I drank a 1 L bottle, 1 mug, and 2 glasses of water.

? How can Pedro measure and keep track of the amount of water he drinks each day?

A. Find a container with each **capacity**.

- less than 250 mL
- about 250 mL
- about 500 mL
- about 1000 mL
- more than 1 L

Use a 250 mL container to measure each capacity.

B. List your containers for part A in order from the greatest to least capacity.

capacity
The amount a container can hold. Capacity is measured in millilitres (mL) and litres (L).
1000 mL = 1 L

C. Estimate the number of mugs of water that would fill a 1 L bottle.

D. Estimate the number of glasses of water that would fill a 1 L bottle.

E. Measure the number of mugs of water it takes to fill a 1 L bottle. About how many millilitres does the mug hold?

F. Measure the number of glasses of water it takes to fill a 1 L bottle. About how many millilitres does the glass hold?

G. About how much water did Pedro drink each day? Complete a chart like this.

Day	Number and container	Capacity (mL or L)
Saturday	4 mugs	
Sunday	4 glasses	

Reflecting

1. On what day did Pedro drink the most water? How do you know?

2. Each of these containers has a capacity of 2 L.

a) 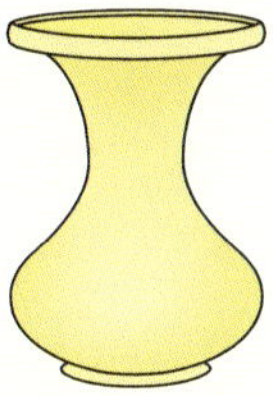

b) 

c) 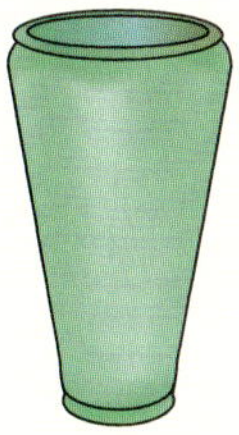

Suppose 1 L of water was poured into each container. Where would the water line be for each container?
- halfway between the top and bottom
- closer to the top than the bottom
- closer to the bottom than the top

Explain your thinking.

3. Does it take more litres or millilitres of water to fill a glass? Explain your thinking.

LESSON

Skills Bank

1

1. Use a 3-D model for this shape.

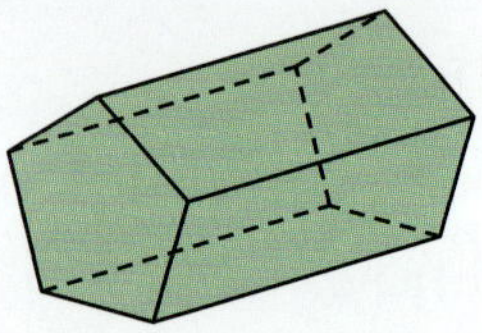

a) Draw a net for the shape. Fold and tape your net to make the shape.

b) Name the shapes of the faces.

c) How many faces are there of each shape?

d) Name the 3-D shape.

3

2. Name the 3-D shape that can be made with each net.

a)

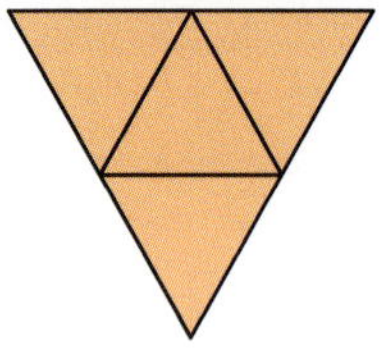

b)

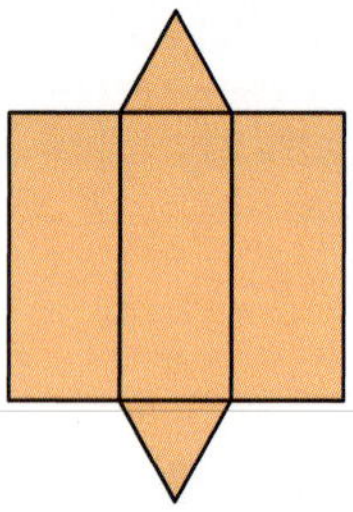

3. Which shapes match each description?

a) There are an even number of edges.

b) The total number of sides of its faces is greater than 25.

c) The total number of sides of its faces is twice the number of edges.

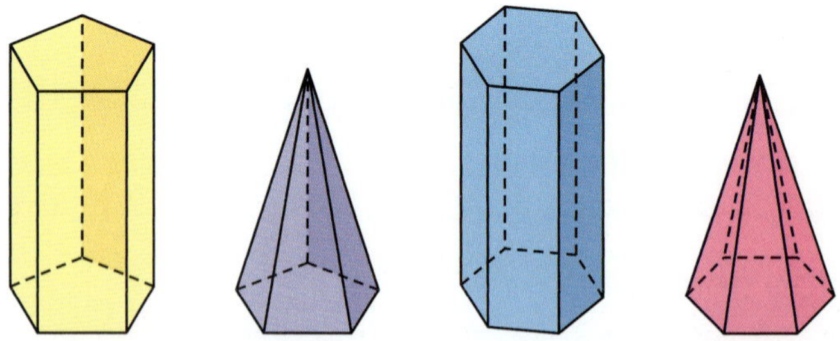

A B C D

4. Which shapes in Question 3 match each description?

a) The same number of faces meet at each vertex.

b) The number of faces is the same as the number of vertices.

3 **5.** Which shapes in Question 3 match each description?

a) There are an odd number of vertices.

b) There are more edges than vertices.

5 **6.** Use linking cubes.
Build each shape.
What is the volume of each shape?

a) 

b) 

c) 

7. Use linking cubes.
For each shape in Question 6, build another shape with the same volume.
Record your work.

8. Use ones blocks.
Build each shape in Question 6 with the same number of cubes.
What is the volume of each shape?

6 **9.** Name something with each mass.

a) about 10 g
b) about 30 g
c) about 100 g
d) about 5 kg
e) about 10 kg
f) about 75 kg

7 **10.** Name something with each capacity.

a) about 5 mL
b) about 5 L
c) about 50 L
d) about 50 mL
e) about 100 mL
f) about 750 mL

11. Which measurements seem unreasonable?
Explain each answer.

A. a 50 kg pencil sharpener
B. a 100 g candy bar
C. a 2 kg radio
D. a 50 mL juice box

LESSON

Problem Bank

1

1. How many different nets can you make for a cube? Sketch the nets.

2. Paulette thinks that any 3-D shape has twice as many nets as it has faces.
Choose a pyramid and a prism.
Is Paulette correct for the pyramid and prism you chose? Use models and diagrams to explain.

3

3. A net has 5 faces.
What 3-D shape might it be when it is folded?
What other 3-D shape might it be?
Use drawings in your explanation.

4. Suppose all the vertices of a triangle-based prism were cut off.
How many faces would the new shape have?
How do you know?

5. I have 8 faces and 12 vertices.
2 of my faces are different from the other 6 faces.
What shape am I?

6. We both have 6 vertices.
We have different numbers of edges.
What shapes are we?

6

7. 1 medium egg has a mass of about 50 g.
An empty egg carton weighs about 25 g.
Denny is holding about 5 kg of eggs, including cartons.
How many full cartons of eggs is he holding?

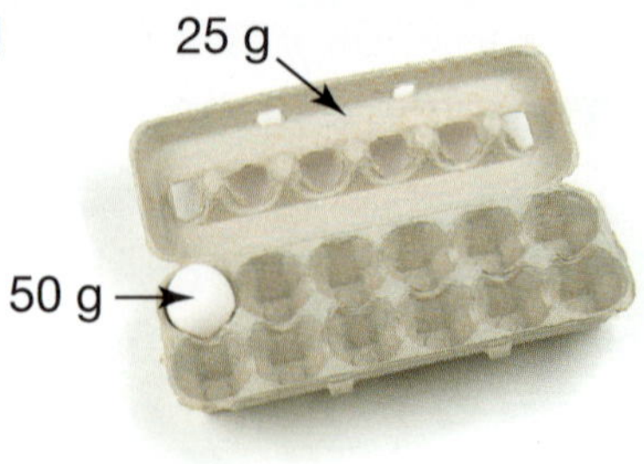

6 8. Which foods might go in each bag?

9. 100 pennies have a mass of 250 g.
Is your own mass closer to the mass of $100 in pennies or $1000 in pennies? How do you know?

10. Would the mass of a hot dog and bun be greater than or less than 50 g?
Show your work.

7 11. A can holds 350 mL of frozen orange juice concentrate. To make juice, 3 cans of water must be added to the concentrate.
Will a pitcher with a capacity of 1 L be large enough to hold the juice?
Show your work.

12. About how many litres of water does your bathtub hold? Explain how you estimated.

13. The tap in the health room drips 1 mL of water every 15 seconds.
a) How many millilitres does the tap drip every minute?
b) How many millilitres does the tap drip every hour?
c) About how many litres of water are wasted each day?

LESSON

Chapter Review

1 **1.** Use a 3-D model for a triangle-based pyramid.
   a) Draw 2 different nets for the pyramid.
   b) How are your nets the same? How are they different?

3 **2.** A 3-D shape has 7 faces. It is either a prism or a pyramid.
   a) What 3-D shapes could it be?
   b) Sketch all the faces of each shape.
   c) How many edges does each shape have? How do you know?

3. **a)** Look at models for a triangle-based prism and a square-based pyramid. Look at each shape from different angles. Which shape has more faces that can be seen at once? Why do you think this is?

   b) Compare other prisms and pyramids. What do you notice?

4. **a)** Complete a chart to organize data about these prisms.

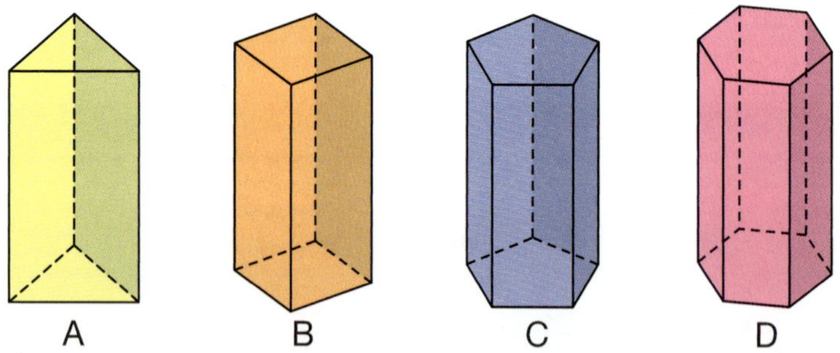

Name of prism	Number of sides of base	Number of faces	Number of edges	Number of vertices

   b) Describe any relationships you see.
   c) Are the relationships the same for a pyramid?

4 5. Write a description of this shape for someone who has never seen it. Do not include a drawing.

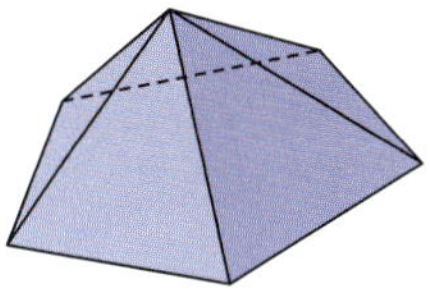

5 6. Miki made a prism using ones blocks. Jon made a prism with the same dimensions using linking cubes.

a) Who used more cubes? Explain your thinking.

b) Jon used the same number of linking cubes to make a different shape. Compare the volumes of his 2 shapes.

6 7. Angela was preparing an order of 500 g of meat. She estimated and then measured 3 times. She measured 462 g, then 521 g, and finally 498 g.

a) Did her estimates get closer each time?

b) Why did she stop measuring after 498 g?

8. Which mass is closest to 2 kg? How do you know?

A. 1950 g B. 2020 g C. 1590 g

7 9. Estimate how many more millilitres need to be added to fill each container. Order your estimates from the least to greatest capacity.

a)

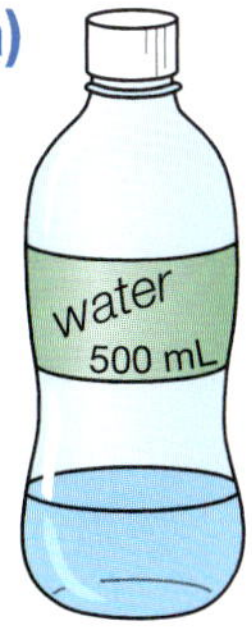

b)

c)

d)

10. Which capacities are greater than 1 L? How do you know?

A. 1200 mL B. 978 mL C. 3005 mL D. 1000 mL

Chapter Task

Cube Creature

Create a Cube Creature following these guidelines.

- Your creature must have at least 1 head, 2 or more arms, a body, and 2 or more legs.
- Each body part of your Cube Creature must be at least 2 cubes wide and 2 cubes long. For example, the smallest head would have a volume of 4 cubes.

? How can you describe your Cube Creature so that someone can make a copy of it?

Task Checklist

- ✓ Is your description clear and organized?
- ✓ Did you include diagrams?
- ✓ Did you use math language?
- ✓ Did you include the mass and volume of your Cube Creature?

Cumulative Review

Cross-Strand Multiple Choice

1. Which list shows areas that should *all* be measured in square metres?
   A. field, parking lot, skating rink, basketball court
   B. basketball court, field, playground, sandwich
   C. napkin, field, playground, skating rink
   D. field, sandbox, skateboard, parking lot

2. Choose another way to write 5×24.
   E. 4×25
   F. $5 \times 20 + 5 \times 4$
   G. $5 \times 2 + 5 \times 4$
   H. 5×2 ones $+ 5 \times 4$ tens

3. 352 slices of cake were sold at the school fun fair. Each cake was cut into 8 slices. How would you calculate the number of cakes that were used?
   A. Multiply 352×8.
   B. Divide 352 by 8.
   C. Add $352 + 8$.
   D. Subtract 8 from 352.

4. Describe the faces of a pentagon-based prism.
   E. pentagons only
   F. pentagon and triangles
   G. pentagons and trapezoids
   H. pentagons and rectangles

5. Choose the measurement that doesn't make sense.
   A. jug of milk: 1 L
   B. flower vase: 500 mL
   C. water bottle: 800 L
   D. kitchen sink: 14 L

6. Choose the measurement that doesn't make sense.
   E. bag of potatoes: 5000 g
   F. truck: 454 g
   G. small paper salt package: 23 g
   H. bag of uncooked spaghetti: 500 g

Cross-Strand Investigation

We Need a Playground

The children had nowhere to play. They wanted a playground that was big enough for all of them.

7. a) 456 children marched to City Hall in 2 arrays. The 1st array had 8 rows. The 2nd array had 5 rows. Draw a diagram to show the arrays. Write the multiplication sentence for each array.
   b) The children walked home in groups of 6. How many groups were there? Show your calculations.

8. Carlitos designed a playground on a centimetre grid.
   a) Is there enough room for the children to walk and run? Compare the areas for walking, running, and playing on equipment. Show your work.
   b) Find 2 parts of the playground that have the same area. Explain how you know the areas are the same.

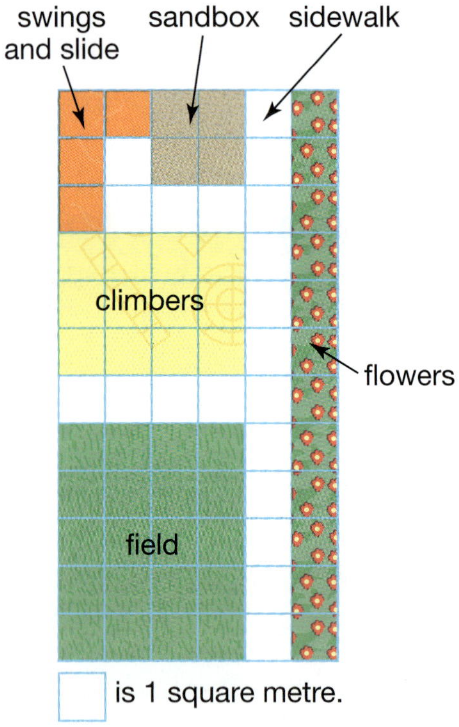

9. What 2-dimensional and 3-dimensional shapes can you see in this playground equipment? Use geometry words.

CHAPTER 12

Fractions and Decimals

Goals

You will be able to

- describe fractions of a region and of a set
- describe and create regions and sets using decimal tenths and decimal hundredths
- use decimals for measurements
- add and subtract decimals

Sea sails

Getting Started

Fractions

? How many ways can you show $\frac{2}{3}$?

2 of 3 students are wearing green shirts.

$\frac{2}{3}$ of the students are wearing green shirts.

2 of 3 three students are boys.

$\frac{2}{3}$ of the students are boys.

A. Look at the picture. Create more descriptions of the students that are examples of the **fraction** $\frac{2}{3}$.

B. Use materials to make as many models of $\frac{2}{3}$ as you can.

You will need

- pattern blocks

- linking cubes

- fraction circles

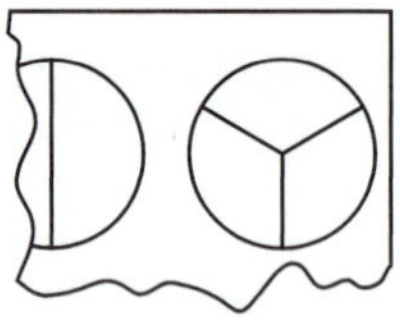

- paper strips

C. Work with a partner. Do your models show $\frac{2}{3}$ of a set, $\frac{2}{3}$ of a region, or $\frac{2}{3}$ of a length?
Sort your models of $\frac{2}{3}$ using a chart like this.

$\frac{2}{3}$ **of a set**	$\frac{2}{3}$ **of a region**	$\frac{2}{3}$ **of a length**
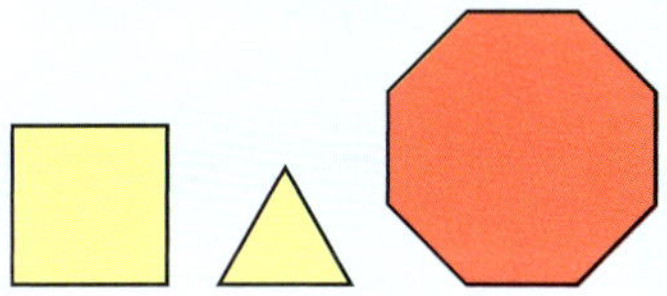	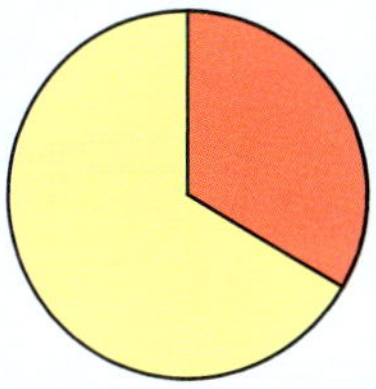	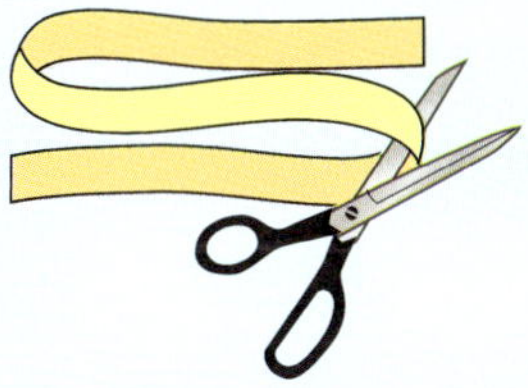

D. Find examples in your classroom that you can describe with fractions.

Do You Remember?

1. Explain what each word means and give an example.
a) numerator **b)** denominator

2. Which picture cannot be described by the fraction $\frac{3}{4}$? Explain your thinking.

A. 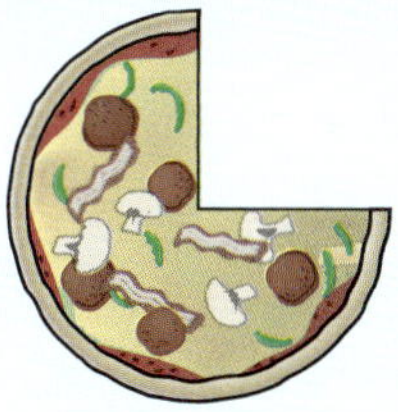

B. 

C. 

3. Use a fraction to estimate how full each glass is. Explain your answer.

a) 

b) 

c) 

d) 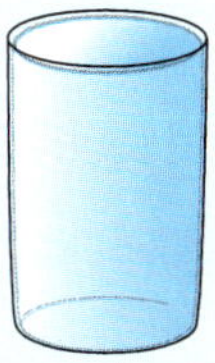

1 Tenths of a Region

Goal **Describe tenths as fractions and decimals for part of a region.**

Chris's school is planning a play day. Each team was asked to make a flag with each part a tenth.

You will need

- pencil crayons

- paper
- scissors

- ruler

? How can flags represent tenths?

Chris's Team's Flag

My team designed these 3 flags and then chose the flag on the left.

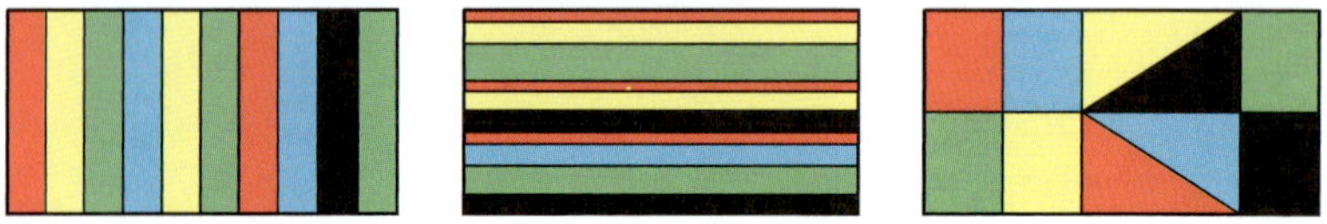

My team's flag is coloured in 10 equal parts and 3 of the parts are green.

That means 3 tenths of the flag is green.

I can write three tenths as a fraction.

$$\frac{3}{10}$$

3 ← green parts

10 ← equal parts in all

I can also write three tenths as a **decimal**.

0.3

Zero means there is less than 1 whole. → 0

3 ← number of tenths

. ← decimal point

A. How do you know the flag Chris's team chose shows tenths?

B. Do the other 2 flags show tenths? Explain how you know.

C. Use fractions, decimals, and words to describe one of the flags that shows tenths so that someone could identify it.

decimal
A way to describe fractions using place value.
A decimal point separates the ones place from the tenths place.

Reflecting

Natalie's team made this flag.

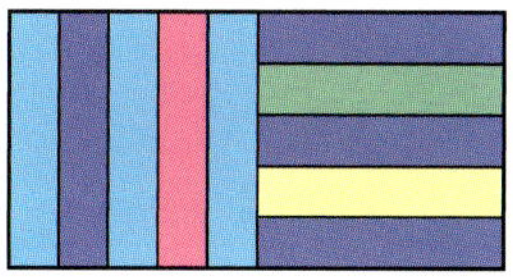

1. Does the flag for Natalie's team show tenths? Explain how you know.
2. Do the parts need to be congruent to describe them as tenths? Explain how you know.

Checking

3. Look at the flag for Chris's team.
   a) Write a fraction, a decimal, and words for the part of the flag that is blue.
   b) Write a fraction, a decimal, and words for the part of the flag that is not blue.

Practising

4. Rami's team made this flag. Write a fraction, decimal, and words to describe the number of tenths for each colour.

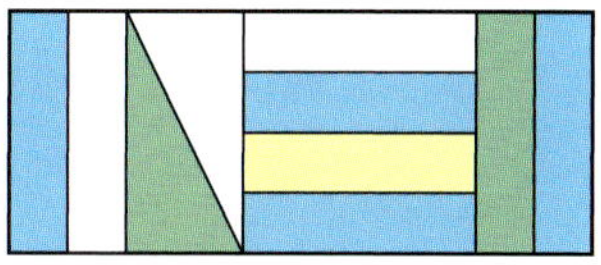

   a) blue b) green c) white d) yellow
5. Explain how you know the flag in Question 4 shows tenths.
6. When tenths are written as a decimal, the digit to the right of the decimal point is the numerator of the fraction. Use 2 examples to explain this.
7. a) Draw a flag that is 0.2 red, 0.5 blue, and the rest yellow.
   b) Write the fraction, decimal, and words for the tenths that are yellow.
   c) Write the fraction, decimal, and words for the tenths that are not yellow.

2 Tenths of a Set

You will need

- pencil crayons

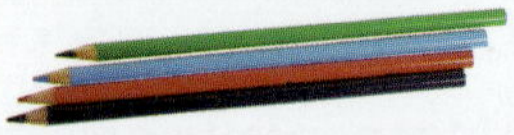

Goal Describe tenths as fractions and decimals for part of a set.

Miki has rings on all of her fingers.

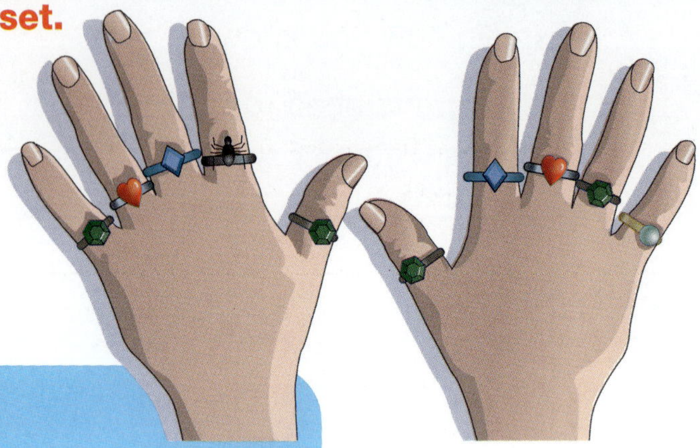

? How can Miki represent the number of blue rings on her fingers?

Miki's Blue Rings

I have blue rings on 2 fingers.
That's two tenths of my fingers.

$\frac{2}{10}$ ← 2 fingers have blue rings; ← 10 fingers in all

0.2 fingers have blue rings

0.0	0.1	0.2	0.3	0.4	0.5	0.6	0.7	0.8	0.9	1.0
$\frac{0}{10}$	$\frac{1}{10}$	$\frac{2}{10}$	$\frac{3}{10}$	$\frac{4}{10}$	$\frac{5}{10}$	$\frac{6}{10}$	$\frac{7}{10}$	$\frac{8}{10}$	$\frac{9}{10}$	$\frac{10}{10}$

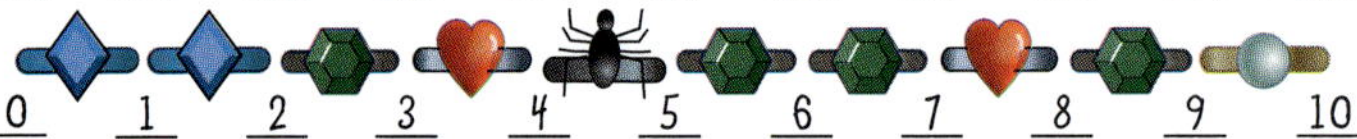

Two tenths is closer to 0 than it is to 1.

Reflecting

1. Why are 2 hands a good model to show decimal tenths?
2. Why is 0.2 different from $\frac{0}{2}$?
3. What decimal is halfway along the row of rings?

Checking

4. a) Describe Miki's other rings using words, fractions, and decimals.
   b) Tell which of your decimals is greatest and which is least.

Practising

5. Write a fraction and a decimal that tell that all of Miki's fingers have rings.

6. Write each as a fraction and a decimal.
   a) one tenth b) zero tenths c) ten tenths

7. Take 10 objects from your desk.
   a) Sort them into 3 or 4 groups. Describe each group.
   b) Write a fraction, decimal, and words to describe each group.

8. Tell whether each decimal is closer to 0, $\frac{1}{2}$, or 1.
   a) 0.9 b) 0.4 c) 0.7 d) 0.1

9. a) Trace your hands and draw rings on your fingers. Leave some fingers without rings. Make less than half of your rings the same colour.
   b) Use words and decimals to describe your rings.

10. Describe each colour of flag using words, fractions, and decimals.

11. Write the decimal that represents the colour for the most flags in Question 10. Explain how you know by comparing the words and by comparing the fractions.

3 Decimal Tenths Greater Than 1

You will need

- a measuring tape

Goal **Model, write, and compare measurements for decimal tenths greater than 1.**

Josef wants to enter his frog in a jumping contest.
Sarah's champion frog can jump 1.6 m.
Sarah says, "My frog jumped one-and-six-tenths metres."

? Do you think Josef's frog can win?

Josef's Frog Jump

I placed 2 metre sticks end-to-end to measure my frog's best jump.

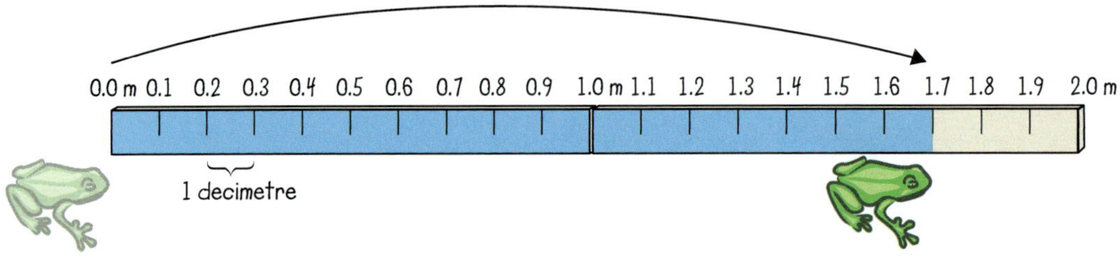

A. Part of the metre stick between 1.0 m and 2.0 m is shaded. Write this part of a metre as a decimal.

B. Why is the jump distance for Josef's frog 1 m 7 dm?

C. Why is the decimal for this distance 1.■ m?

D. Write the jump distance of Josef's frog as a decimal.

E. Whose frog can jump farther, Sarah's or Josef's? Show how you know.

F. Write each jump distance in **decimetres**.

Reflecting

1. a) Explain how you know 1.7 m is longer than 1.2 m.
   b) Explain how you know 0.9 m is shorter than 1.2 m.

Checking

2. A frog jumped 2.3 m.
   a) What does the 2 tell you?
   What does the 3 tell you?
   b) Write the jump distance in decimetres.
   c) Which jump distance is greater, 2.3 m or 2.1 m?

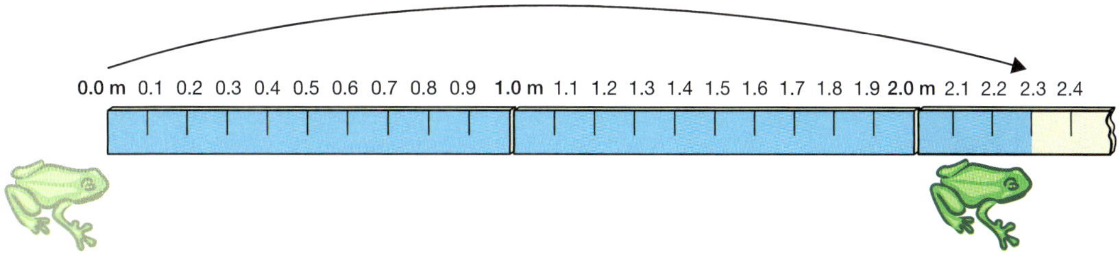

Practising

3. A frog jumped three-and-two-tenths metres.
   a) Use a tape measure or draw a picture to show where the frog landed.
   b) Use a decimal to write the jump distance.
   c) Write the jump distance in metres and decimetres.

4. Complete.
   a) 1.6 m = ■ m 6 dm
   b) 3.1 m = ■ m ■ dm
   c) 2.9 m = 2 m ■ dm
   d) 1.7 m = ■ m ■ dm

5. Write each measurement in metres. Use decimals.
   a) one metre four decimetres
   b) twenty three decimetres
   c) three metres one decimetre
   d) two-and-four-tenths metres

6. Arrange these measurements from longest to shortest.
   2.1 m, 3.5 m, 0.9 m, 1.8 m, 2.6 m, 2.0 m

4 Adding Decimal Tenths

You will need

- a measuring tape

Goal **Add decimals in tenths.**

Brian's shadow is 1.9 m long.
Jennifer's shadow is 0.3 m longer than Brian's shadow.

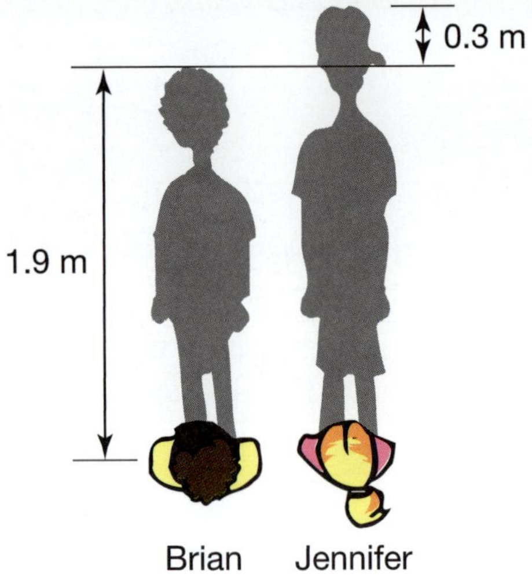

? How long is Jennifer's shadow?

A. Suppose Jennifer's shadow were only 0.1 m longer than Brian's shadow. How long would her shadow be?

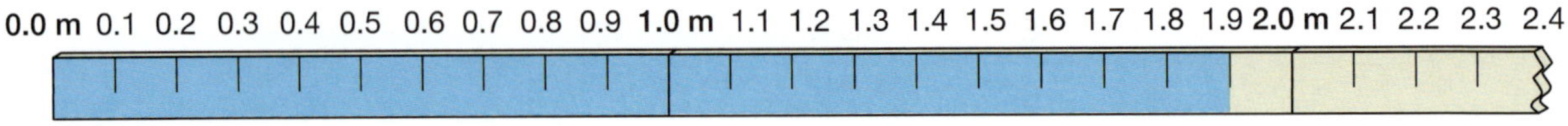

$1.9 + 0.1 = \blacksquare$

B. How much more is 0.3 than 0.1?

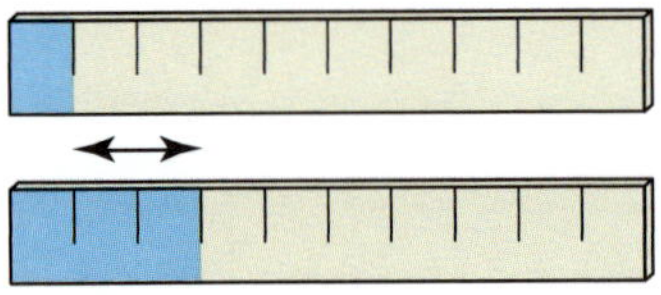

C. How can you use your answer to part B to find the length of Jennifer's shadow?

D. Suppose the shadow of a light pole is 1.8 m longer than Jennifer's shadow. How long is the light pole's shadow?

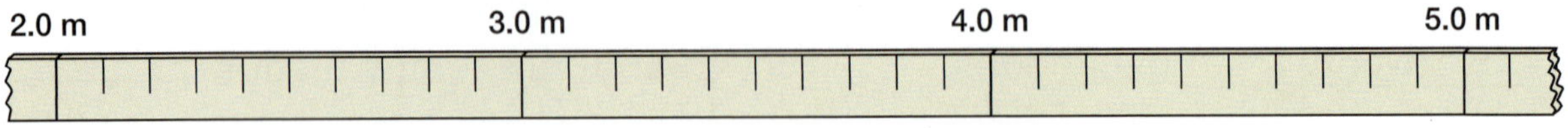

Reflecting

1. How do you know your answers are reasonable?
2. How could you use decimetres to check the addition?

Checking

3. Later in the day, Jennifer's shadow is 4.4 m long. Brian stands at the end of her shadow. His shadow is 3.8 m long.
   a) Why would the length of their shadows together be less than 8.4 m?
   b) What is the length of their shadows together? Use a picture to explain your answer.

Practising

4. Late in the day, Jennifer's shadow is 6.4 m long. Her shadow grows longer, first by 2.3 m and then by another 2.5 m. How long is Jennifer's shadow? Use a model or picture to find the answer. Show your work.

5. Use models or pictures to add the lengths.
   a) 0.5 m and 0.7 m
   b) 1.6 km and 2.4 km
   c) 3.2 cm and 1.8 cm
   d) 2.6 dm and 1.5 dm

6. Use models or pictures to find each sum. Show your work.
   a) 1.9 + 7.2
   b) 0.4 + 5.8
   c) 6.1 + 1.9
   d) 4.5 + 2.6

7. A shadow with a length between 1 m and 2 m grows to be twice as long.
   a) What is the longest the shadow could be?
   b) What is the shortest the shadow could be?

8. How many different pairs of decimal tenths add together to make 1.0? Use pictures to show your work.

5 Subtracting Decimal Tenths

You will need

- a measuring tape

Goal **Subtract decimals in tenths.**

A baby boa constrictor is about 0.6 m long at birth.
An adult boa constrictor might be 2.1 m long.

? How much does the boa constrictor grow?

Chantal's Method

The answer will be between 1 m and 2 m.

I started with the length of the baby boa constrictor and counted on to the length of the adult.

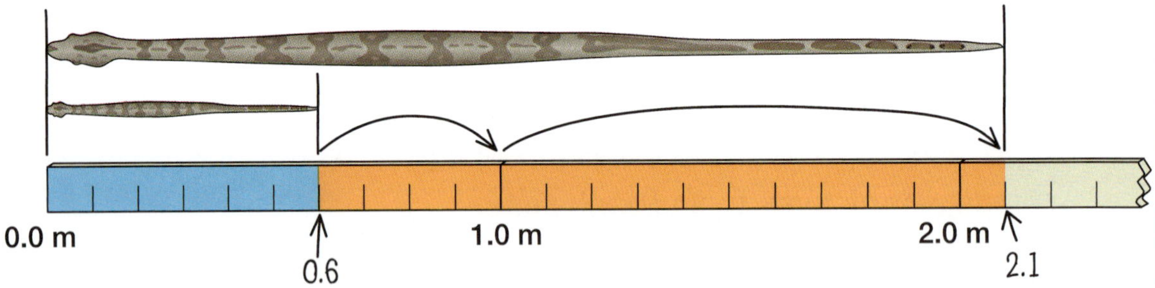

I counted on 0.4 m to get from 0.6 m to 1.0 m.

Then it's another 1.1 m to get from 1.0 m to 2.1 m.

$0.4 + 1.1 = 1.5$

The boa constrictor grew by 1.5 m.

Reflecting

1. Why do you think Chantal knew the answer would be between 1 m and 2 m?
2. How could you use decimetres to do the subtraction?

Checking

3. A bull snake can grow to be 1.5 m long. If a baby bull snake is about 0.4 m long, how much will it grow?

Practising

4. A giraffe's neck can be about 2.5 m long. If a giraffe is about 5.4 m tall, how tall is the rest of its body? Show your work.
5. Use models or pictures to find the difference between these lengths.
   a) 5.5 m and 1.4 m   b) 4.3 cm and 3.4 cm
6. Use models or pictures to find each difference.
   a) 1.6 − 0.9   b) 2.0 − 1.4   c) 2.8 − 1.5
7. a) Which difference is greatest in Question 6?
   b) How much greater is this difference than the least difference?
8. A snake grew 1.7 m over its lifetime. The snake was shorter than 1 m when it was born. What could its birth length and its final length have been? Give 2 possibilities. Draw a picture to explain your thinking.

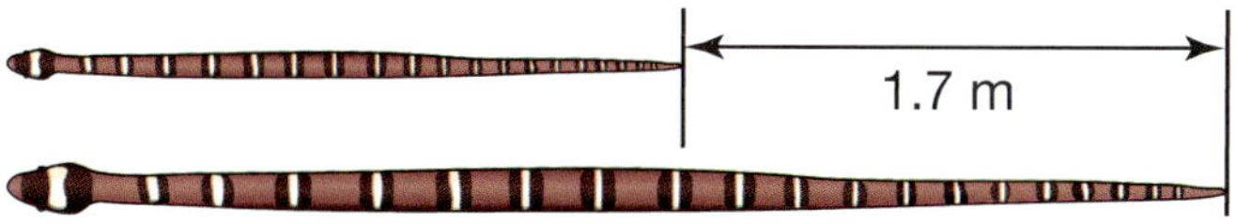

CHAPTER 12

6 Communicate About Decimal Operations

You will need

- base ten blocks

Goal **Use a model to explain how to add and subtract decimals.**

You can spot an American kestrel in flight because its wingspan is more than double its body length.

American kestrel

? How can Vinh explain how to find the difference between the American kestrel's wingspan and its body length?

Vinh's Explanation

I'll use blocks to explain.

This rod is 1 dm long.

This cube is 0.1 dm long.

First, I model the wingspan and the body length.
The wingspan is 5.8 dm.
The body length is 2.7 dm.

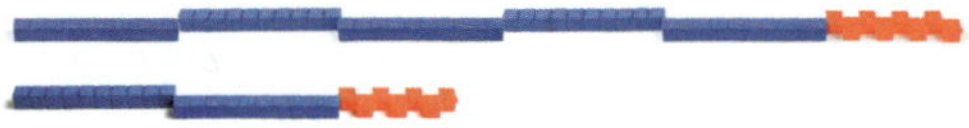

The difference is how much longer the wingspan is than the length.
I count the whole decimetres in the difference first.

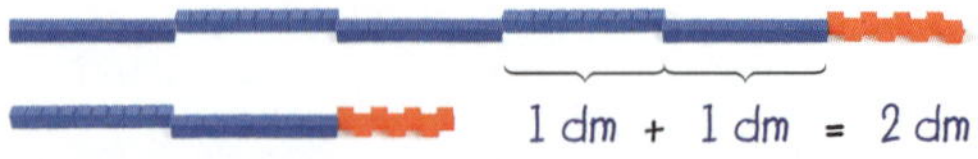

Then I count tenths of a decimetre.
I look for 10 tenths, or 1 dm.

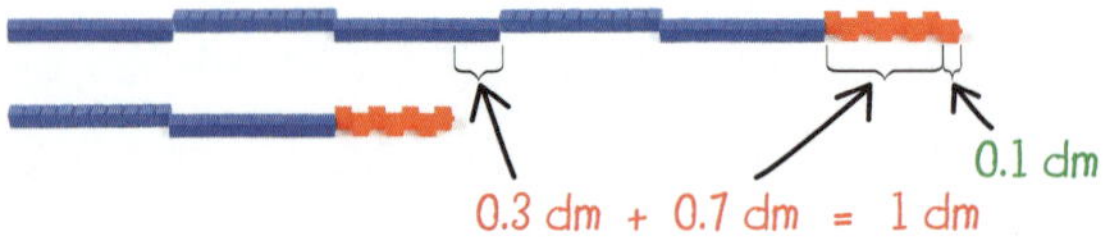

The difference is 2 dm + 1 dm + 0.1 dm = 3.1 dm.

A. How can you tell that Vinh considered all the questions in the Communication Checklist?

B. How does the block model help to make Vinh's words easier to understand?

C. Explain why Vinh could add 2.7 dm + 3.1 dm to check his answer.

D. Use blocks to explain to a classmate how you would add 2.7 dm + 3.1 dm.

Communication Checklist

- ✓ Did you use a model?
- ✓ Did you show all your steps?
- ✓ Did you put the steps in order?
- ✓ Did you explain your thinking?
- ✓ Did you use math language?

Reflecting

1. Use the Communication Checklist.
   a) What strengths can you identify in the way you explained your addition steps in part D?
   b) Suggest 1 way to improve your explanation.

Checking

2. Use blocks to show how you would subtract to find the difference between the least bittern's wingspan and its body length.

least bittern
body length 3.3 dm
wingspan 4.4 dm

Practising

3. The least bittern has a wingspan of 4.4 dm. The American kestrel has a wingspan of 5.8 dm. Use blocks to explain how you would subtract to find the difference between these 2 wingspans.

broad-winged hawk
body length 4.1 dm

4. The broad-winged hawk is 4.1 dm long. Its wingspan is 4.5 dm longer than its body length. Use blocks to explain how you would add to find its wingspan.

LESSON

Mid-Chapter Review

1 **1.** Use a 2 by 5 grid to make a flag.

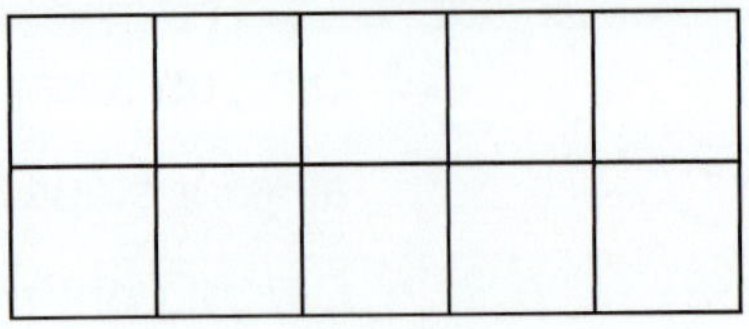

a) Colour one tenth of the flag **yellow**.
b) Colour 0.5 of the flag **red**.
c) Colour $\frac{2}{10}$ of the flag **green**.
d) What part of the flag is not coloured? Write the decimal.

2 **2.** Calvin had 10 buttons. He sewed 7 on a shirt.

a) What fraction of his buttons did Calvin sew on the shirt?
b) Write this fraction as a decimal.
c) What fraction of his buttons did Calvin not sew on the shirt?
d) Write this fraction as a decimal.

3 **3.** Complete by using >, =, or <. Explain your thinking. Use pictures or a tape measure to help you.
a) 0.8 m ■ 1.3 m **b)** 2.6 m ■ 2.7 m **c)** 3.0 m ■ 0.3 m

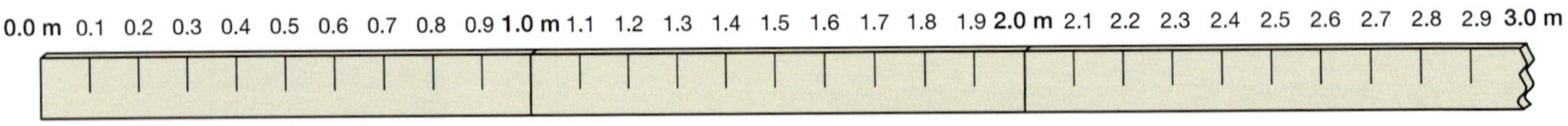

4 **4.** Use models or pictures to add these lengths.
a) 0.5 m and 1.4 m **b)** 1.5 dm and 0.7 dm **c)** 1.0 cm and 1.8 cm

5 **5.** Use models or pictures to find the difference between these lengths.
a) 1.3 m and 0.5 m **b)** 2.1 cm and 1.9 cm **c)** 2.8 dm and 2.3 dm

Decimal Cakes

Number of players: 2 or more
How to play: Take turns spinning and taking square tiles for tenths of cakes. Try to get enough tenths for 3 cakes.

Step 1 Spin the spinner. Take the same number of tiles as the number of tenths you spin. Each square tile is 1 tenth of a cake.

Step 2 Take turns spinning and taking tenths of a cake. After your first turn, add the tenths to your total. Record the addition sentence.

The first player with exactly 3 cakes wins.

You will need

- square tiles

- spinners

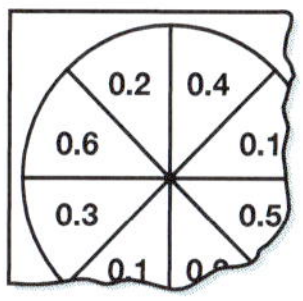

- paper clip

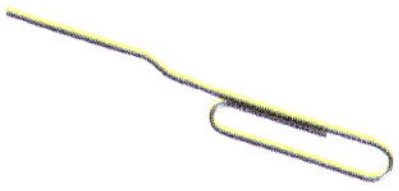

Rey's Turn

My total before this spin was 1.5.

I will take 6 squares for 6 tenths of a cake.

My addition sentence is $1.5 + 0.6 = 2.1$.

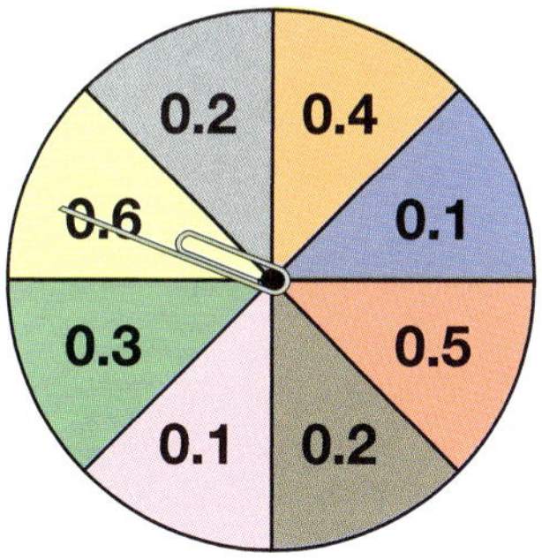

Another way to play: Each player starts with 3 whole decimal cakes. Subtract the same number of tiles as the number of tenths you spin. The first player with no cake left wins.

CHAPTER 12

7

Hundredths Less Than or Equal to 1

Goal **Write hundredths as fractions and decimals using words and symbols.**

You will need

- a 100 grid

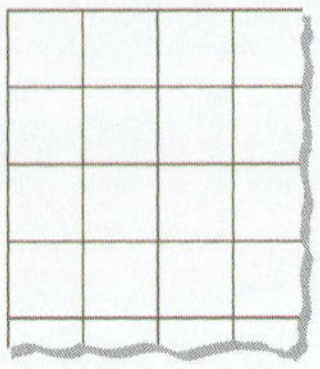

- pencil crayons

Pedro drew a design on a 100 grid using different colours.

? How can Pedro use decimals to describe his design?

Pedro's Grid

My grid is divided into 100 parts.

I coloured 1 part red.

One hundredth of the grid is red.

I write the fraction $\frac{1}{100}$ as the decimal 0.01.

Pedro's whole grid is represented by 1, 1.00, or $\frac{100}{100}$.

A. Which colour covers $\frac{23}{100}$ or 0.23 of the grid?

B. Write a fraction and a decimal in hundredths for every other colour in the grid.

C. Arrange your decimals in order from least to greatest. Explain how you know your answer is right.

Reflecting

1. Is 0.07 closer to 0 or to 1? How do you know?
2. Use a 100 grid to explain how 0.07 is different from 0.70.

Checking

3. Write a fraction and a decimal for the shaded part.

a) 

b) 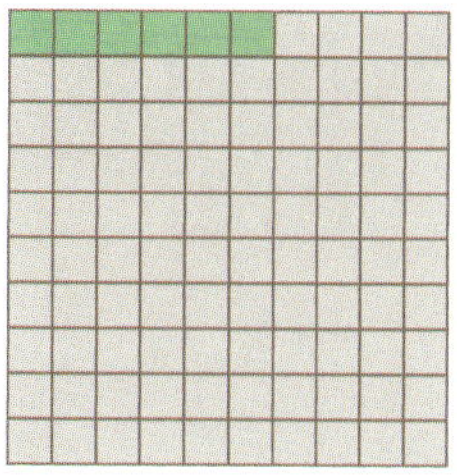

Practising

4. Write a fraction and a decimal for the shaded part. Then write a fraction and a decimal for the part that is not shaded.

a) 

b) 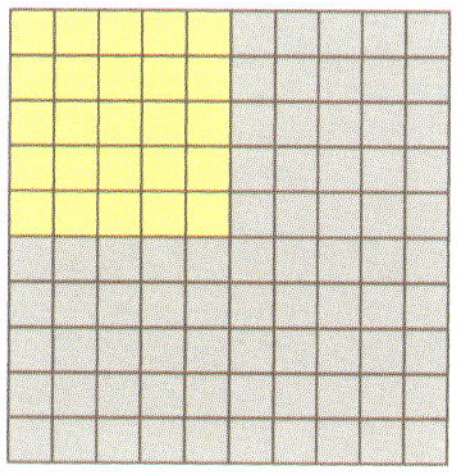

5. Write each fraction as a decimal.

a) $\frac{89}{100}$ b) $\frac{7}{100}$ c) $\frac{30}{100}$ d) $\frac{12}{100}$

6. Write each decimal as a fraction.

a) 0.67 b) 0.29 c) 0.40 d) 0.04

7. Arrange these measurements from shortest to longest.
0.45 m, 0.80 m, 0.54 m, 0.08 m, 0.05 m, 0.48 m

8. a) What decimal describes half of Pedro's 100 grid?
b) Is the blue part of Pedro's design closer to $\frac{1}{2}$ or 1? Explain how you know.

9. a) Create your own design on a 100 grid. Use at least 4 colours.
b) Describe your design using words, fractions, and decimals.
c) Which colour is closest to half of your design? Explain how you know.

8 Add and Subtract Hundredths

Goal **Add and subtract decimal hundredths using grids.**

You will need

- a 100 grid

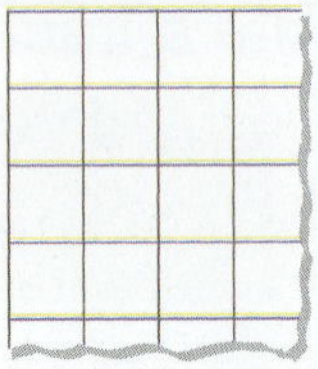

- pencil crayons

A city plans to develop 1 square kilometre of land.
The city will use 0.45 of a square kilometre for homes and 0.32 of a square kilometre for a park.
The rest of the land will be used for a shopping mall.

? How much land will be used for homes and parks?

Carmen's Solution

I can use a 100 grid to represent the land.

Each small square is 0.01 of a square kilometre.

I can colour 45 hundredths yellow to represent 0.45 of a square kilometre for homes.

I can colour the park green.

A. Colour the regions for homes and the park on your grid.

B. How many small squares are coloured in all? How much land will be used for homes and the park together?

C. How much land will be used for the mall? Write a subtraction sentence.

D. Which covers the most land: the homes, the park, or the mall? Which covers the least land? Show how you know.

Reflecting

1. How can you use a 100 grid to show the addition of 2 different decimals like 0.37 + 0.54?
2. How can you use a 100 grid to show the subtraction of 2 decimals like 0.63 − 0.15?

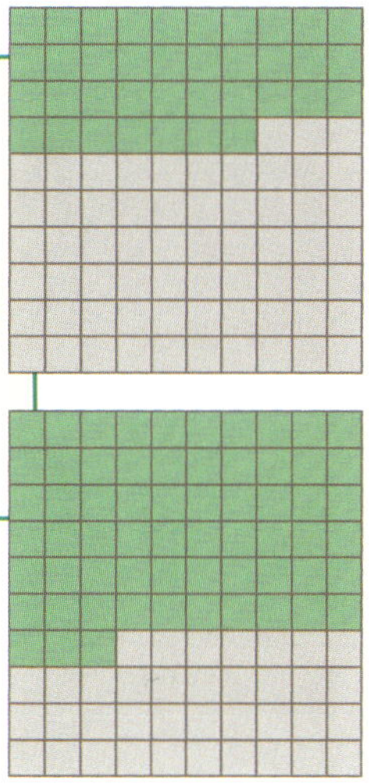

Checking

3. Use a 100 grid.
   a) Add 0.15 over and over.
   Use a different colour for each 0.15.
   Stop when you reach the sum of 0.75.
   b) How many different colours did you use?

4. Use a 100 grid.
   a) Start with 1.00 and subtract 0.09 over and over.
   After each subtraction, record the decimal for the part of the grid that is blank.
   Use a different colour each time.
   Stop when 0.10 of the grid is still blank.
   b) How many colours did you use?
   c) Describe the patterns in the digits of the decimals you wrote.

Practising

5. Add or subtract using a 100 grid.

a) 0.43 + 0.26	c) 0.23 + 0.07	e) 0.37 − 0.16
b) 0.35 + 0.47	d) 1.00 − 0.54	f) 0.45 − 0.09

6. a) Find 2 different pairs of decimals that add to 1.
   Show your work on 100 grids.
   b) Find 2 different pairs of decimals that have a difference of 0.35.
   Show your work on 100 grids.

9 Relating Decimals and Measurements

You will need

- a measuring tape

Goal **Estimate, measure, and compare measurements using decimals.**

? How can you use decimals with measurements?

A. Find an object you estimate is slightly longer than 1 m. Measure the length to the nearest tenth of a metre.

B. Find an object you estimate is not as high as 1 cm. Measure the height to the nearest tenth of a centimetre.

C. Find an object you estimate has a perimeter between 80 dm and 100 dm. Measure the perimeter to the nearest tenth of a decimetre.

D. Find an object you estimate has a perimeter slightly less than 4.5 m. Measure the perimeter to the nearest hundredth of a metre.

E. Find a round object. Choose a unit for estimating the circumference. Estimate the circumference. Measure the circumference to the nearest tenth of the same unit.

Reflecting

1. **a)** Explain how you estimated for part A and for part C.
   b) Explain how you chose the unit for part E.
2. **a)** How can you write the measurement for part B in centimetres and millimetres: ■ cm ■ mm?
   b) How can you write the measurement for part D in metres and centimetres: ■ m ■ cm?
3. Why are decimals useful for writing measurements?

Mental Math

Quarters and Dimes

Jon spends \$0.75 for a bus ticket.
He pays with a \$2 coin.
How much change will Jon receive?

Jon's Change

\$0.75 is 3 quarters and \$2 is 8 quarters.

8 quarters − 3 quarters = 5 quarters

I will get \$1.25 in change.

A. How did Jon know that \$2 is the same as 8 quarters?

B. What other way could Jon count coins to calculate his change mentally?

Try These

1. Use Jon's method to find the change.
   a) spend \$0.75 from a \$5 bill
   b) spend \$0.30 from a \$1 coin

2. Find the change.
   a) \$2.00 − \$0.25
   b) \$1.50 − \$0.75
   c) \$5.00 − \$3.75
   d) \$2.00 − \$0.70

3. Find the sum.
   a) \$2.00 + \$1.75
   b) \$1.50 + \$0.75
   c) \$2.75 + \$1.25
   d) \$1.70 + \$0.60

4. How can thinking about money help you to add decimals like 1.75 + 3.50?

LESSON

Skills Bank

1

1. Write fractions, decimals, and words for the shaded parts of the fraction rectangles.

a) 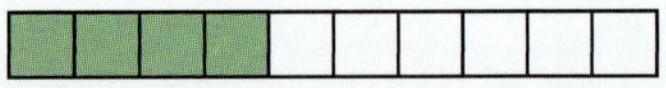

c) 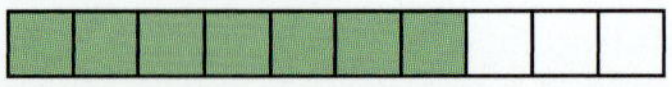

b) 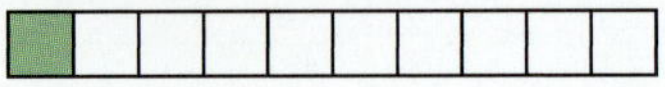

d) 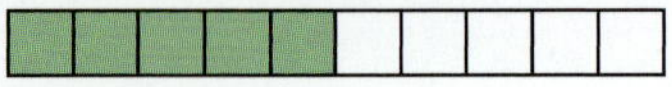

2

2. Write each as a fraction and a decimal.
a) three tenths b) six tenths c) five tenths

3. Miki has 10 bracelets: 4 red bracelets, 2 purple bracelets, 3 green bracelets, and 1 orange bracelet. Write words, fractions, and decimals to describe each colour of Miki's bracelets.

4. Use diagrams to show 0.4 in 3 different ways.

5. Tell whether each decimal is closer to 0, $\frac{1}{2}$, or 1.
a) 0.2 b) 0.3 c) 0.6 d) 0.8

3

6. Complete.
a) 1 m 4 dm = 1.■ m
c) 2 m 5 dm = ■.■ m
b) 27 dm = ■.7 m
d) 36 dm = ■.■ m

7. Complete by using $>$, $=$, or $<$.
a) 2.1 m ■ 0.2 m b) 1.8 cm ■ 1.9 cm c) 2.4 dm ■ 1.7 dm

8. Arrange each set of measurements from shortest to longest.
a) 1.9 m, 2.5 m, 0.8 m, 2.9 m, 2.3 m
b) 2.3 m, 3.5 m, 1.9 m, 0.8 m, 4.1 m
c) 1.5 m, 3.2 m, 0.6 m, 0.1 m, 2.4 m

4

9. Use models or pictures to add the lengths. Show your work.
a) 5.9 m and 3.4 m
d) 3.0 m and 5.9 m
b) 2.9 cm and 2.1 cm
e) 4.2 dm and 3.3 dm
c) 2.5 cm and 1.8 cm
f) 1.5 km and 2.1 km

4

10. Use models or pictures to find each sum.
Show your work.

a) $\begin{array}{r} 2.9 \\ +\ 3.6 \\ \hline \end{array}$ **b)** $\begin{array}{r} 0.3 \\ +\ 2.9 \\ \hline \end{array}$ **c)** $\begin{array}{r} 1.4 \\ +\ 1.9 \\ \hline \end{array}$ **d)** $\begin{array}{r} 0.6 \\ +\ 3.4 \\ \hline \end{array}$

11. Brian and Karyn built Lego towers.
Brian's tower is 1.7 dm tall.
Karyn's tower is 2.8 dm taller.
Use pictures or models.
How tall is Karyn's tower? Show your work.

5

12. Use models or pictures to find the difference between these lengths.
Show your work.

a) 3.7 m and 2.1 m
b) 6.2 cm and 4.8 cm
c) 4.6 m and 2.2 m
d) 6.9 cm and 3.7 cm

13. Use models or pictures to find each difference.
Show your work.

a) $\begin{array}{r} 7.9 \\ -\ 3.5 \\ \hline \end{array}$ **b)** $\begin{array}{r} 6.3 \\ -\ 5.9 \\ \hline \end{array}$ **c)** $\begin{array}{r} 8.3 \\ -\ 1.7 \\ \hline \end{array}$ **d)** $\begin{array}{r} 4.6 \\ -\ 3.8 \\ \hline \end{array}$

14. Kylie is running a 10.0 km race.
She stops for a drink at 6.5 km.
How much farther does she need to run?
Use models or pictures to show your work.

15. Zola had a roll of wrapping paper 3.2 m long.
She cut off 1.9 m to wrap some presents.
How long was the piece Zola had left?
Use models or pictures to show your work.

6

16. A black-billed magpie has a length of 4.8 dm.
Its wingspan is 1.5 dm longer than its length.
Use blocks to explain how you would add to find its wingspan.

6 **17.** A red-winged blackbird has a length of 2.3 dm and a wingspan of 3.2 dm. Use blocks to explain how you would subtract to find the difference between its wingspan and length.

red-winged blackbird
body length 2.3 dm
wingspan 3.2 dm

18. A blue jay has a wingspan of 4.1 dm. A gray jay has a wingspan that is 0.5 dm longer. Use blocks to explain how you would add to find the wingspan of a gray jay.

7 **19.** Write a fraction and a decimal for the shaded part.

a) 

b) 

c) 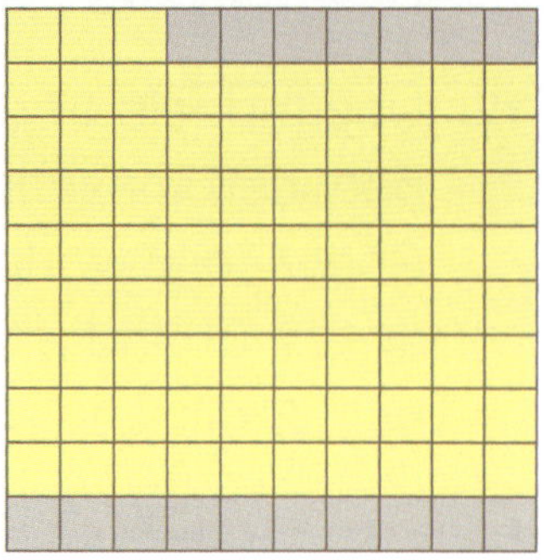

20. Write each fraction as a decimal.

a) $\frac{50}{100}$ b) $\frac{9}{100}$ c) $\frac{63}{100}$ d) $\frac{90}{100}$

21. Write each decimal as a fraction.

a) 0.67 b) 0.38 c) 0.06 d) 0.70

8 **22.** Add or subtract.
Use a grid.

a) 0.21 + 0.34
b) 0.09 + 0.80
c) 0.89 + 0.02
d) 0.62 + 0.17
e) 0.99 − 0.02
f) 1.00 − 0.32
g) 0.87 − 0.45
h) 0.32 − 0.12
i) 0.56 + 0.34

9 **23.** Manitok erased the units of measurement.
Write each unit.

a) My calculator is 8 ____ 6 ____ or 8.6 ____ wide.
b) A computer screen is 3 ____ 2 ____ or 3.2 ____ wide.
c) The window in my bedroom is 6 ____ 4 ____ or 0.64 ____ wide.

LESSON

Problem Bank

1 **1.** If $\frac{3}{10}$ of a circle is red and the rest is green, which colour would cover more area? Show how you know.

2 **2.** If a red trapezoid represents 0.3, how many green triangles would represent 0.7? Draw a picture to show how you know.

3. A junk drawer has 5 pencils, 2 erasers, 1 pencil sharpener, and 2 crayons. Write decimals and fractions that tell about the things in the junk drawer.

3 **4.** Allison is thinking of a decimal. It is 1 tenth greater than the decimal that is halfway between 2.8 and 3.4. What is the decimal?

4 **5.** Which of these decimals add to 5.0?

1.2   0.8   4.2   1.0   2.2   1.7   0.3   0.6

Find more than 1 group of decimals that add to 5.0. Show how you know.

5 **6.** The sum of 2 decimals is 0.8. Their difference is 2 tenths. What are the decimals? Show how you know.

6 **7.** A baby deer leaps 1.4 m. Its mother leaps 4 m. How much farther did the mother leap than the baby? Show how you know.

8 **8.** The sum of 2 decimals is 0.33. One decimal is represented by twice as many squares on a 100 grid as the other decimal. What are the decimals? Show how you know.

LESSON

Chapter Review

1 **1.** 10 friends went to the Square Pizza Shop. Show how the friends could share 1 pizza equally.

2. Write fractions, decimals, and words for the shaded parts of the fraction rectangles.

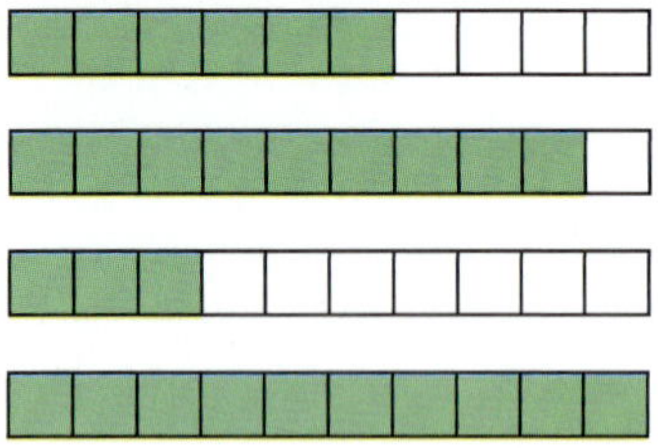

a)
b)
c)
d)

2 **3.** Tell whether each number is closer to 0, $\frac{1}{2}$, or 1.

a) 0.4 b) $\frac{3}{10}$ c) 0.9 d) $\frac{1}{10}$

4. This nursery rhyme has been popular for more than 2 centuries.
Suppose the old lady had 3 rings on each hand and 2 bells on each foot.

a) Describe the lady's rings using words, fractions, and decimals.

b) Describe the lady's bells using words, fractions, and decimals.

Ride a cock-horse
to Banbury Cross
To see an old lady
get up on her horse.
Rings on her fingers
and bells on her toes,
She will make music
wherever she goes.

3 **5.** a) Draw a picture to show a decimal number that is between 3 and 4.

b) Explain why more than 1 decimal is possible.

6. Complete by using >, =, or <. Explain your thinking. Use pictures or a tape measure to help you.

a) 2.7 m ■ 2.0 m b) 0.4 cm ■ 4.0 cm c) 3.1 dm ■ 2.3 dm

4 **7.** Use models or pictures to add these lengths.

a) 0.6 m and 1.5 m **b)** 2.0 dm and 1.4 dm

8. A rabbit hops 2.4 m to the right.
Then the rabbit hops another 1.1 m to the right.
How far is the rabbit from where it began?
Use models or pictures to show your work.

5 **9.** Use models or pictures to find the difference between these lengths.

a) 1.8 m and 0.9 m **b)** 3.0 dm and 2.3 dm

10. What are the missing digits? Show how you know.
Use models or pictures.

a)
$$\begin{array}{r} 2.9 \\ +\ \blacksquare.7 \\ \hline 4.6 \end{array}$$

b)
$$\begin{array}{r} 4.\blacksquare \\ -\ 3.6 \\ \hline 0.5 \end{array}$$

eastern screech owl
body length 2.2 dm
wingspan 5.1 dm

6 **11.** An eastern screech owl is 2.2 dm long and has a wingspan of 5.1 dm. Use blocks to explain how you would subtract to find the difference between its wingspan and length.

12. A crow is 4.5 dm long. Its wingspan is 5.4 dm longer than its length. Use blocks to explain how you would add to find its wingspan.

crow
body length 4.5 dm

7 **13.** Write each as a fraction and a decimal.

a) thirty three hundredths **b)** nine hundredths

8 **14.** Add or subtract using a 100 grid.

a) 0.54 + 0.09 **b)** 0.85 − 0.66 **c)** 1.00 − 0.37

9 **15.** Write each measurement with a decimal point so that the sentence is reasonable.

a) My pen is 135 cm long.
b) My father is 18 m tall.
c) The width of this page is about 20 dm.
d) The perimeter of my bedroom is 155 m.

Chapter Task

Decimal Kites

Shani's Kites

I designed these 4 kites using different shapes and colours.

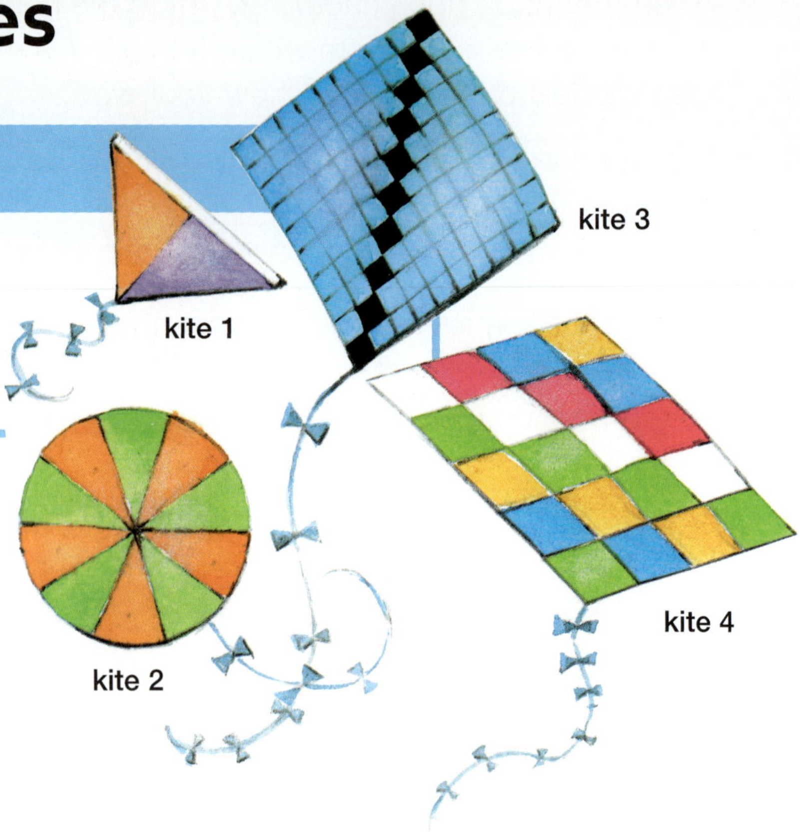

? How can you design and describe a kite?

Part 1

A. For each of Shani's kites, describe the shape and express each color as a fraction.

B. Express each colour fraction as a decimal.

Part 2

C. Draw your own kite.

D. Describe your kite completely using words and numbers.

Task Checklist

- ✓ Did you include a diagram?
- ✓ Did you describe your kite using fractions and decimals?
- ✓ Did you use math language?

CHAPTER 13

Probability

Goals

You will be able to

- **describe the probability of events using mathematical language**
- **compare the probability of events**
- **conduct probability experiments**
- **predict the probability of events**
- **solve probability problems using tree diagrams**

Spinning

Getting Started

What's Likely?

? How can you describe the probabilities of events?

A. Who is likely to be right? Explain your thinking.

B. Who is likely to be wrong? Explain your thinking.

C. Describe the probability of each **event**.
Use a word from 1 of the cards below.

a) I will have a sandwich for lunch tomorrow.
b) The teacher will be away from school next week.
c) Tomorrow will be cloudy.
d) It will rain in the first week of May.
e) We will go on a class trip soon.
f) Every day will be pizza day in school.
g) Tomorrow will be warmer than today.
h) It will be sunny in the morning.
i) It will rain on sports day.
j) Our school volleyball team will be the champions.

likely	unlikely	probably	expected

Do You Remember?

1. a) Describe 2 events that are likely to happen at school this week.
   b) Describe 2 events that are unlikely to happen at school this week.
   c) Describe 2 events that you expect to happen at school this week.
2. Describe 2 other events that might happen at school. Which of the events is more likely to happen?
3. When you toss a coin, what 2 events are equally likely to happen?

1 Probability Lines

Goal **Use a probability line to compare the probability of events.**

? Where do Mandy's events go on the probability line?

You will need

- letter cards

- chart paper

Mandy's Events

I made a list of events in a chart with the probability of each event.
I will probably phone Rey this week, so event A is likely.

Events	Probability
A I will use the phone this week.	likely
B I will go to the North Pole this year.	very unlikely
C I will stay home or go away for vacation.	equally likely
D I will go to school on Monday.	very likely
E I will have a birthday this year.	certain
F I will sit and run at the same time.	impossible
G I will miss the bus after school.	unlikely
H I will watch television tonight.	possible
I I will go to the park tomorrow.	uncertain

I can place each event on a **probability line** between impossible and certain. Event A is likely, so I put it closer to certain than impossible.

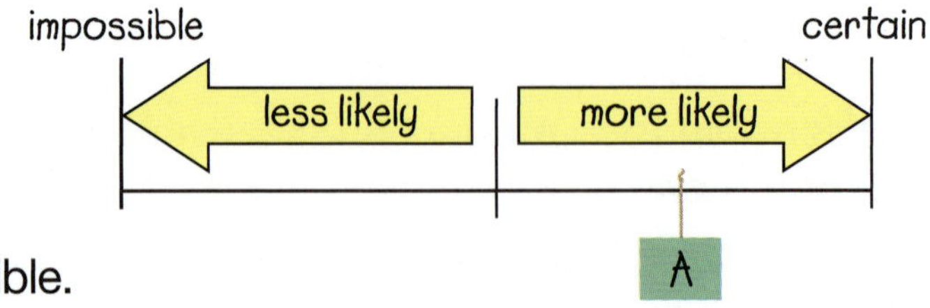

A. Finish placing Mandy's events.

Reflecting

1. a) Which events are closer to **impossible** than **certain**?
   b) Which events are closer to certain than impossible?
2. a) Which events are **possible**?
   b) Which events are **uncertain**?
3. Look at 2 events that are beside each other. Discuss which event is more unlikely.

Checking

4. Use probability words to describe the probability of each event.
   a) I will have eggs for breakfast.
   b) I will be allowed to stay up late tonight.
   c) I will have a party this year.
   d) A tree will talk to me today.
   e) I will go swimming this summer.
5. Draw a probability line. Label it with the letters of the events from Question 4.

Practising

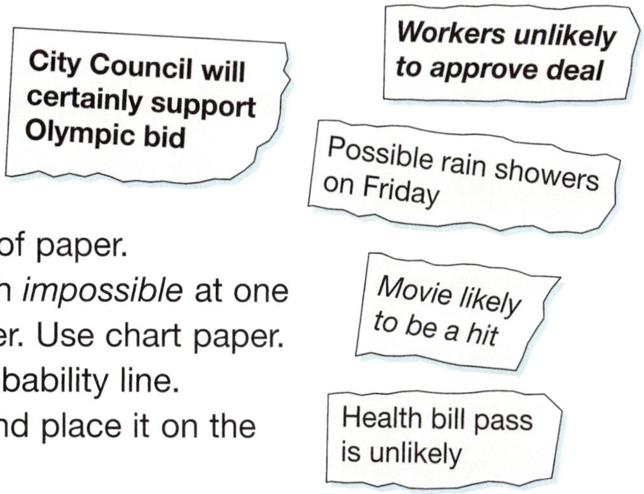

6. a) Use these headlines or go through newspapers to find events with probability words.
   Write the events on strips of paper.
   b) Make a probability line with *impossible* at one end and *certain* at the other. Use chart paper. Glue the events on the probability line.
   c) Write your own headline and place it on the probability line.

CHAPTER 13

2 Experimenting with Spinners

You will need

- spinners

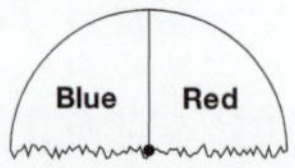

- a paper clip

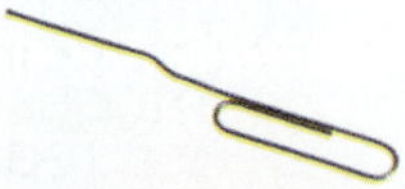

Goal **Make predictions and experiment with spinners with equal sections.**

Jon got a board game for his birthday.
The game has 2 spinners in it.
A player who spins **red** scores a point.

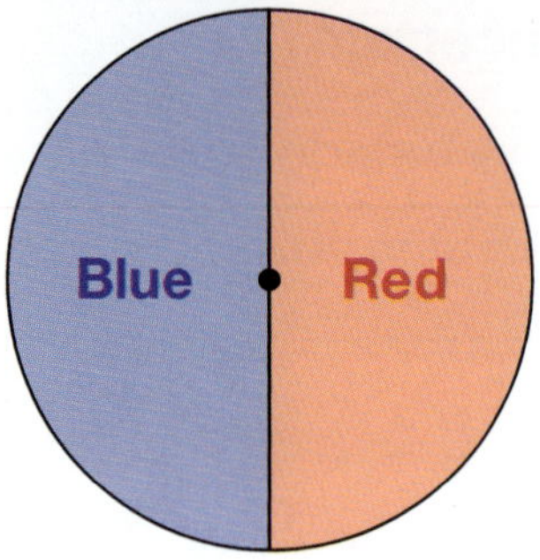

Spinner X

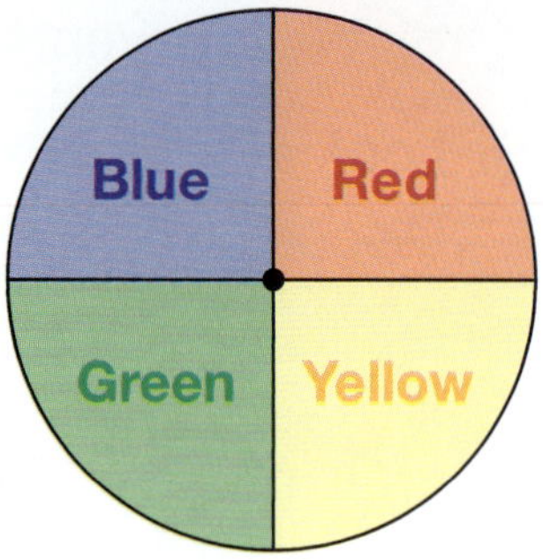

Spinner Y

? Which spinner should Jon choose?

Jon's Experiment

I spun Spinner X 20 times.
I recorded the result of each spin.
Then I made a tally chart.

B, R, B, B, B, R, B, R, R, R,
R, B, B, B, R, B, R, R, B, B

Red	Blue
𝍸 ||||	𝍸 𝍸 |

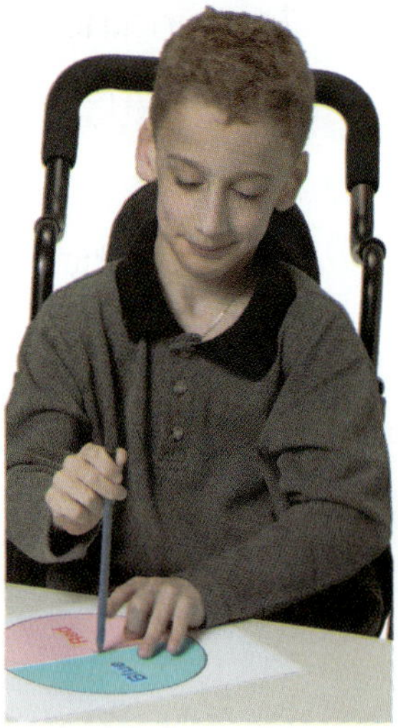

A. Jon says that **red** and **blue** are **equally likely** on Spinner X. Explain why he thinks so.

B. Spin Spinner Y 20 times.
Record the results in a tally chart.

C. Which spinner is more likely to spin **red**?
Give a reason for your answer.

Reflecting

1. Use probability words to describe the probability of spinning **green** with Spinner X.
2. Which spinner should Jon choose to score points? Give a reason for your choice.

Checking

3. Use Spinner Z.
   a) Predict the number of times you will spin **red** in 20 spins on Spinner Z.
   b) Spin 20 times and record the results.
   c) Compare the results with your prediction.
   d) Compare the probabilities of spinning **red** and **blue**.
   e) Is spinning **red** more likely on Spinner X or Spinner Z? Give a reason for your answer.

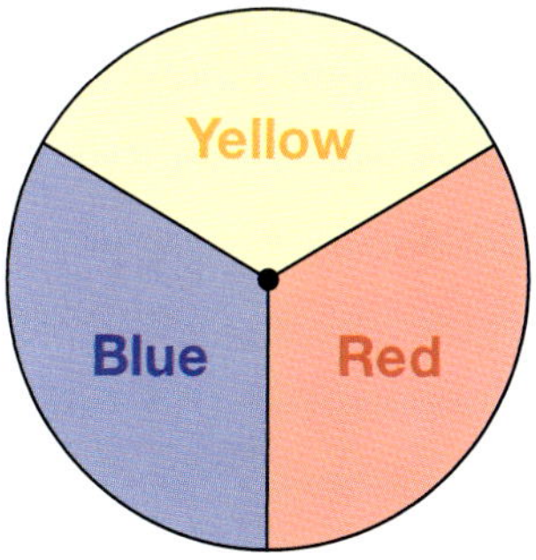

Spinner Z

Practising

4. a) Predict the number of times you will spin an even number in 20 spins on Spinner P.
   b) Spin 20 times and record the results. Compare the results with your prediction.

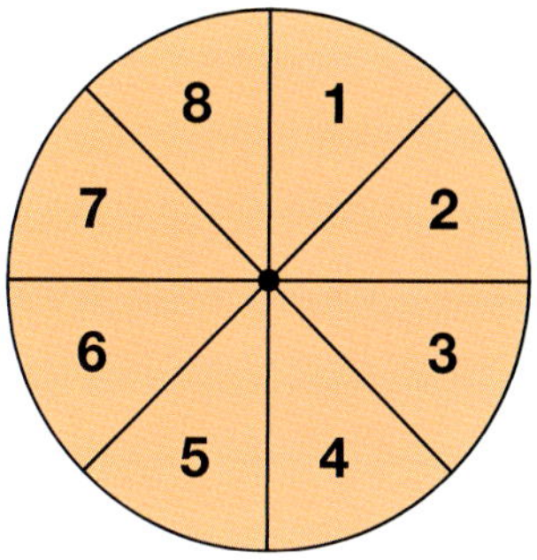

Spinner P

5. Use probability words to describe the probability of each spin on Spinner Q.
   a) spinning an even number
   b) spinning an odd number

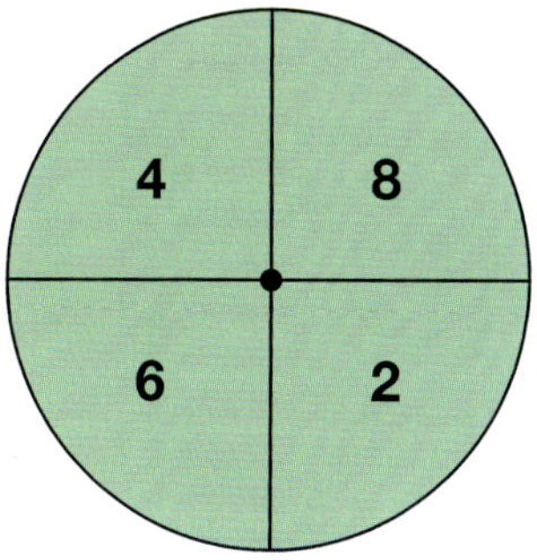

Spinner Q

6. a) Make up your own question about spinning Spinner P, Q, or Z.
   b) Experiment to answer your question. What happened?

3 Making Predictions

You will need

- a paper bag

- red and blue tiles

Goal **Make predictions and design and carry out experiments.**

Suppose you put 2 colours of tile in a paper bag.

? What colour tile are you more likely to draw from the bag?

Experiment 1

A. Put 5 **red** tiles and 5 **blue** tiles in a paper bag.

B. Predict the number of tiles of each colour you will pick in 20 draws.

C. Draw a tile from the paper bag. Record its colour in a tally chart like this. Replace the tile.

D. Repeat the experiment 20 times.

E. Compare the results to your prediction.

Prediction: ____ Red	____ Blue
Red	**Blue**
Result: ____ Red	____ Blue

Experiment 2

F. Sketch a bag that contains 10 tiles. Design the bag so that drawing a **blue** tile is more likely than drawing a **red** tile.

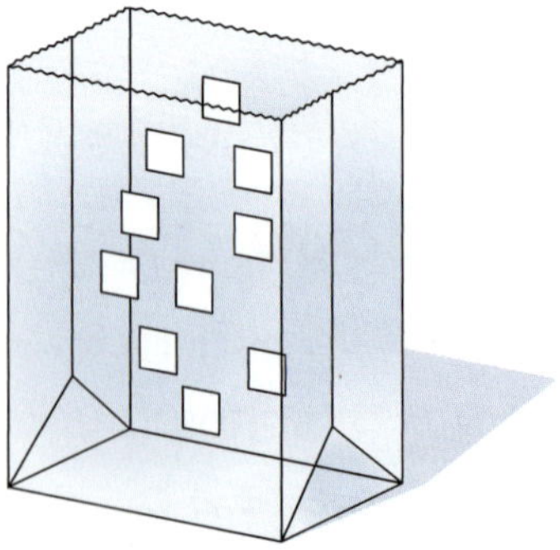

G. Explain your strategy for designing the bag.

H. Predict the result of 20 draws from your bag.

I. Put the tiles in the bag and draw from the bag 20 times. Replace the tile after each draw. Record your results in a tally chart.

J. Compare the results to your prediction.

Reflecting

1. a) When you draw a tile from the bag, what is certain about its colour?
   b) What is uncertain about its colour?
   c) What colour is impossible to draw?
2. Suppose you put 3 more red tiles in the bag for Experiment 1. How might the **outcomes** change? Explain why this might happen.
3. Make up a question about drawing tiles from a bag. Explain how you could experiment to answer your question.

Math Game

Predicting Tiles

You will need

- a paper bag
- red and blue tiles

Number of players: 2
How to play: Predict the colour of the tile you will draw.

Step 1 Put 7 red tiles and 3 blue tiles in a bag.

Step 2 Both players predict what colour tile will be drawn.

Step 3 Draw a tile from the bag.

Step 4 Score 1 point if the colour of the tile matches a prediction of red.
Score 2 points if the colour of the tile matches a prediction of blue.

Step 5 Replace the tile.

Play until a player has 10 points.

Mid-Chapter Review

LESSON

1 **1.** Draw a probability line and place these words along it.

impossible	unlikely	likely
certain	very likely	

2. Use a probability word to describe each event.

a) There will be a fire drill today.
b) The next time I am in a car, I will wear a seat belt.
c) A dog will come to school today.
d) I will grow to be 3 m tall.

2 **3.** A board game has these 2 spinners.

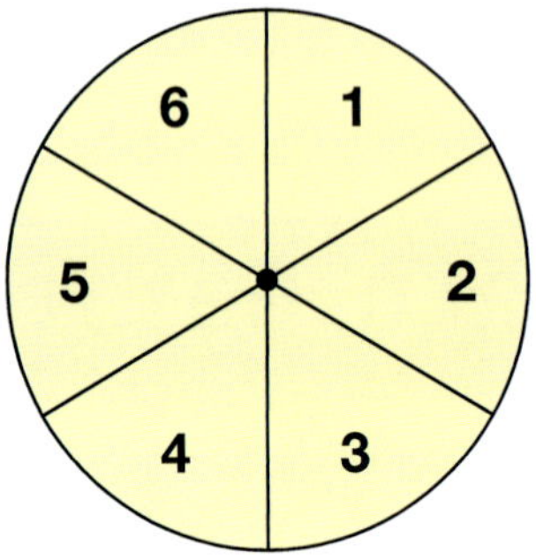

Spinner E

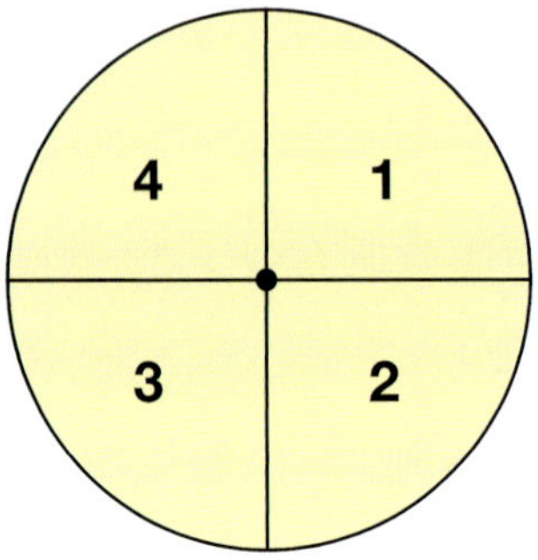

Spinner F

a) Use probability words to describe the probability of spinning 2 on each spinner.
b) Predict the number of times you would spin 2 in 20 spins of each spinner. Explain your prediction.

3 **4.** Sketch a draw bag to match each description. Each bag should contain 10 tiles.

a) **Red** is certain to be drawn.
b) **Black** is impossible to be drawn.
c) 2 colours are equally likely to be drawn.
d) 1 colour is more likely to be drawn than another colour.

Choose Your Spinner

Number of players: 2 to 4
How to play: Take turns spinning and moving your marker toward **Finish**.

Start	Green	Red	Blue	Green
Red	Blue	Green	Red	Blue
Green	Red	Blue	Green	Red
Blue	Green	Red	Blue	Green
Red	Blue	Green	Red	Finish

Step 1 All players begin by placing a marker on **Start**.

Step 2 Choose a spinner and spin it.

Step 3 Move to the next square only if you spin its colour. You can move vertically or horizontally, but not diagonally.

The first player to arrive at **Finish** wins.

You will need

- a coloured marker for each player

- a game board
- 3 spinners
- paper clips

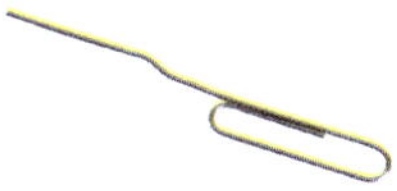

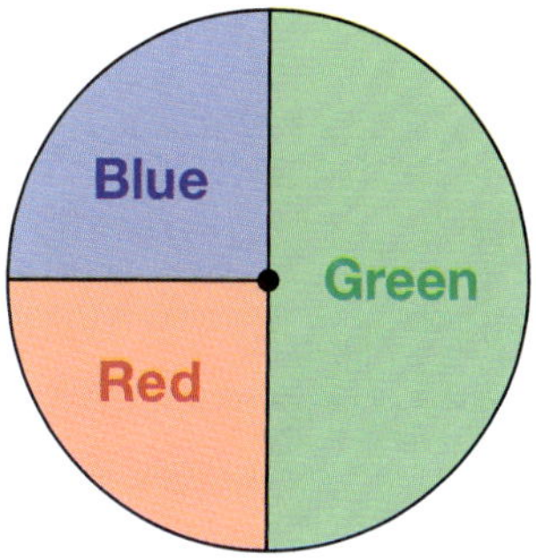

Spinner 1

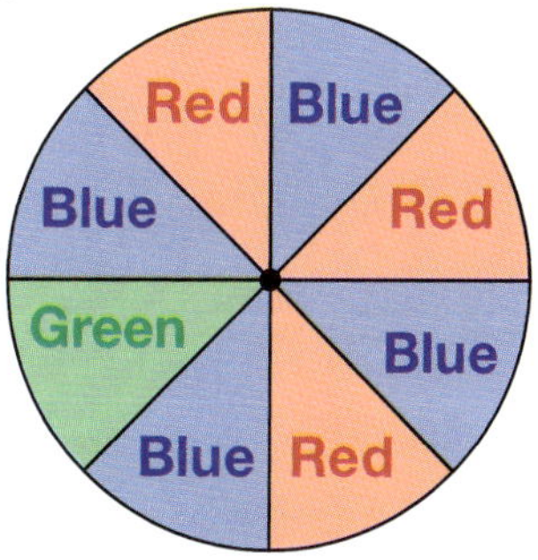

Spinner 2

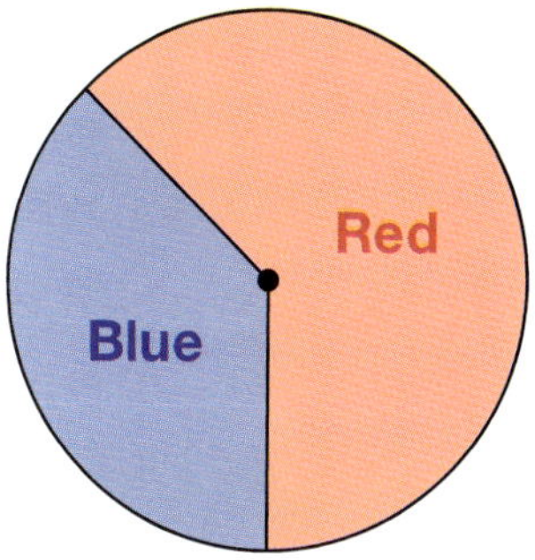

Spinner 3

4 Comparing Probabilities

You will need

- spinners

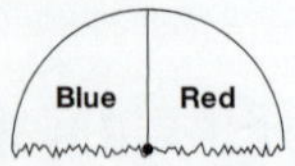

- a paper clip

Goal **Make predictions and experiment with spinners with unequal sections.**

Sarah can win a stuffed bear by spinning **blue** at a Playland booth.

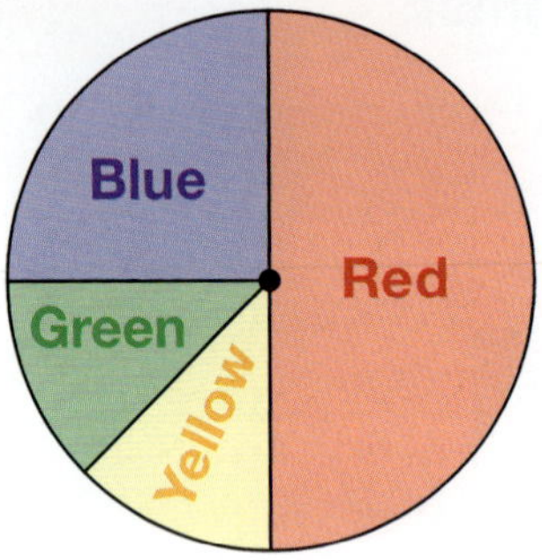

Playland Spinner

? How likely is it that Sarah will win with 1 spin?

A. Predict the number of times you will spin **blue** in 40 spins on the Playland Spinner.

B. Spin 40 times. Record the results in a tally chart.

C. Compare the results with your prediction.

D. Is spinning **red** more likely than spinning **green**? Explain your thinking.

E. Is spinning **blue** more likely than spinning **green**? Explain your thinking.

F. What 2 spins are **equally likely**?

G. Place the colours (**R**, **Y**, **G**, **B**) on a probability line. Explain your placement of each colour.

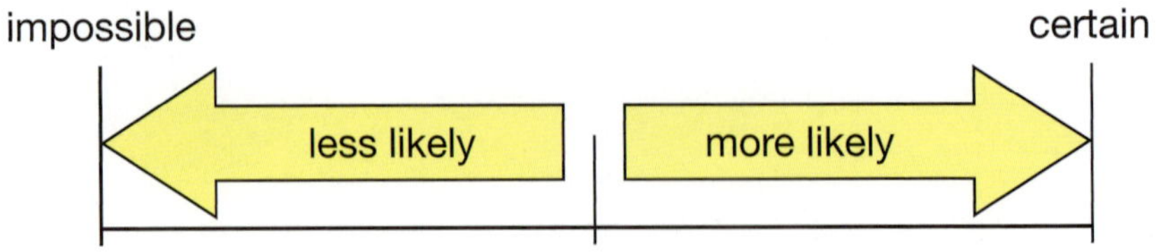

Reflecting

1. How likely is it that Sarah will win with 1 spin?
2. Suppose Vinh wins a stuffed bear by spinning **red**. Is this game fair when Sarah wins by spinning **blue**? Explain your thinking.

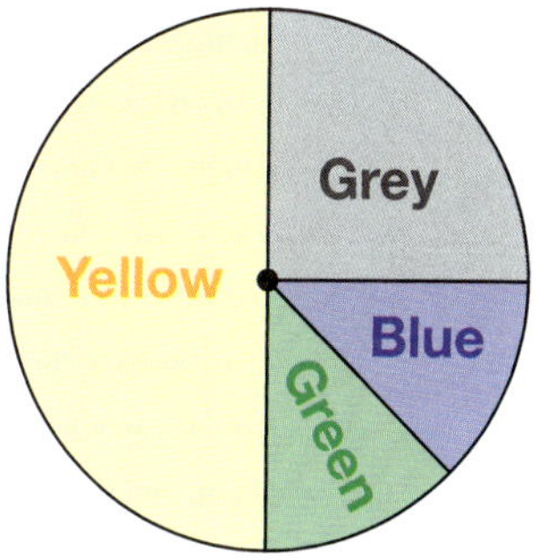

Funland Spinner

Checking

3. Sarah can also win a stuffed bear by spinning **blue** at a Funland booth.
   a) Make a probability line for spinning the colours.
   b) Which spin is most likely?
   c) Which spins are equally likely?
   d) Which Playland colour is impossible to spin at Funland? Explain your thinking.
   e) Suppose Sarah plays the game at Playland and Vinh plays at Funland. Is this fair? Explain your thinking.

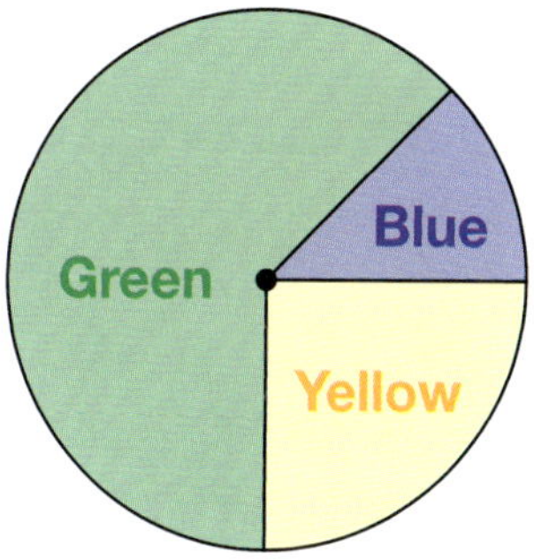

Spinner S

Practising

4. A game has 2 spinners, S and T. You get 1 spin. Explain your thinking for each question.
   a) Which colours on Spinner T are impossible to spin on Spinner S?
   b) Which colour is equally likely on Spinner S and Spinner T?
   c) Make a probability line for Spinner S and for Spinner T.

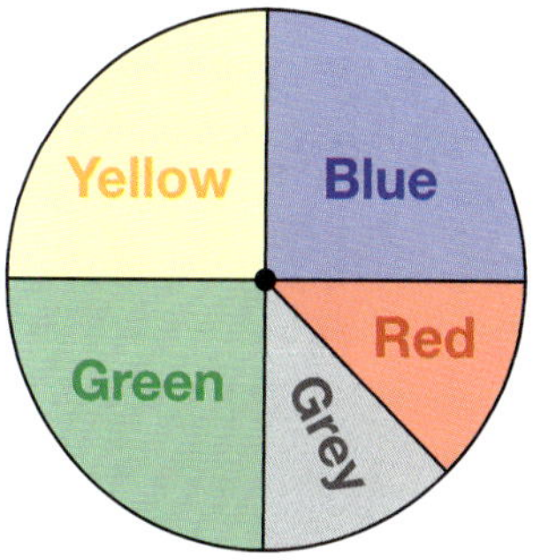

Spinner T

5. Suppose you are playing a game. You get a point for spinning **blue** and another player gets a point for spinning **green**.
   a) Is the game fair with Spinner S? Why or why not?
   b) Is the game fair with Spinner T? Why or why not?

5 Creating Spinners

Goal Design spinners to meet given conditions and test the spinners.

You will need

- spinners

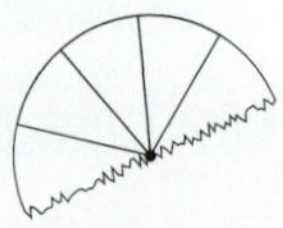

- a paper clip

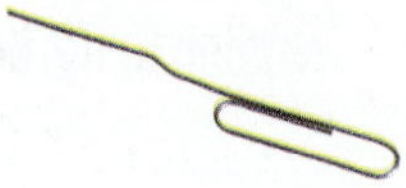

? How can you design a spinner to match given probabilities?

Understand

The spinner should match 1 of the cards.

A. If 1 colour is more likely than another, what do you know about the sections?

Make a Plan

B. Choose 1 card.

C. How many colours should the spinner have? Decide on your colours.

D. Decide what size each colour section will be.

Carry Out the Plan

E. Create your spinner.

Look Back

F. Test your spinner by spinning it 20 times. Record the results in a tally chart.

G. Choose another card. Design and test another spinner.

2 colours
1 colour is more likely than the other.

3 colours
Blue is very likely. Red is impossible.

3 colours
2 colours are equally likely. Another colour is less likely.

4 colours
All colours are equally likely.

Yellow is certain.

Reflecting

1. Could you have designed your spinners differently? Explain your thinking.
2. Can you design 1 spinner to match more than 1 card? If so, explain how.

Curious Math

Fair Game

1 Vinh, Miki, and Pedro are using this spinner for a game. Vinh scores 1 point for spinning **green**. Miki scores 1 point for spinning **yellow**. Pedro scores 1 point for spinning **blue**. Is the game fair? Explain your thinking.

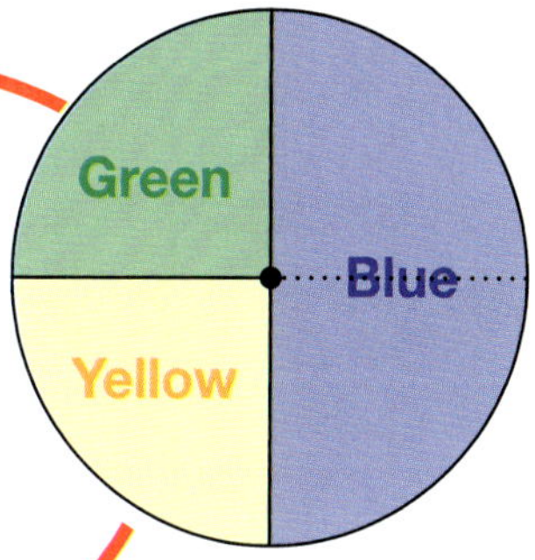

2 Allison and Josef are playing a game. Allison scores 2 points for spinning **red**, and Josef scores 1 point for spinning **blue**.

a) Sketch 2 spinners that are fair for this game. Explain how your 2 fair spinners are the same.

b) Sketch 2 spinners that are unfair for this game. Explain how your 2 unfair spinners are different.

3

a) Explain when this spinner would not be fair for a game.

b) Suppose half the spinner was changed to **blue**. Explain when it would not be fair for a game.

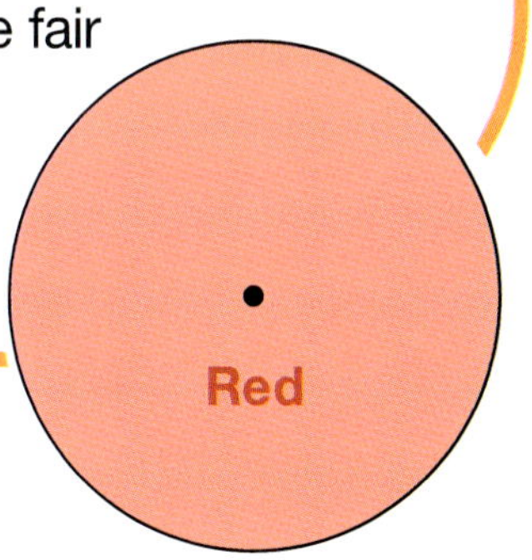

4 Explain when tossing a coin would be fair for a game.

6 Solve Problems Using Tree Diagrams

Goal **Use tree diagrams to find all possible combinations.**

Calvin's class brought white bread and brown bread, turkey, tuna, and cheese to make sandwiches for a trip.

? How many types of sandwiches are possible?

Calvin's Tree Diagram

Understand

I assume that each sandwich is made with 1 type of bread and 1 filling.

tree diagram
A way to record and count all combinations of events

Make a Plan

I will draw a **tree diagram** to find all the combinations of breads and fillings.

Carry Out the Plan

This is my diagram.

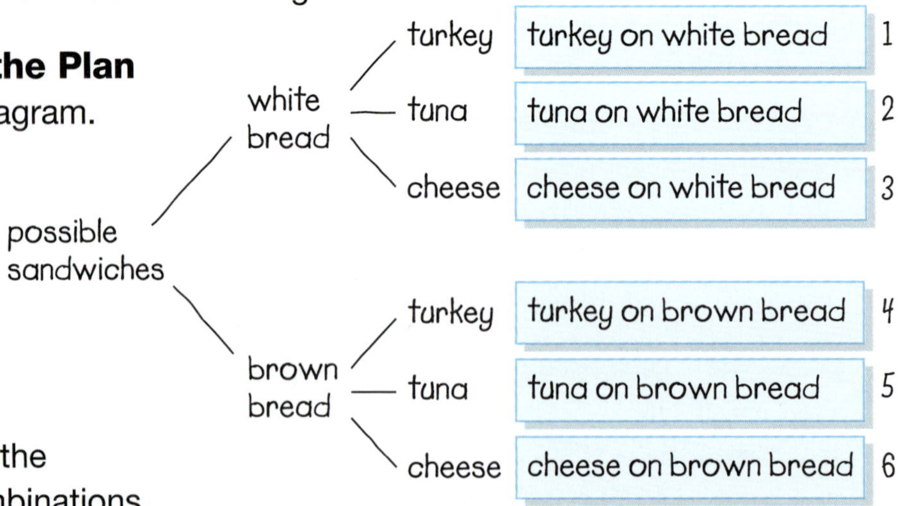

I counted all the possible combinations.

There are 6 possible sandwiches.

Reflecting

1. How did the tree diagram help Calvin solve the problem?
2. What kind of sorting must you do before you draw the branches of a tree diagram?
3. Suppose Calvin ate one of the sandwiches. Is it more likely that he ate a sandwich on white bread, or a turkey sandwich on brown bread? Explain using the tree diagram.

Checking

4. The students brought muffins and icing.
Muffins: carrot, banana
Icing: cream cheese, chocolate, strawberry, vanilla
   a) Make a tree diagram to show all the possible combinations of muffins and icing.
   b) How many combinations are possible?
   c) A student eats a muffin with icing.
   Is it more likely that the muffin was
   • a banana muffin with any icing, or
   • a carrot muffin with cream cheese icing?
   Explain your answer.

Practising

5. The students brought chips and dips.
Chips: corn, salted potato, plain potato
Dips: salsa, sour cream, chick pea
   a) How many combinations are possible?
   b) A student dips and eats a chip.
   Is it less likely that the chip and dip were
   • a corn chip with salsa, or
   • any potato chip with sour cream?
   Explain your answer.

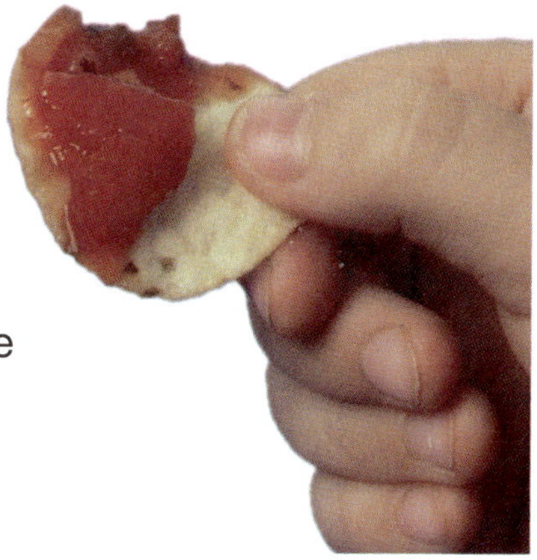

Spinning Decimals

1. **Blue** is 0.10 of this spinner.
   a) How can you tell that **yellow** is greater than 0.50?
   b) What is the sum of these 3 decimals?
   c) Estimate the decimals for **yellow** and **red**.

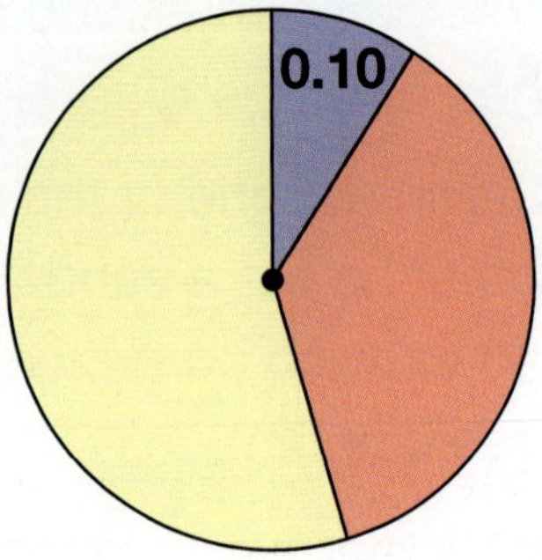

2. **Blue** is 0.50 of this spinner. Estimate the decimals for the other 2 colours.

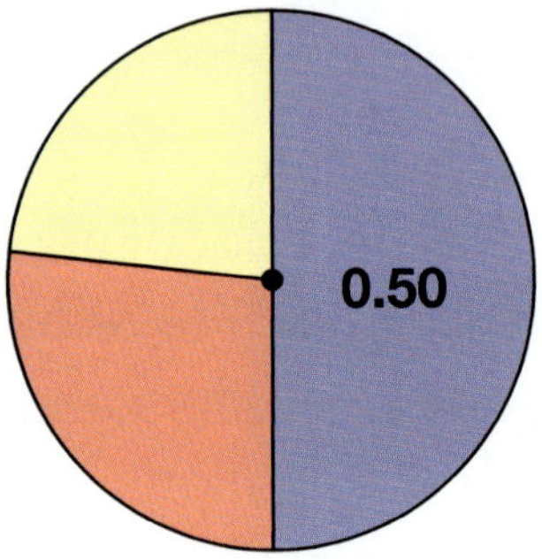

3. a) Trace this circle. Divide it into 3 or more parts.
   b) Estimate the decimal for each part of the circle.

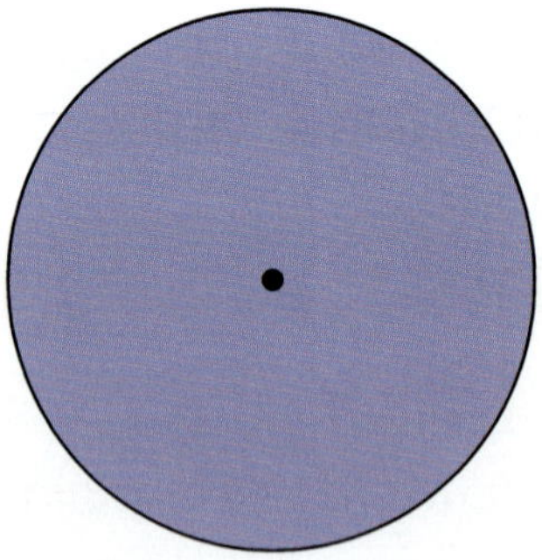

Skills Bank

LESSON

1

1. a) Draw a probability line.
Label it with the letters of these events.

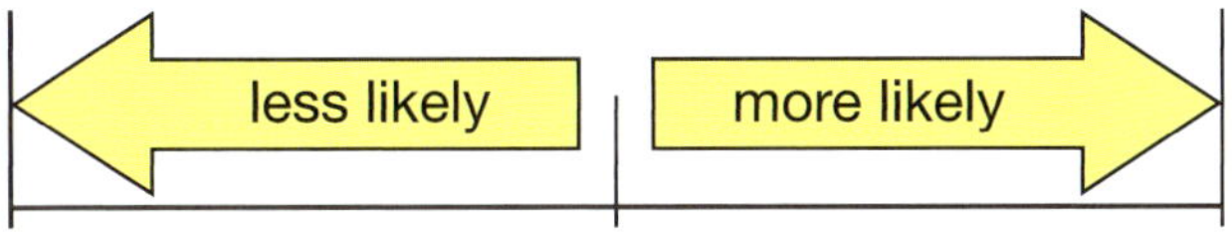

A. Everyone who lives in Calgary will ski this winter.
B. It will rain tomorrow.
C. At least 1 teacher in Vancouver will give homework tonight.
D. Most grade 4 students in Canada are in school today.

b) Make up an event to match each probability word. Add the new events to the probability line from part a).
E. very likely
F. certain
G. unlikely
H. uncertain

2

2. Predict the number of times you will spin a number greater than 5 in 20 spins.

a)
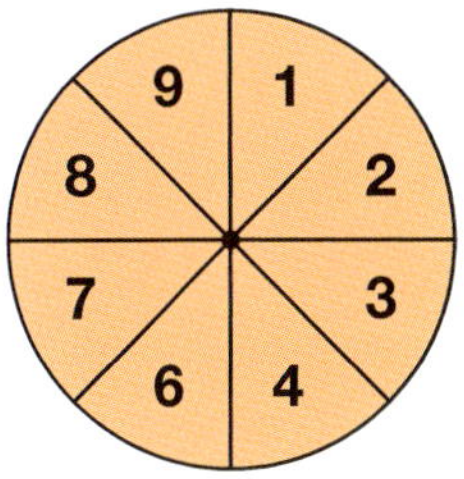

b)
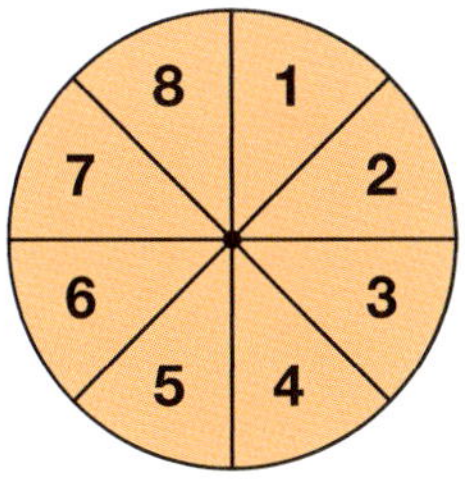

c)
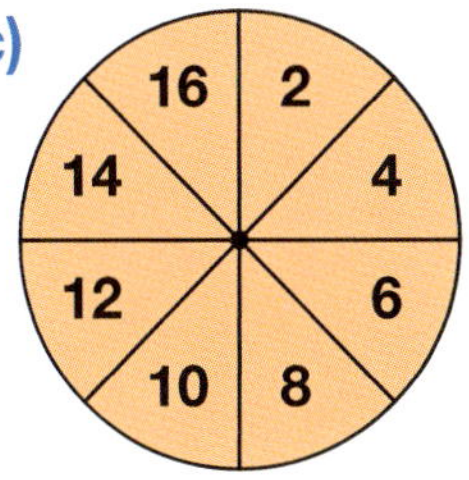

3. Use a probability word to describe the probability of each spin.

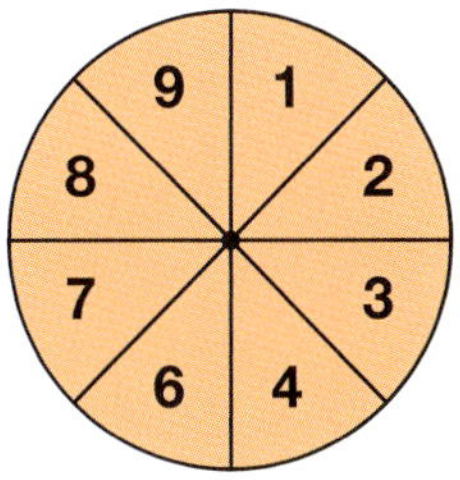

a) a number less than 9
b) a number less than 6
c) 6
d) a number greater than 9
e) an even number
f) a multiple of 3

3

4. Predict the number of tiles of each colour you will get in 20 draws.

a) 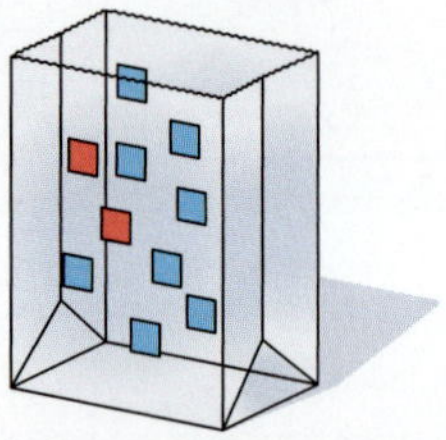

b) 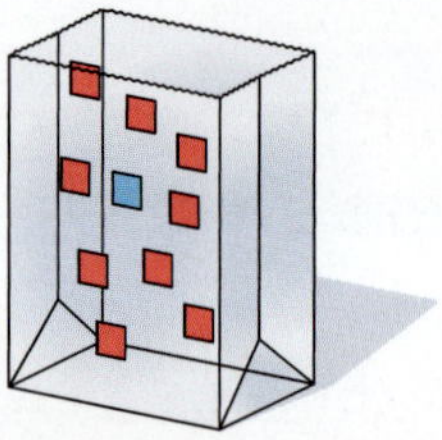

c) 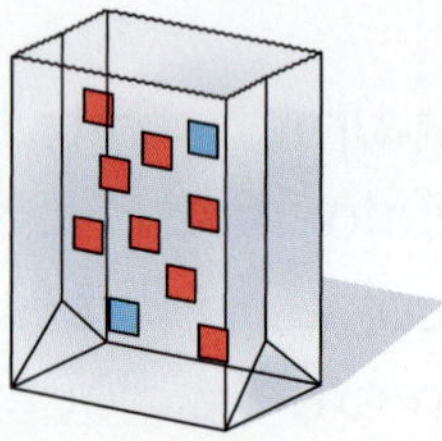

4

5. A game has 2 spinners, S and T.
Each player gets 1 spin.

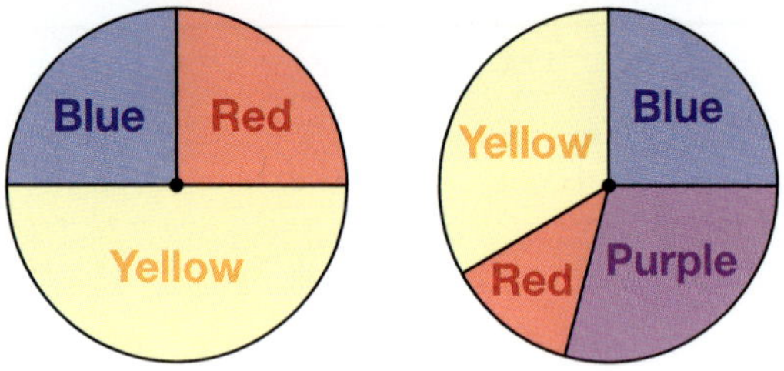

a) Is spinning **red** more likely on Spinner S or Spinner T?
b) On which spinner is **yellow** less likely?
c) Which colour on Spinner T is impossible to spin on Spinner S?
d) Make a probability line for each spinner.
e) Which colour is equally likely on Spinner S and Spinner T?

5

6. Draw 4 spinners, one for each description.
a) On this spinner, **blue** is certain.
b) This spinner has 2 colours.
1 colour is less likely than the other.
c) This spinner has 3 colours.
Red, **blue**, and **green** are equally likely.
d) This spinner has 3 colours.
Yellow is impossible. **Red** is very likely.
Blue is possible. **Green** is less likely than blue.

Problem Bank

LESSON

3

1. Sketch a draw bag to match each description.
   a) The colour **green** is certain to be drawn.
   b) The colour **red** is impossible to draw.
   c) There are 3 colours. All 3 colours are equally likely to be drawn.

2. a) Is it possible to design a draw bag to match all 3 descriptions in Question 1?
   Explain your thinking.
   b) Is it possible to design a draw bag to match 2 of the descriptions in Question 1?
   Explain your thinking.

5

3. The planners of this year's school fun fair want to include a fish pond booth.
   The pond will have 40 fish in it.
   There will be 4 colours of fish.
   Design the pond so that 3 colours of fish are equally likely to be caught.
   The 4th colour of fish wins a big prize, and must be very unlikely.

6

4. In a 3-legged race, each team must have 1 boy and 1 girl.
   a) List all the possible teams.
   b) The names are put in 2 bags, 1 bag for boy names and 1 bag for girl names.
   A boy's name and a girl's name are drawn from the bags.
   Describe the probability that both names start with R.

LESSON

Chapter Review

1 **1.** Describe an event to match each probability.

a) impossible
b) unlikely
c) very likely
d) certain
e) uncertain
f) possible

2 **2.** A game has 2 spinners, G and H.

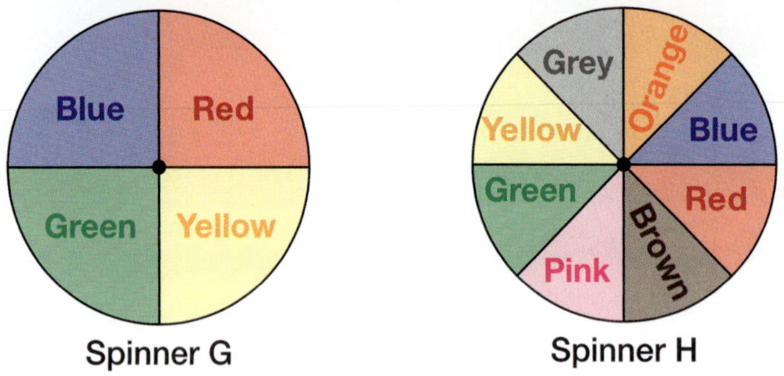

a) Which colours are equally likely on Spinner G?

b) Predict the number of times you will spin **red** in 20 spins of each spinner. Explain each prediction.

c) If you want to spin **green**, which spinner would you choose? Explain your thinking.

3 **3.** A draw bag contains 7 **green** marbles and 3 **black** marbles.

a) Is it more likely to draw a **black** marble or a **green** marble from the bag? Explain.

b) Suppose 4 more **black** marbles are put in the bag. Use probability words to describe the probability of drawing a **green** marble.

4. **a)** Sketch a draw bag with 2 colours of tile. Design the bag so that it is less likely to draw 1 colour than the other colour.

b) Explain why there is more than 1 way to colour the tiles.

4

5. A game booth has 2 spinners, O and P.

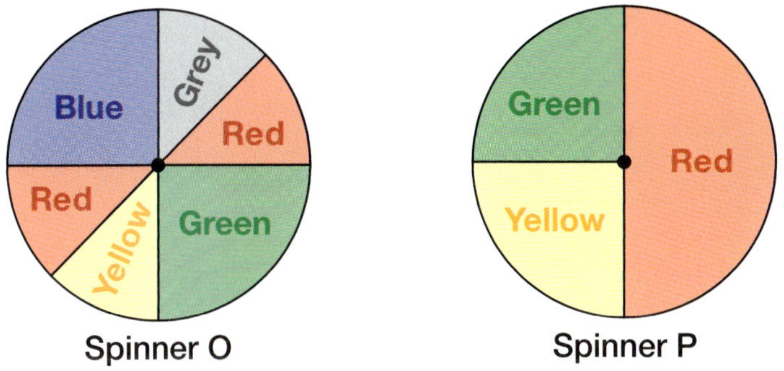

a) On which spinner is spinning **red** more likely?

b) Make a probability line for each spinner. Label each line with the colours of each spinner.

c) Suppose Manitok and Miki are playing a game with Spinner P. Manitok gets a point for spinning **red** and Miki gets a point for spinning **yellow**. Is this fair? Explain reasons for your answer.

5

6. Design a spinner to match each description.

a) There are 3 numbers. All the numbers are equally likely.

b) There are 2 colours. Spinning 1 colour is more likely than spinning the other colour.

c) There are 3 colours. **Red** and **black** are equally likely. **Blue** is less likely.

6

7. Vanessa is making sundaes at her party.
Ice cream: vanilla, chocolate
Toppings: strawberry, pineapple, chocolate, caramel

a) Make a tree diagram of the sundaes that can be made with 1 type of ice cream and 1 topping.

b) Vanessa gives you the sundae she has just made. Which is more likely:
- the sundae has chocolate ice cream and chocolate topping, or
- the sundae has vanilla ice cream and fruit topping?

Chapter Task

Probability Prizes

Paulette wants to design a spinner for her booth at the fun fair.

Paulette's Prizes

I want visitors to my booth to win the prize that they spin.

I drew this probability line to show how the prizes will be won.

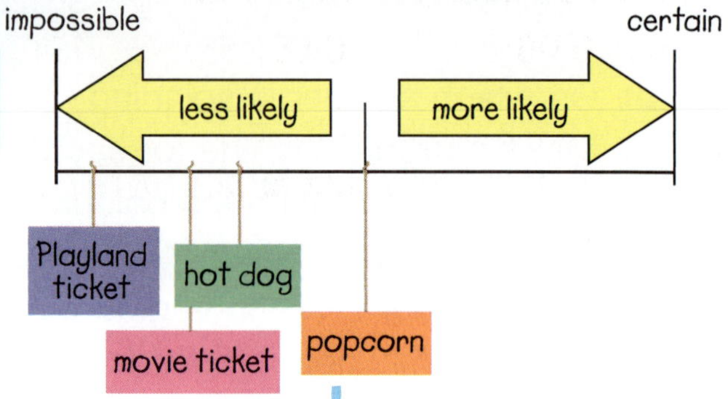

? What spinner would you design for the booth?

Understand

A. Which is more likely, a hot dog or a movie ticket?

Make a Plan

B. How many different sections does the spinner need? Which section should be the biggest?

Carry Out the Plan

C. Gather the materials and make the spinner.

D. Make a prediction. If you spin 40 times, how many times will each prize be won?

E. Conduct the experiment. Compare the results with your prediction.

Look Back

F. Explain how your spinner shows the probabilities on Paulette's probability line.

Task Checklist

- ☑ Did you include diagrams?
- ☑ Did you include the data from your experiment?
- ☑ Did you explain your thinking?
- ☑ Did you use math language?

Cumulative Review

Cross-Strand Multiple Choice

1. Which measurement set is ordered from longest to shortest?
   - A. 2.2 m, 0.2 m, 1.2 m, 2.0 m
   - B. 2.0 m, 1.2 m, 2.2 m, 0.2 m
   - C. 0.2 m, 1.2 m, 2.0 m, 2.2 m
   - D. 2.2 m, 2.0 m, 1.2 m, 0.2 m

2. Which decimal represents $\frac{60}{100}$?
   E. 0.06 F. 0.66 G. 0.60 H. 0.67

3. Which is the sum for the decimal represented by the green squares plus 0.19?

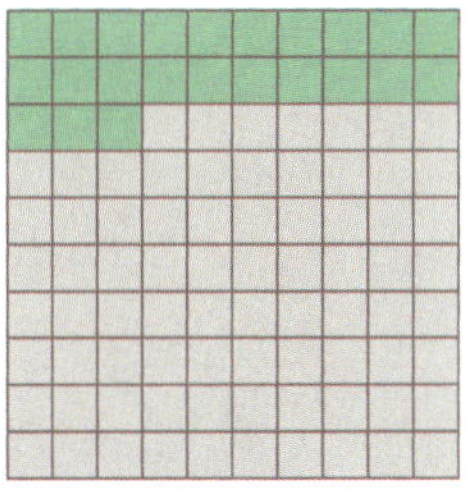

   - A. 0.23
   - B. 0.42
   - C. 0.52
   - D. 4.2

4. Jon spun 20 times and got the results in the chart. Which spinner did he likely use?

Section	Spins
blue	3
red	10
yellow	3
green	4

E.
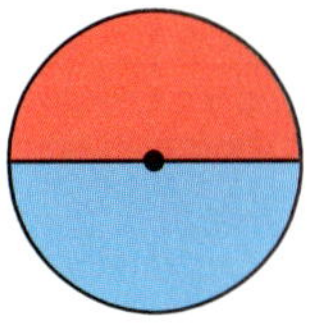
F.
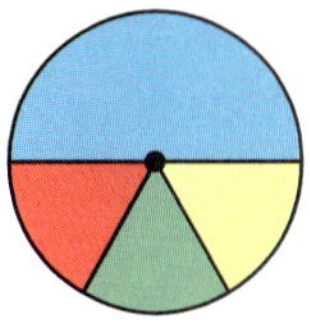
G.
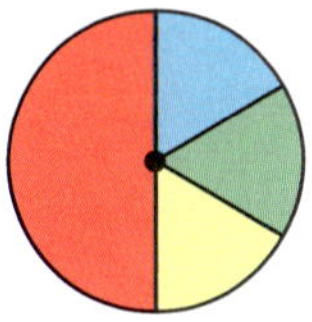
H.
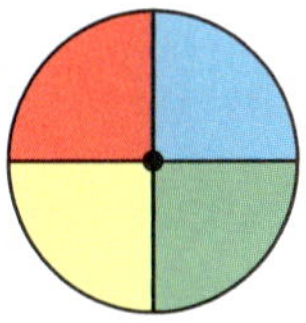

5. Which sentence describes probabilities for this spinner?
   - A. Spinning red is more likely than spinning grey.
   - B. Spinning blue is equally likely to spinning green.
   - C. Spinning yellow is less likely than spinning blue.
   - D. Spinning red is more likely than spinning yellow.

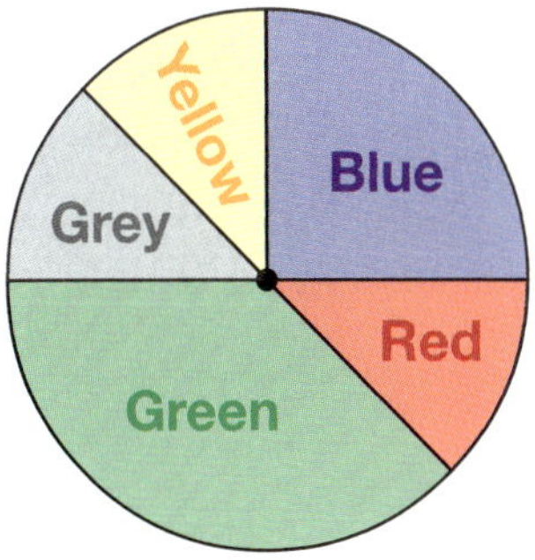

6. How many possible combinations are there for 1 pancake and 1 syrup flavour?
   Pancakes: plain, blueberry
   Syrup: maple, raspberry, blueberry
   E. 6 F. 5 G. 9 H. 2

Cross-Strand Investigation

A Butterfly Conservatory

giant owl butterfly

blue morpho butterfly

monarch butterfly

The number of giant owl butterflies at a conservatory is the same as the number of blue morpho butterflies. There are half as many monarch butterflies.
You want to find out if visitors are more likely to see a monarch butterfly first or a blue morpho butterfly. You will use an experiment to find out.

7. a) Create a spinner with a section for each kind of butterfly. The sections should represent how likely it is to see each kind of butterfly.
   b) First, predict what you will discover. Then spin 20 times and record the results. Compare the results with your prediction.
   c) Pose a different problem about comparing 2 spinner sections. Predict the solution and then spin 20 times. What happened?

This stone tile walkway is incomplete.

8. a) Copy the diagram onto a grid with 5 columns of 20 squares. Create a symmetrical design to fill the empty columns.
   b) Your diagram has 5 columns of 20 squares or 100 squares in all. Write a fraction and a decimal to describe each part.
      i) number of dark brown stones
      ii) number of grey stones
      iii) total number of light brown and dark brown stones
      iv) difference between the number of light brown and dark brown stones

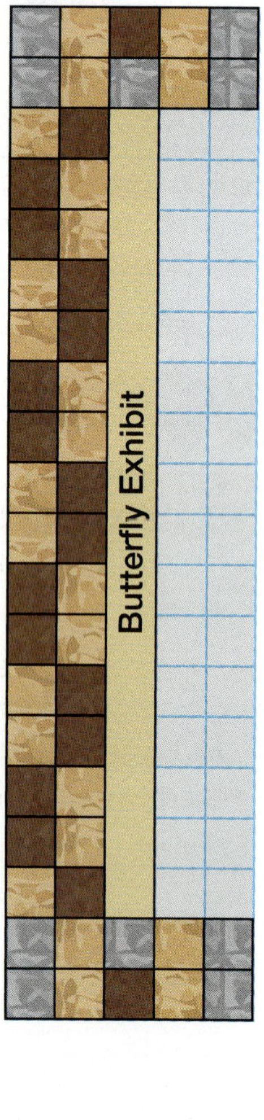

Glossary

Instructional Words

C

calculate: Figure out the number that answers a question; compute

clarify: Make a statement easier to understand; provide an example

classify: Put things into groups according to a rule and label the groups; organize into categories

compare: Look at 2 or more objects or numbers and identify how they are the same and how they are different (e.g., Compare the numbers 6.5 and 5.6. Compare the size of the students' feet. Compare 2 shapes.)

construct: Make or build a model; draw an accurate geometric shape (e.g., Use a ruler to construct a line.)

create: Make your own example

D

describe: Tell, draw, or write about what something is or what something looks like; tell about a process in a step-by-step way

draw: 1. Show something in picture form (e.g., Draw a diagram.)
2. Pull or select an object (e.g., Draw a card from the deck. Draw a tile from the bag.)

E

estimate: Use your knowledge to make a sensible decision about an amount; make a reasonable guess (e.g., Estimate how long it takes to cycle from your home to school. Estimate how many leaves are on a tree. What is your estimate for 3210 + 789?)

evaluate: Determine if something makes sense; judge

explain: Tell what you did; show your mathematical thinking at every stage; show how you know

explore: Investigate a problem by questioning, brainstorming, and trying new ideas

extend: 1. In patterning, continue the pattern
2. In problem solving, create a new problem that takes the idea of the original problem further

J

justify: Give convincing reasons for a prediction, an estimate, or a solution; tell why you think your answer is correct

L

list: Record thoughts or things one under the other or one after the other

M

measure: Use a tool to describe an object or determine an amount (e.g., Use a ruler to measure the height or distance around something. Use balance scales to measure mass. Use a measuring cup to measure capacity. Use a stopwatch to measure time in seconds or minutes.)

model: Show an idea using objects and/or pictures (e.g., Model a number using base ten blocks.)

P

predict: Use what you know to work out what is going to happen (e.g., Predict the next number in the pattern 1, 2, 4, 8, … .)

R

reason: Develop ideas and relate them to the purpose of the task and to each other; analyze relevant information to show understanding

record: Set down work in writing or pictures

relate: Show a connection between objects, drawings, ideas, or numbers

represent: Show information or an idea in a different way that makes it easier to understand (e.g., Draw a graph. Make a model.)

S

show (your work): Record all calculations, drawings, numbers, words, or symbols that make up the solution

sketch: Make a rough drawing (e.g., Sketch a picture of the field with dimensions.)

solve: Develop and carry out a process to find a solution to a problem

sort: Separate a set of objects, drawings, ideas, or numbers according to an attribute (e.g., Sort 2-D shapes by the number of sides.)

V

verify: Work out an answer or solution again, usually in another way to show that the original answer is correct; show evidence of

visualize: Form a picture in your mind of what something is like; imagine

Glossary

Mathematical Words

A

addend: A number that is added to another number

algorithm: A series of steps to carry out a procedure (e.g., add, subtract, multiply, or divide)

a.m.: The time between midnight and noon

analog clock: A clock that measures time using rotating hands

angle: The space between 2 arms that meet at a **vertex** (e.g., The **arc** shows the angle.)

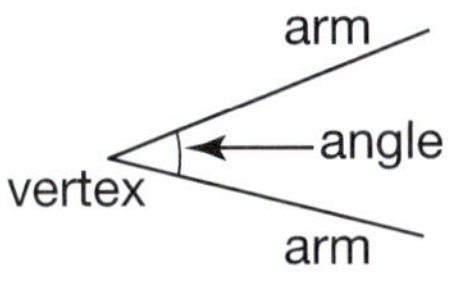

arc: Curved mark

area: The number of square units needed to cover a surface

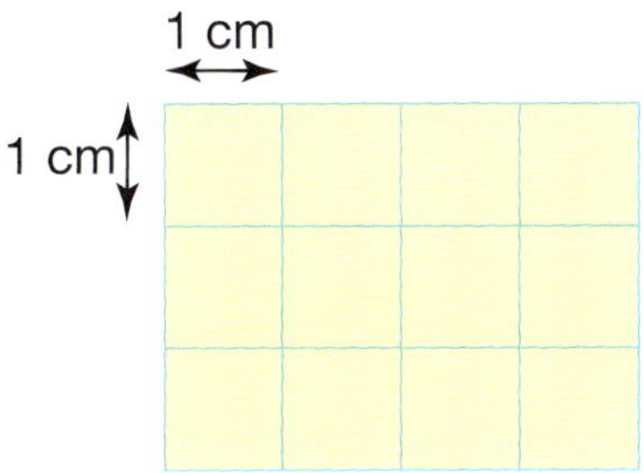

area = 12 square centimetres

array: A rectangular arrangement of items or pictures in **rows** and **columns** (e.g., An array can show why 2 × 3 and 3 × 2 have the same product.)

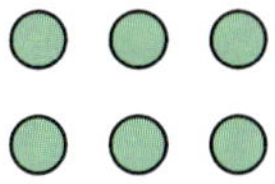

This array shows 2 rows of 3 or 2 × 3.
It also shows 3 columns of 2 or 3 × 2.
In both cases, the product is 6.

attribute: A characteristic or quality, usually of a pattern or geometric shape (e.g., Some common attributes of shapes are size, colour, texture, and number of edges.)

axis (plural is axes): A horizontal or vertical line in a graph, labelled with words or numbers to show what the bars or pictures in the graph mean

B

bar graph: A way to show data that uses horizontal or vertical bars

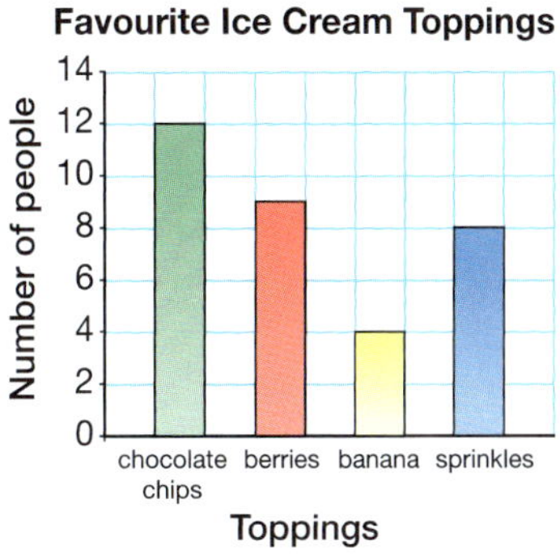

base: In a 3-D shape, the **face** on which it is resting

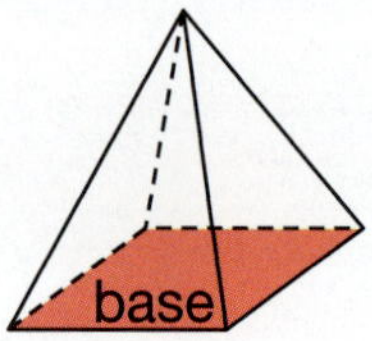

base ten blocks: Blocks that represent numbers as ones, tens, hundreds, and thousands

Thousands	Hundreds	Tens	Ones
2	3	5	4

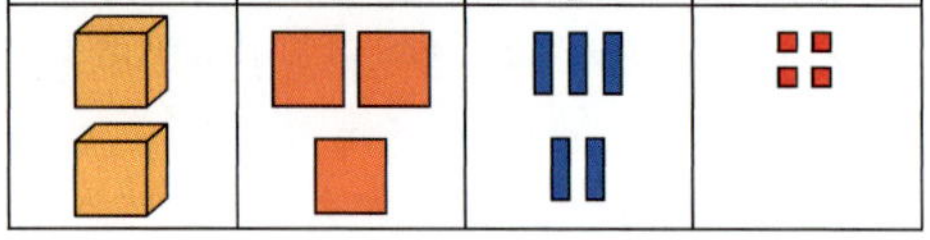

C

capacity: The amount a container can hold (e.g., Capacity is measured in millilitres (mL) and litres (L).)

Carroll diagram: A diagram that uses rows and columns to show relationships

	Less than 12	Greater than or equal to 12
Odd	1, 3, 5, 7, 9	13, 15, 27, 119
Even	2, 4, 6, 8, 10	12, 14, 20, 140

centimetre (cm): A unit of measurement for **length**; one hundredth of a metre (e.g., A fingertip is about 1 cm wide.) 1 cm = 10 mm, 100 cm = 1 m

century: A unit of measurement for time; 100 years

certain outcome: A result that will always happen (e.g., If you roll a die with a 3 on every face, rolling a 3 is a certain outcome.)

chance: The probability or likelihood that something will happen

circumference: The distance around a circle or circular object; the **perimeter** of a circle

column: A set of items lined up vertically; a **vertical** set of squares in a grid (See also **row**.)

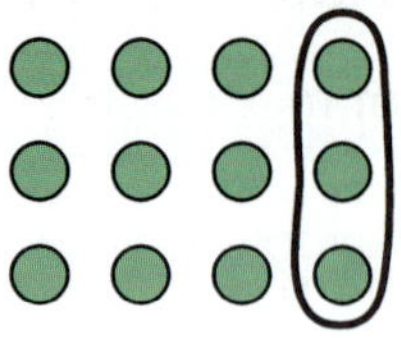

congruent: Identical in size and shape

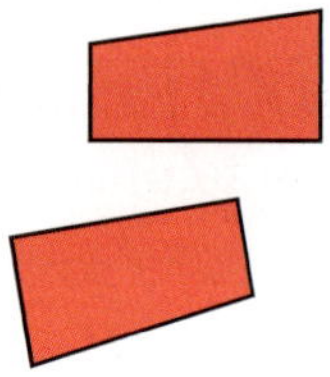

coordinates: Numbers or letters that describe a location on a **grid** (e.g., The counter is on C4.)

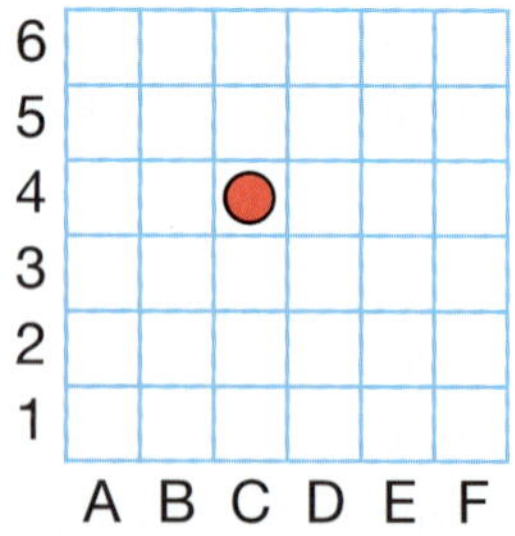

cross-section: The region formed by a **diagonal** of a 3-D shape

cube: A 3-D shape with 6 **congruent** square faces

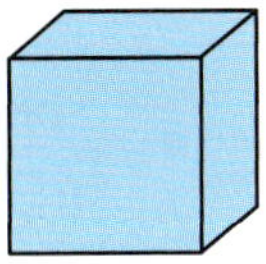

D

data: Information gathered in a survey, in an experiment, or by observing (e.g., Data can be in words like a list of students' names, in numbers like quiz marks, or in pictures like drawings of favourite pets.)

decade: A unit of measurement for time; 10 years

decimal: A way to describe fractional amounts using place value when the **denominator** is 10, 100, … (e.g., 0.6 is a decimal for $\frac{6}{10}$ or 6 tenths.)

decimal point: A dot used to separate the ones place from the tenths place in a decimal

decimetre (dm): A unit of measurement for **length**; one tenth of a metre (e.g., A tens rod from a set of **base ten blocks** is 1 dm long.); 1 dm = 10 cm, 10 dm = 1 m

degree Celsius (°C): A unit of measurement for temperature (e.g., Water freezes at 0°C and boils at 100°C.)

denominator: The bottom number in a fraction that tells how many are in the whole set, or how many parts the whole or set has been divided into. (See also **numerator**.) (e.g., In $\frac{3}{4}$ the fractional unit is fourths.)

$\frac{3}{4}$ ⟵ denominator

diagonal: In a 2-D shape, a diagonal can join any 2 **vertices** that are not next to each other.
In a 3-D shape, a diagonal can join any 2 vertices that are not on the same **face**.

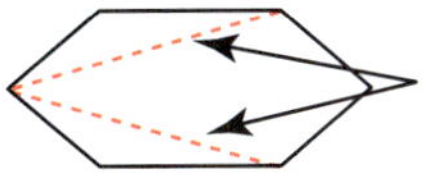

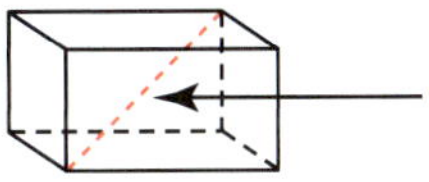

difference: The result when you subtract; the amount by which one number is greater than or less than another number

$$\begin{array}{r} 93 \\ -45 \\ \hline 48 \end{array}$$ ⟵ difference

digit: A symbol used to write a numeral (e.g., The digits are 0, 1, 2, 3, 4, 5, 6, 7, 8, and 9. For 5608, the digits are 5, 6, 0, and 8. For 7.9, the digits are 7 and 9.)

dimension: Measurements such as length and width (e.g., A line has only length, so it is 1-dimensional.
Area is 2-dimensional (2-D).
Volume is 3-dimensional (3-D).)

dividend: The number that is divided into equal parts in division

$9 \div 3 = 3$
↑ dividend (9)

divisible: Can be divided with no remainder (e.g., 30 is divisible by 6 because you can make exactly 5 groups of 6 from 30.)

divisor: The number you divide by in a division sentence

$24 \div 3 = 8$

↑ divisor

double: To multiply a number by 2

E

edge: The **line** formed where 2 **faces** meet on a 3-D shape

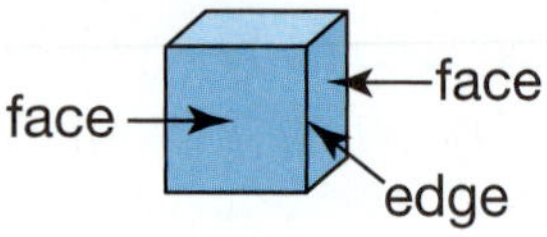

equally likely outcomes: Results that have an equal chance of happening (e.g., In flipping a coin, heads and tails are equally likely outcomes.)

estimate: A reasoned guess about a measurement or answer

even number: A number that is **divisible** by 2 (e.g., 12 is even because $12 \div 2 = 6$.)

event: 1 **possible outcome** of a **probability** experiment (e.g., When rolling a die, you could decide that rolling an even number, such as 2, 4, or 6, is an event.)

expanded form: A way to write a number that shows the value of each digit (e.g., In expanded form, 2365 is 2000 + 300 + 60 + 5 or 2 thousands + 3 hundreds + 6 tens + 5 ones.)

F

face: A 2-D shape that forms a flat surface of a 3-D shape

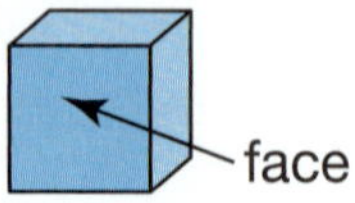

fact: An addition, subtraction, multiplication, or division sentence (e.g., In grade 4, we learn multiplication facts to $9 \times 9 = 81$.)

fact

fact family: Addition and subtraction or multiplication and division facts; each fact uses the same numbers

$3 \times 2 = 6 \quad 6 \div 3 = 2$

$2 \times 3 = 6 \quad 6 \div 2 = 3$

factor: Any one of the numbers you multiply

$2 \times 6 = 12$

factor factor

fraction: Numbers used to name part of a whole or part of a set (See also **numerator** and **denominator**.)

G

gram (g): A unit of measurement for **mass** (e.g., 1 mL of water has a mass of 1 g.); 1000 g = 1 kg

graph: A way of showing information so it is more easily understood. A graph can be concrete (e.g., boys in one line and girls in another), pictorial (e.g., pictures of boys in 1 row and girls in another), or abstract (e.g., 2 bars on a bar graph to show how many students are boys and how many are girls)

greater than (>): A sign used when comparing 2 numbers (e.g., 10 is greater than 5, or $10 > 5$.)

grid: A pattern with **rows** and **columns** for spaces, dots, or lines

H

halve: To divide a number by 2

heptagon: A **polygon** with 7 straight sides and 7 angles

hexagon: A **polygon** with 6 straight sides and 6 angles

horizontal line: A **line** that goes left and right; a line across a page that is **parallel** to the bottom edge

I

impossible outcome: A result that cannot happen (e.g., If you roll a die with a 3 on every face, rolling a 5 is an impossible outcome.)

intersecting lines: Lines that cross or meet

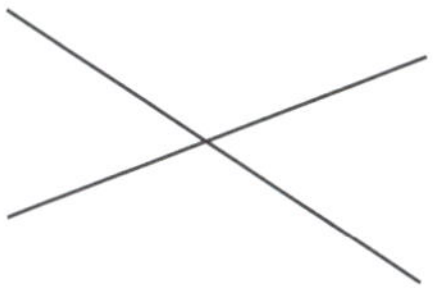

interval: The distance between 2 values on a graph scale; intervals on a graph should be equal (e.g., If the **scale** is numbered 0, 5, 10, 15, …, then the intervals are 5.)

K

kilogram (kg): A unit of measurement for **mass** (e.g., A math textbook has a mass of about 1 kg.); 1 kg = 1000 g

kilometre (km): A unit of measurement for **length**; one thousand metres; 1 km = 1000 m

kite: A **quadrilateral** with 2 pairs of equal sides and no sides parallel

L

legend: A feature on a graph that explains what symbols mean

length: The distance between 2 **points**

less than (<): A sign used when comparing 2 numbers (e.g., 5 is less than 10, or 5 < 10.)

likely outcome: A result that can easily happen (e.g., If you roll a die with a 3 on all the faces but one, rolling a 3 is a likely outcome.)

line: A straight path going in both directions

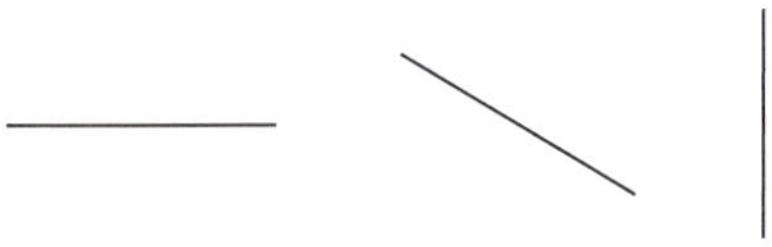

line of symmetry: The fold line that divides a 2-D shape into matching halves

litre (L): A unit of measurement for **capacity** (e.g., A container the size of a thousands cube from a set of **base ten blocks** has a capacity of 1 L.); 1 L = 1000 mL

M

mass: The amount of matter in an object (e.g., Mass is measured in grams (g) and kilograms (kg).)

metre (m): A unit of measurement for **length** (e.g., 1 m is about the distance from a doorknob to the floor.); 1000 mm = 1 m; 100 cm = 1 m; 1000 m = 1 km

millennium: A unit of measurement for time; 1000 years

millilitre (mL): A unit of measurement for **capacity** (e.g., A container the size of a centimetre cube from a set of **base ten blocks** has a capacity of 1 mL.); 1000 mL = 1 L

millimetre (mm): A unit of measurement for **length** (e.g., A dime is about 1 mm thick.); 10 mm = 1 cm, 1000 mm = 1 m

multiple: The product when a whole number is multiplied by any whole number except 0 (e.g., When you multiply 10 by the whole numbers 0 to 4, you get the multiples 0, 10, 20, 30, and 40.)

N

net: A 2-D pattern that can be folded into a 3-D shape

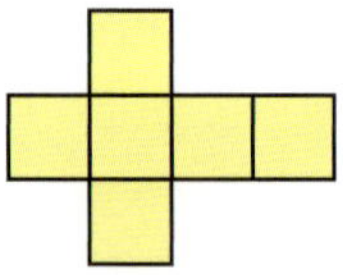

This is a net for a cube.

nonstandard unit: A unit of measurement that is not part of an accepted system (e.g., A desk is about 5 juice cans wide.)

number line: A diagram that shows ordered numbers or points on a line

number sentence: A mathematical statement that shows how 2 quantities are related (e.g., $3 \times 8 = 24$)

numeral: The written symbol for a number (e.g., 148, $\frac{3}{4}$, and 2.8)

numerator: The top number in a **fraction** that shows how many parts of a given size the fraction represents. (See also **denominator**.)

$\frac{3}{4}$ ← numerator

O

octagon: A **polygon** with 8 straight sides and 8 angles

odd number: A number that has a remainder of 1 when it is divided by 2 (e.g., 15 is odd because $15 \div 2 = 7$ R1.)

ordinal number: A way of describing an item's place in a numbered order (e.g., 1st, third, 15th)

organized list: The problem-solving strategy of following an order to find all possibilities

outcome: A single result (e.g., If you roll a die, the possible outcomes are 1, 2, 3, 4, 5, and 6; 7 is an impossible outcome.)

P

p.m.: The time between noon and midnight

parallel lines: Lines that never cross or meet, always the same distance apart

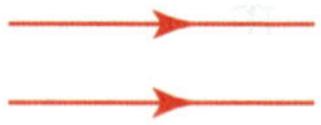

parallelogram: A **quadrilateral** with equal and **parallel** opposite sides (e.g., A **rhombus**, a **rectangle**, and a **square** are parallelograms.)

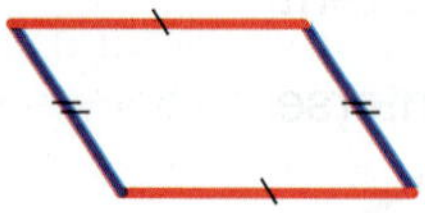

pattern: Something that follows a rule while repeating or changing

pattern rule: A description of how a pattern starts and how it can be extended

pentagon: A **polygon** with 5 straight sides and 5 angles

perimeter: The total length of the sides of a shape

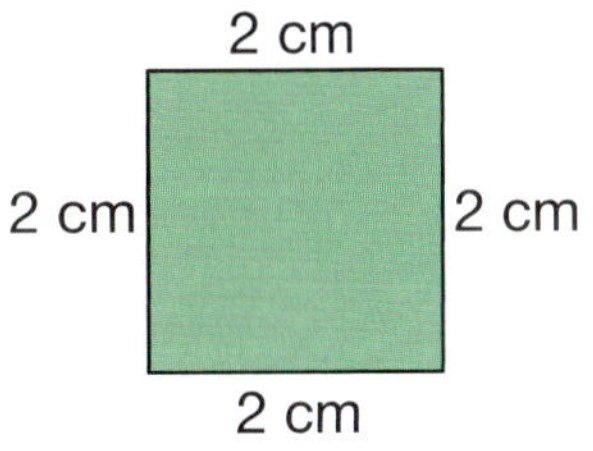

The perimeter of this square is 8 cm.

perpendicular lines: Lines that cross or meet at right angles

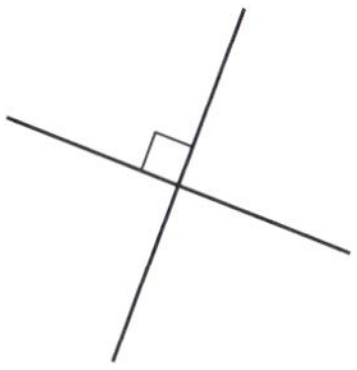

pictograph: A **graph** that uses pictures or symbols to represent quantities

place value: The value given to a digit based on its position in a number (e.g., The 3 in the number 237 represents 3 tens, while in the number 5.3 it represents 3 tenths.)

point: Where intersecting lines cross or meet

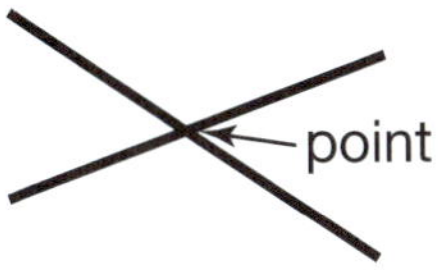

polygon: A 2-D shape with sides made from straight lines

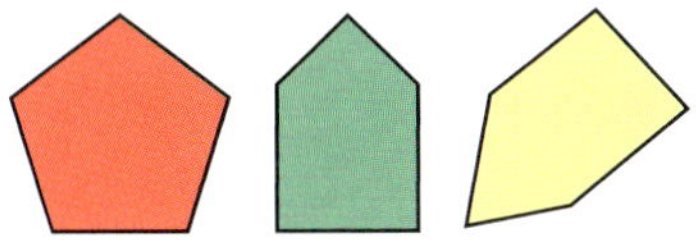

population: The whole group

possible outcome: Any result that can happen (e.g., If you roll a die, the possible outcomes are 1, 2, 3, 4, 5, and 6.)

precision: A way to compare tools and measurements (e.g., A measurement made with a ruler divided in millimetres is more precise than a measurement made with a ruler divided in centimetres.)

prism: A 3-D shape with opposite **congruent bases**; the other faces are rectangles (e.g., a triangle-based prism)

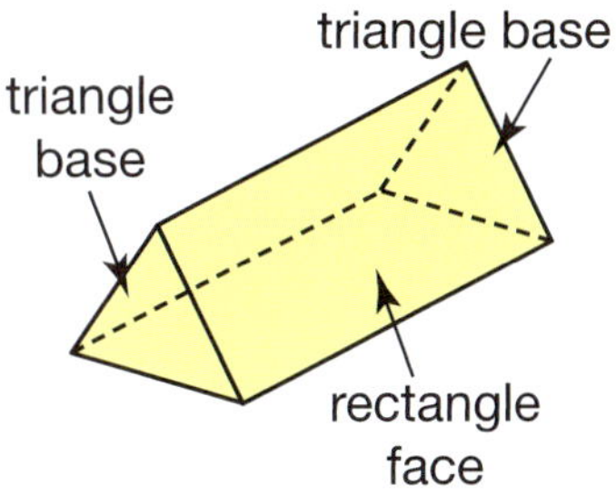

probability: How likely it is that a particular outcome will happen

probability line: A way to show probabilities of several outcomes

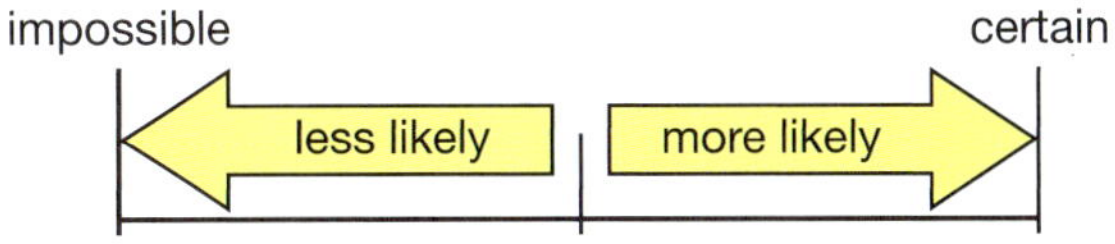

product: The result when you multiply

$2 \times 6 = 12$

product

pyramid: A 3-D shape with a **polygon** for a base; the other faces are triangles that meet at a single **vertex** (e.g., a rectangle-based pyramid)

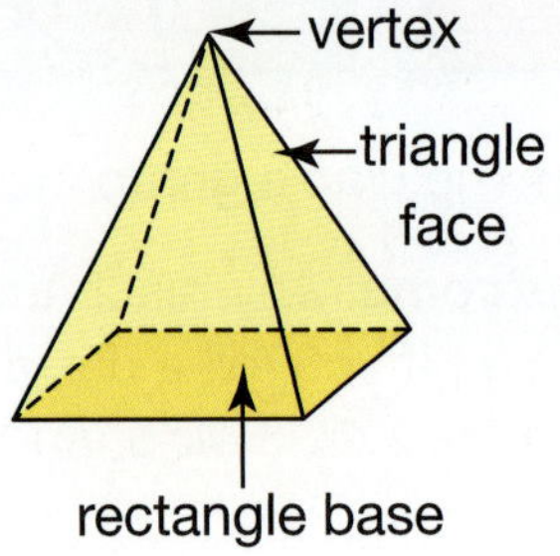

Q

quadrilateral: A **polygon** with 4 straight sides and 4 angles (See also **kite**, **parallelogram**, **rectangle**, **rhombus**, **square**, **trapezoid**.)

quotient: The result when you divide, not including the **remainder**

12 ÷ 5 = 2 R2

↑ quotient

R

rectangle: A **parallelogram** with 4 **right angles**

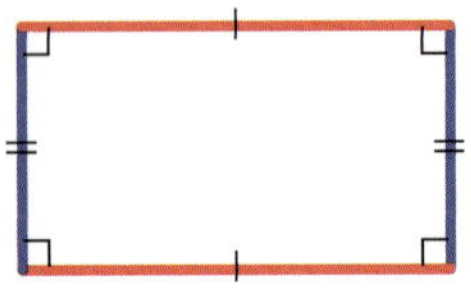

region: An area inside a shape

regroup: Trade 10 smaller units for 1 larger unit, or 1 larger unit for 10 smaller units

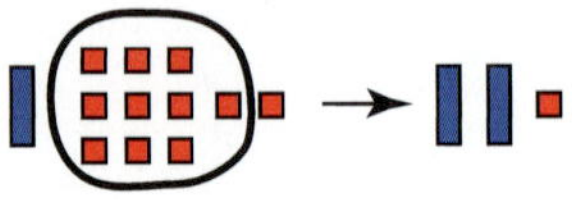

1 ten 11 ones = 2 tens 1 one

regular polygon: A straight-sided 2-D shape with equal sides

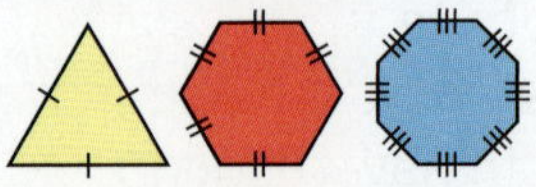

remainder: The number left over after an amount is divided into equal parts

14 ÷ 4 = 3 R2

↑ remainder

rhombus: A **parallelogram** with 4 equal sides

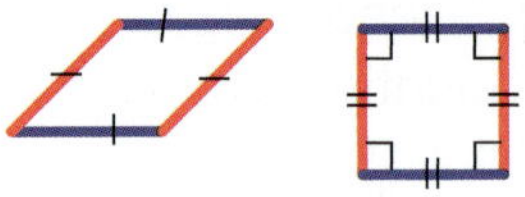

right angle: An **angle** that forms a square corner

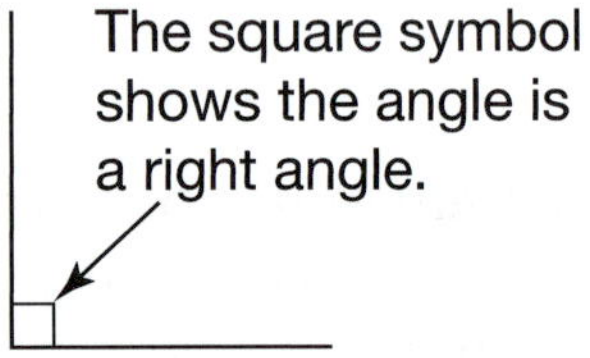

round: To approximate a number to a given place (e.g., 8327 rounded to the nearest hundred is 8300.)

row: A set of items lined up horizontally; a horizontal set of squares in a **grid** (See also **column**.)

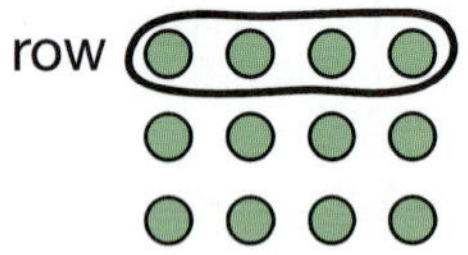

S

sample: A part of the **population** or group

scale: Numbers and marks arranged at regular intervals that are used for measurement or to establish position (e.g., the markings on the side of a measuring cup or on the **axis** of a graph)

set: A collection of items or numbers

shape: 1. A geometric object (e.g., A square is a 2-D shape. A cube is a 3-D shape.) 2. The **attribute** that describes the form of a geometric object (e.g., Circles and spheres both have a round shape.)

side: A **straight line** that forms part of a polygon

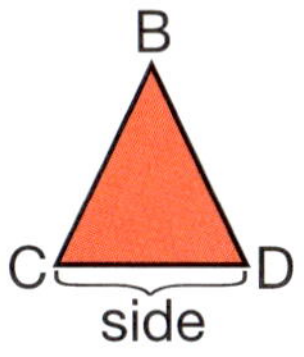

skip count: To count without using every number, but according to a pattern or rule (e.g., counting to 100 by 5s)

sorting rule: A rule used for sorting

square: A **parallelogram** with 4 equal sides and 4 right angles

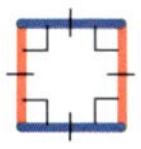

square centimetre: A unit of measurement for **area**; the area covered by a square with sides that are 1 cm long

square corner: A **vertex** of a shape that forms a square (e.g., Squares and rectangles have 4 square corners or **right angles**.)

square metre: A unit of measurement for **area**; the area covered by a square with sides that are 1 m long

square unit: A nonstandard unit of measurement for **area**

standard form: The usual way we write numbers (e.g., 2365 is written in standard form. See also **expanded form** and **word name**.)

standard unit: A unit of measurement that is part of an accepted measurement system (e.g., metres, kilograms, litres, and square metres are all standard units.) (See also **nonstandard unit**.)

sum: The result when you add

$$14 + 37 = 51$$

↑ sum

survey: 1. A set of questions planned to obtain information directly from people 2. To ask a group of people a set of questions

symmetrical: A shape or object that has symmetry

T

2-D shape: A shape that has the **dimensions** of length and width

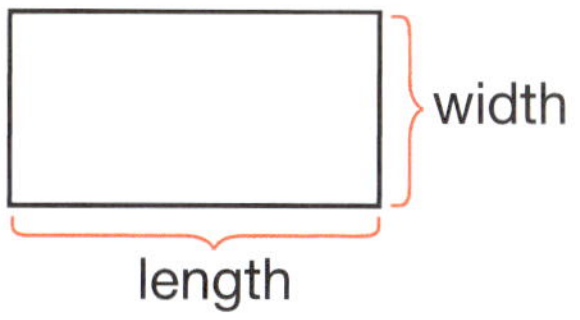

3-D shape: A shape that has the **dimensions** of length, width, and height

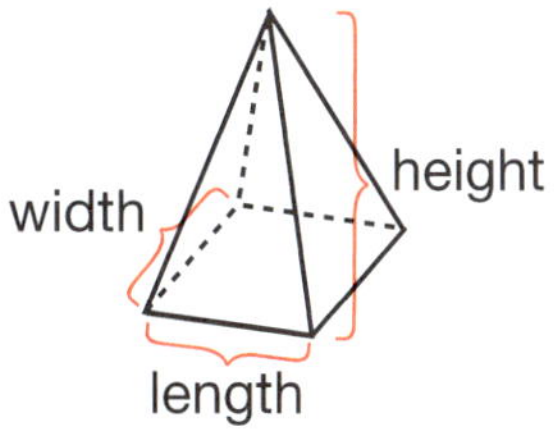

tally: A way to keep track of data using marks

~~////~~ ~~////~~ /

tally chart: A way to record tallies (e.g., If you are surveying students about favourite flavours of ice cream you could use a tally chart like this.)

Favourite Ice Cream Flavours	
vanilla	~~////~~ ~~////~~ /
chocolate	~~////~~ ~~////~~ ~~////~~ ~~////~~ ///
strawberry	~~////~~

t-chart: A way to organize information; both sides of the T are labelled

Weeks	Days
1	7
2	14
3	21

tetrahedron: A 3-D shape with 4 **faces** that are triangles

trapezoid: A **quadrilateral** with only 1 pair of **parallel** sides

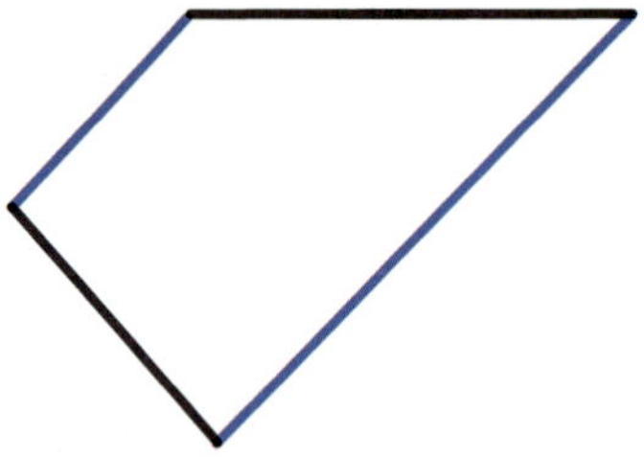

tree diagram: A way to record and count all combinations of **events** (e.g., This tree diagram shows all the 3-digit numbers that can be made from the digits 1, 2, and 3, if 1 must be the first digit and each digit is used only once.)

1 — 2 — 3 (123)
  \ 3 — 2 (132)

triangle: A **polygon** with 3 straight sides and 3 angles

U

unlikely outcome: A result that has little chance of happening (e.g., If you roll a die with a 3 on all the faces but one, rolling a number other than 3 is an unlikely outcome.)

V

Venn diagram: A diagram that uses circles and a rectangle to show relationships

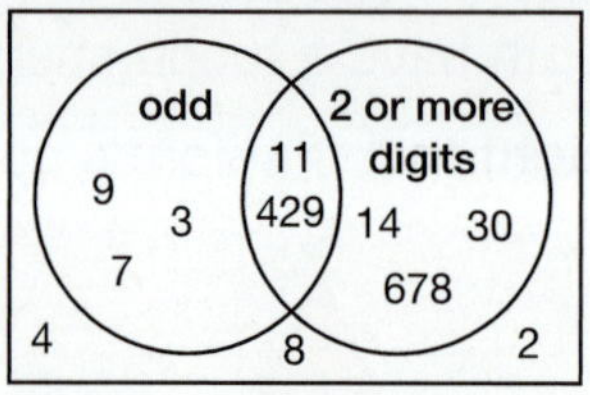

This Venn diagram shows each of 11 and 429 is an odd number and has 2 or more digits.

vertex (plural is **vertices):** The point at the corner of an angle or a shape (e.g., A cube has 8 vertices. An angle has 1 vertex.)

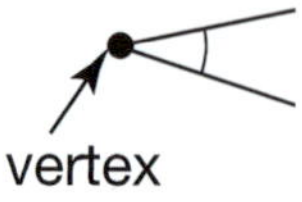

vertical line: A line that goes up and down; a line on a page **parallel** to the side edge, or straight up and down from the floor

volume: The amount of space taken up by an object

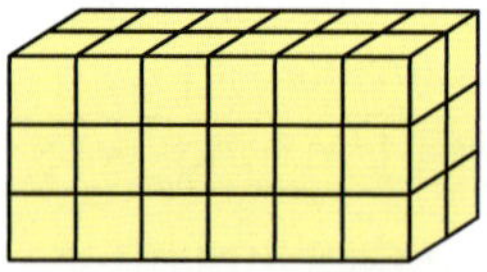

The volume of this box is 36 cubes.

W

whole numbers: The counting numbers that begin at 0 and continue forever; 0, 1, 2, 3, … .

word name: The way a number is presented in words (e.g., The word name for 365 is three hundred sixty-five.)

Index

D

N

O

P

S

T

U

V

Y

Z

Photo Credits

Cover Image © Corbis/Magma

Table of Contents Page xiv left: © James Davis; Eye Ubiquitous/Corbis/Magma, right: © Gunter Marx/Corbis/Magma; Page xv left: Digital Vision/Getty Images, right: © John B. Boykin/ Corbis/Magma

Chapter 2 Opener Page 27: © Gabriel Jecan/ Corbis/Magma; Page 30: © Ken Wilson; Papilio/Corbis/Magma; Page 31: First Light; Page 33 top: © Jerry Cooke/Corbis/Magma, bottom: David Madison/Stone/Getty Images; Page 34: © AFP/Corbis/Magma; Page 40: Mike Johnson; Page 49: © Wolfgang Kaehler/ Corbis/Magma

Chapter 3 Page 62: Gotta Find A Footprint excerpted from BONE POEMS. Text copyright © 1997 by Jeff Moss. Used by permission of Workman Publishing Co., Inc., New York. All rights reserved; Page 75: © Corbis/Magma

Chapter 4 Opener Page 91: © Kevin Fleming/ Corbis/Magma; Page 99: © Michael Newman/ Photo Edit; Page 93: First Light

Chapter 5 Page 146: Corel

Chapter 6 Opener Page 155: Eyewire/Getty Images; Page 176: Ryan McVay/PhotoDisc/ Getty Images; Page 179: Corbis/Magma

Chapter 7 Opener Page 189: Gary Conner/ Index Stock; Page 190: Corbis/Magma; Page 196: © Jose Luis Pelaez, Inc./ Corbis/Magma

Chapter 8 Page 219: © David Zimmerman/ Corbis/Magma

Chapter 9 Opener Page 243: © Canadian Museum of Civilization, photographer Harry Forster, 1988, image no.. S89-1873; Page 249: © David Lees/Corbis/Magma; Page 254: Corbis/Magma; Page 256: Infocus International/ Image Bank/Getty Images; Page 257 top: Corel, bottom: © Jim Winkley; Ecoscene/ Corbis/Magma

Chapter 10 Page 284: Viktor Pivovarov/CP Picture Archive; Page 290 top: Jeff Greenberg/ Photo Edit, bottom: © Christie's Images/ Corbis/Magma; Page 291: © Wolfgang Kaehler/Corbis/Magma; Page 294: Jules Frazier/PhotoDisc/Getty Images; Page 295: Russell Illiq/PhotoDisc; Page 297: David Young-Wolff/Photo Edit

Chapter 11 Opener Page 299: Courtesy of Levitt Architect Limited; Page 302 left: Corbis/Magma, right: © Gunter Marx Photography/Corbis/Magma; Page 304: © James Davis; Eye Ubiquitous/Corbis/Magma; Page 309 both: Courtesy of Levitt Architect Limited; Page 313 left: Ryan McVay/PhotoDisc/ Getty Images, right: © David Young-Wolff/ Photo Edit; Page 321: HIRB/Index Stock

Chapter 12 Opener Page 325: © Ron Watts/ Corbis/Magma; Page 336: Oxford Scientific Films; Page 337 top: W. Perry Conway/ Corbis/Magma, bottom: Digital Vision/Getty Images; Page 338: © W. Perry Conway/ Corbis/Magma; Page 339 top: S. Charles Brown, Frank Lane Picture Agency/Corbis/ Magma, bottom: © Ron Austing; Frank Lane Picture Agency/Corbis/Magma; Page 340: First Light; Page 344: © John B. Boykin/Corbis/ Magma; Page 350: © D. Robert & Lorri Franz/ Corbis/Magma; Page 353 upper: © Joe MacDonald/Corbis/Magma, lower: First Light

Chapter 13 Page 380 top: © Buddy Mays/ Corbis/Magma, middle: Gail Shumway/Taxi/ Getty Images, bottom: First Light